Structuralist Theory of Science

Perspektiven der Analytischen Philosophie

Perspectives in Analytical Philosophy

Herausgegeben von

Georg Meggle und Julian Nida-Rümelin

Band 6

Walter de Gruyter · Berlin · New York

1996

Structuralist Theory of Science

Focal Issues, New Results

Edited by
Wolfgang Balzer and C. Ulises Moulines

Walter de Gruyter · Berlin · New York

1996

∞ Printed on acid-free paper which falls within the guidelines of the ANSI to ensure permanence and durability.

Library of Congress Cataloging-in-Publication Data

Structuralist theory of science : focal issues, new results / edited by Wolfgang Balzer and C. Ulises Moulines.
p. cm. – (Perspektiven der analytischen Philosophie ; Bd. 6 = Perspectives in analytical philosophy)
Includes bibliographical references and index.
ISBN 3-11-014075-6
1. Structuralism. 2. Science – Philosophy. I. Balzer, Wolfgang, 1947– . II. Moulines, Carlos Ulises. III. Series: Perspectives in analytical Philosophy ; Bd. 6.
Q175.32.S78S78 1996
501–dc20 95-45839
CIP

Die Deutsche Bibliothek – Cataloging-in-Publication Data

Structuralist theory of science : focal issues, new results / ed. by Wolfgang Balzer and C. Ulises Moulines. – Berlin : de Gruyter, 1996
(Perspektiven der analytischen Philosophie ; Bd. 6)
ISBN 3-11-014075-6
NE: Balzer, Wolfgang [Hrsg.]; GT

Printed in Germany
Printing: Werner Hildebrand, Berlin
Binding: Lüderitz & Bauer, Berlin
Cover Design: Rudolf Hübler, Berlin

To Wolfgang Stegmüller

in memoriam

Foreword

The aim of this volume is to provide a survey of the latest theoretical developments and results of the structuralist program in the philosophy of science. Since the appearance of *An Architectonic for Science,* (a compendium on structuralism by Balzer, Moulines and Sneed) in 1987, a considerable number of contributions to the program, both on general topics and on reconstructions of particular theories, have been published. However, they appeared scattered throughout many different journals, countries, and languages. For the general public interested in philosophy of science, no overview about the state of the art was available. So, the idea occurred to the editors of this volume to bring together a number of outstanding 'structuralists' and ask them to lay out what they have been so far doing on the research front. A conference took place in Munich in February 1994 (actually the very first conference exclusively devoted to the structuralist approach), where each contribution was amply discussed and, to some extent, 'tuned in' to the rest. The result was the present volume. All essays contained in it have been exclusively written for this project.

As will be explained in the first chapter of this book, the structuralist program has two sides: it deals with general problems of the philosophy of science, and it also provides a methodology to reconstruct particular scientific theories. However, the contributions to the present volume have deliberately been chosen so as to address general issues only. We reserve for another book the exposition of case studies, that is, of new applications of the structuralist program to actual science. The reason for this selfrestraint is just that, otherwise, the great number of recent applications of the program would break the limits of any reasonable work.

Since the appearance of Joseph D. Sneed's *The Logical Structure of Mathematical Physics* (the first fully structuralist writing *avant la lettre*) in 1971, the program has been quite successful, we think, in providing a metatheoretical model of the structure and development of science, and in applying this model to the reconstruction of a great number of concrete theories. Nevertheless, several 'classical' topics in general philosophy of science, most notably those of explanation, confirmation, problem solving, the hypothetico-deductive method, and logical foundations, had scarcely been addressed at all. We believe that this was not an accidental feature of the program's development but rather something inherent to the subject matter. In order to deal with those questions in an appropriate way, it appeared necessary to have a most clear notion of what a scientific theory is and also to have as many reconstructed examples as possible. Indeed, it would be bad strategy to tackle those general topics without first having solid ground in the two latter respects. This much we can learn from the overall frustrating development of general philosophy of science in the fifties and sixties, as well as, we dare to say, from much of what is going on today.

Meanwhile, after two decades of research, structuralism's metascientific conception and numerous examples of application have stabilized to a sufficient degree in order to provide a sound basis for attacking at least some of the subjects above-mentioned. Consequently, the present volume lays out recent structuralist research on some classical problem areas which had not been in the program's focus up to now. Thus, Thomas Bartelborth's present contribution deals with the concept of explanation, Ulrich Gähde's with holism, Theo Kuipers' with the hypothetico-deductive method and truthlikeness, Bernhard Lauth's with confirmation, Felix Mühlhölzer's with symmetry principles, and Matti Sintonen's with the notion of a scientific problem. On the other hand, Peter Hinst makes clear, in his paper, that contrary to previous repeated claims by critics of structuralism the informal use of set theory doesn't lead to any additional foundational problems besides those already known from the philosophy of mathematics, while Thomas Mormann shows how to fruitfully use (mathematical) category theory as a formal foundation for the program. The chapters by Wolfgang Balzer, Werner Diederich, and C. Ulises Moulines & Marek Polanski elaborate on structuralist 'core' themes: theoretical terms, scientific development, and intertheoretical relations, respectively.

It is not the purpose of this book to provide a full introduction to the structuralist program and its methods, much less to offer a comprehensive exposition of all its aspects. The most comprehensive exposition of structuralism up to now (also containing a number of concrete applications and case studies) is *An Architectonic for Science*. Since this work will be quite often cited or used here, it will be referred to simply as *Architectonic* throughout the text. The reader should take it as the basic reference work for many concepts and constructs more or less taken for granted or not explained in detail in the present book. In a sense, the present collection of essays leaves off the ground settled in *Architectonic* in order to explore new directions. Therefore, we must presuppose that the reader already has some acquaintance with the spirit and the letter of the structuralist research program.

Nevertheless, to facilitate the reading of the essays, especially those of a more technical kind, it is convenient to offer a brief reminder of structuralism's basic notions and theses. This is done by Moulines in his first essay.

Several persons and institutions deserve grateful mention for having contributed to the preparation of this volume. The Munich conference on structuralism above-mentioned, which paved the way for the compilation of the volume, was financially supported by the *'Gesellschaft von Freunden und Förderern der Universität München'*. We are very much indebted also to Dorothea Lotter for her invaluable editorial assistance, as well as to Margit Barrios and Dr. Jamel Tazarki for their equally invaluable help in producing the final version of the typescript. Last but not least, we thank the publisher, Walter de Gruyter & Co., and especially Dr. Hans-Robert Cram and Professor Heinz Wenzel, for their encouragement and patience, which allowed for the timely publication of this book.

The Editors
Munich, May 1995

Table of Contents

Chapter 1: Structuralism: The Basic Ideas

C. Ulises Moulines

The term 'structuralism' will be employed in this book to refer to a particular theory of science. This label was first introduced by Wolfgang Stegmüller in the late seventies, following a suggestion by Y. Bar-Hillel. In previous writings, Stegmüller had used the expression 'non-statement view', while Joseph D. Sneed (the other pioneer of this conception) had described it as an 'emended Ramsey view'. None of these labels is particularly fortunate. They are all too restrictive; taken literally, they point out only some features of the theory. 'Structuralism' may be the less inadequate because of its generality. At any rate, it is the label most used in the literature. So, we will keep it throughout this book.

Other schools of thought have been described as 'structuralistic'. We find 'structuralisms' in the social sciences (psychology, linguistics, ethnology), in (French) philosophy, and in the foundations of mathematics (mainly the Bourbaki program). Only of the latter can we say that it has a substantial connection with 'our' structuralism. With respect to the other uses of the term, the relationships are quite remote, if present at all.

Structuralism is a school of thought only in a rather loose sense. 'Structuralists' may be described as a group of people using the concepts and methods of this particular theory of science, though they may greatly differ on many other philosophical issues. They use these concepts and methods mainly for two kinds of purposes: first, to clarify some general epistemological, methodological and metatheoretical questions about science; second, to reconstruct particular scientific theories. This group of people includes not only philosophers in the strict sense but also a number of practising scientists from very different disciplines, who are interested in applying the general theory to their particular discipline. A considerable portion of structuralist literature deals with case studies[1].

The range of disciplines where structuralism has been applied is very wide: from physics, chemistry, and biology through psychology, economics, and sociology up to linguistics, the theory of literature, and the theory of accounting. It is crucial for the structuralist metatheory to have

[1] The great number of scientific theories reconstructed in detail within structuralism up to the year 1989 may be gathered from the bibliography set up by W. Diederich, A. Ibarra, and Th. Mormann. Since then, many other case studies have been published. An up-to-date version of their 'Bibliography of Structuralism', including the reconstruction of particular theories up to 1994, appeared in Erkenntnis, vol 44 (1994).

this wide range of applications, not only because it shows its methodological potential, but also because it gives some empirical support to one of the theory's central claims, viz. that there are some common features in the deep structure of all empirical disciplines and that these features can actually be captured by the metatheory. This is what we should require, anyway, from any general theory of science which aims at a serious treatment of its subject matter. It should be *checked* against many examples.

So, structuralism essentially is a theory about science. But, of course, it is *not* a theory about *every* aspect of science. For example, it is not a theory about the ethical or political aspects of science, nor about its technological implementation, nor about scientific experimental techniques. Essentially, it is a theory about scientific theories - about their identity, structure, mutual relationships, and evolution. This is not to say that, according to structuralism, the other aspects of science are negligible or completely alien to scientific theories. Rather, the assumption is that theories are quite essential for an adequate understanding of the nature of science in general - that you won't know what science is unless you know what a scientific theory is and how it works.

Since structuralism's main object of analysis are theories, it has sometimes been characterized as a 'metatheory' of science. Now, one of the main results of structuralistic analysis is that the presystematic use of the term 'theory' turns out to be polisemic. That is, an adequate formal explication of the intuitive notion of a theory leads to at least *three* different exact concepts. They are mutually related but they are located, so to speak, on three different structural layers. At the inferior (simpler) level, we have 'theory-elements'; at the medium level, we get 'theory-nets'; and the most complex units are 'theory-holons'. (More about these notions will be said below.) The concept of a theory-net is what comes closest to the most frequent intuitive usage of the term 'theory'. However, in some contexts, people using 'theory' mean something closer to the 'smaller' unit, a theory-element, or to the 'bigger' unit, a theory-holon. When we speak of 'a theory' in this book, we usually mean the intuitive notion corresponding to our theory-nets.

Another reason why the common usage of 'theory' is polisemic, stems from the fact that, frequently, no clear distinction is made between the synchronic and the diachronic perspective. When we speak of a scientific theory, it should always be clear whether we refer to a unit of scientific knowledge considered in a synchronic 'cut' or rather as a diachronic 'worm'. A theory of science is part of the sciences of culture since science is a part of culture. And we know from other disciplines of culture how important it is to make the synchronic/diachronic distinction. Philosophy of science should be no exception to this. The structures and criteria of identity of a cultural entity are not necessarily the same in diachronic and in synchronic perspective. We may not fully grasp the precise nature of a cultural entity as seen diachronically by just considering it synchronically and then letting it 'move' through historical time. In particular, a scientific theory as a diachronic entity is not just a theory-net associated with a time interval. It has a related, but different, structure. For this reason, structuralism takes the synchronic/diachronic distinction very seriously and it has introduced a specific label for a theory in the diachronic sense: 'a theory-evolution'.

As any other scientific theory, structuralism has its own 'metaphysics' and 'methodology'. That is, it makes some general metaphysical, or rather: ontological, assumptions, and it chooses some specific methods of analysis. In structuralist literature, these assumptions and choices have usually been made implicitly, that is, they have just been taken for granted. Let us make at least some of them more explicit here.

Structuralism's basic *ontological* assumptions are as follows.

(1) *There are* scientific theories (in at least three different senses of the term 'theory').

(2) Scientific theories are *cultural* objects of a rather *abstract* kind in the sense that they are not spatiotemporally localized the way macroscopic physical objects are. Their ontological status is similar to that of other abstract cultural objects like languages (in the sense of Saussure's *langue*, not of his *parole*), symphonies, computer programs, and the like.

(3) In a way similar to other abstract cultural objects, scientific theories have a 'deep structure'. This means that their identity criteria and essential components cannot be detected in a straightforward manner by just 'looking at' the surface appearance of the concrete objects (e.g. scientific texts) instantiating them. To display them is a non-trivial task.

(4) Scientific theories are genidentical entities. They have a 'life' of their own, like persons or nations do.

(5) Scientific theories are not 'monads'. They are essentially related to things outside themselves. At least part of this outside world consists of other scientific theories. This means that there are intertheoretical relations and that they belong to the 'essence' of scientific theories.

This is not the place to argue for these ontological commitments. Suffice it to note that denying (1) - (3) is the common fault of crude empiricists and of sociologists of science who are nominalists perhaps without knowing it. The common fault of a vast majority of present-day philosophers of science (critical rationalists, scientific realists, constructive empiricists, and what not) consists in *volens nolens* accepting (1) and (2) but ignoring or directly denying (3). Ignoring (4) was the common mistake of *pre*-Kuhnian philosophy of science. Ignoring or trivializing (5) is the common fault of a great deal of post-Kuhnian philosophy of science. Structuralism is an attempt to provide some remedy to all these mistakes. However, the aim of the collection of essays in this volume is not to discuss assumptions (1) - (5) in detail, nor to argue for them; at most, indirect hints for them will be found. Consequently, those readers who are stubbornly persuaded that the mistakes just listed are not really such but that, on the contrary, the *negations* of (1) - (5), respectively, represent the 'real truth' about science, will probably feel that they are losing their time reading one more page of this book.

As for structuralism's specific *methodology*, its choice is, in part, prompted by the previous ontological assumptions.

(6) The best way to reveal the deep structure of a scientific theory as an abstract entity is by means of *formal analysis*. As far as possible, formal techniques of analysis and reconstruction should be preferred to explications in ordinary language. The reason is simple. Formal (or 'semi-formal') techniques of analysis lead to more precise and controllable constructs than their ordinary language counterparts. This is hardly disputed in any other scientific field; there is no reason why a theory of science should be an exception to this. Of course, it may be the case that the formal methods required are still not available, and in this situation one should be content with informal analysis. This is actually the case for some of the issues examined by the structuralist metatheory. But it is no excuse for not employing formal methods whenever possible.

(7) More specifically, given the kind of conceptual analysis required by structuralism, the most adequate formal tool, because of its universal character, appears to be set theory. This is why structuralism has sometimes been described as a 'set-theoretical approach'. However, it should be clear that set theory is only the method, not the 'substance' of the theory. In principle, other formal (as well as non-formal) methods of analysis and reconstruction could be used, and have actually been used; for example, methods from metalogic and from category theory. Some essays in this book illustrate this pluralistic methodology.

Structuralism owes its name to the 'hermeneutic' claim that the most adequate way to interpret the 'essence' of scientific theories is to view them not as sets of statements or propositions (as in more traditional theories of science) but rather as different kinds of *complex structures* consisting, in turn, of several simpler structures. The term 'structure' is understood here in the technical sense of set theory, or more precisely, in the sense of Bourbaki.

The simplest structural units a theory consists of are its *models* (in the sense of formal semantics). They are sequences of the form

$$\langle D_1, \ldots, D_m, R_1, \ldots, R_n \rangle,$$

where the D_i are so-called 'basic sets' and the R_i are relations constructed over (some of) these sets. The D_i settle the theory's 'ontology', i.e. they contain the objects assumed by the theory as 'real' - be they empirically detectable or purely mathematical objects. As for the R_j, they often are functions; in quantitative disciplines, they usually are functions from empirical objects to real numbers or vectors. In a first step, we may say that the identity of a scientific theory is given by a class of models so conceived. The choice of the particular axioms to be satisfied by the models of this class, is considered by structuralism as a relatively unimportant question. It is just a matter of convenience. The really important matter is that the set of axioms chosen exactly determine the class of models we need to represent a certain field of phenomena we are interested in for some reason.

Though the particular choice of axioms is not very important as long as they lead to the same class of models, nevertheless it is quite important that we distinguish two kinds of axioms among those chosen. We have to distinguish between 'frame conditions' on the one hand and 'substantial laws' on the other. The first don't 'say anything about the world' but just settle the formal properties of the concepts we want to use; the second group of axioms 'say something about the world' by means of the concepts previously fixed. The class of structures of which we only require that they satisfy the first kind of conditions we call 'the class of *potential models*' of the theory; let's symbolize it by 'M_p'. Those structures which, in addition, satisfy the substantial laws we call 'actual models'; we symbolize their class by 'M'. In principle, any means to settle M_p and M are good as long as we actually get the classes of structures identifying the theory. For example, we could use an adequate formal language to express the frame conditions and the substantial laws and then define M_p and M in the usual way known from formal semantics as interpretations of the formal language. However, in most developed scientific theories this procedure would be very clumsy if not fully impracticable. As soon as you employ a bit of higher mathematics in your theory, a complete formalization of the axioms becomes an extremely tedious task. Consequently, for *practical-methodological* reasons, structuralists prefer to use a more expedient tool to determine M_p and M precisely: informal set theory instead of a formal language. The particular method to do this has been popularized by Patrick Suppes as 'axiomatization by means of a *set-theoretical predicate*'. It consists in defining, within set theory, a 'second-order' predicate summarizing the corresponding axiomatic conditions. Suppose the sentences $\alpha_1, \ldots, \alpha_m$ are the set-theoretical version of the theory's frame conditions (= the axioms describing the potential models) and the sentences $\beta_1, \ldots, \beta_n$ are the set-theoretical version of the substantial laws (= the axioms for the actual models). Then, we can define two predicates for structures, call them 'A' and 'B', in the following way:

1) For any structure $x, A(x)$ iff $\alpha_1[x] \wedge \ldots, \wedge \alpha_m[x]$ where $\alpha_i[x]$ means that the sentence α_i is true of structure x.

2) For any structure $x, B(x)$ iff $A(x) \wedge \beta_1[x] \wedge \ldots \wedge \beta_n[x]$

Then, we define $M_p =: \hat{A}$ and $M =: \hat{B}$, where $\hat{A}$ and $\hat{B}$ are the extensions of the predicates 'A' and 'B'[2]

Clearly, different set-theoretical predicates may have the same extension; this is why the identity of a theory is not given by a particular set-theoretical predicate but by the corresponding class of structures. Thus, in a first step, we may say that the identity of a scientific theory is given by an ordered pair of the form $\langle M_p, M \rangle$. Let us call this construct a 'model-element'. Model-elements represent the most basic unit to grasp the 'essence' of a theory. However, this is only the first step. A central thesis of structuralism is that 'normal' theories (i.e. theories which are not too primitive or underdeveloped) will consist of more essential components than just model-elements. The structure of an empirical theory of a discipline which has attained a certain degree of conceptual and methodological development will typically be determined by four additional components at least. This conclusion may be reached after a careful metatheoretical analysis of the following facts:

1) The models of one and the same empirical theory don't appear isolated; they are mutually related by certain second-order conditions.

[2] The reader not well-acquainted with the method of set-theoretical predicates should consult the last chapter of Suppes (1957) or any general exposition of structuralism, in particular Ch. I of *Architectonic*. In these works, many examples of axiomatizations of concrete theories by this method are provided as well.

Structuralism has introduced a special label for them: *'constraints'*. The symbol for the class of all constraints of a given theory is 'C'.

2) The theories themselves are not isolated units; this means that models of different theories are mutually connected. These intertheoretical relationships are called *'links'*. The corresponding symbol for all links of a theory is 'L'.

3) Two different conceptual and methodological levels usually are distinguishable within one and the same theory: the level of concepts which are specific of the theory in question and which can *only* be determined by assuming that theory; and the level of concepts coming from 'outside'. Typically, the latter are determined by other, 'underlying', theories. If T is the theory in question, we describe the first set of concepts as *'T-theoretical'* and the second set as *T-non-theoretical'*. The class of substructures satisfying only the axioms for the 'T-non-theoretical' concepts represents the (relative) basis of data for T. These substructures receive a special name: *'partial potential models'*. Their class is symbolized by 'M_{pp}'.

4) No empirical theory which is supposed to be really applicable to this not-too-Platonic world, will work without accepting a certain degree of approximation. The approximation may be qualitative or quantitative, and it can vary according to the kind of application we have in mind. The theory's models will then have to be 'blurred' to some extent. But we should not go too far in this blurring; some limits are to be set, otherwise the theory becomes completely useless. The most general way of characterizing such blurs is to take them as elements of a so-called *'uniformity'* U, in the sense of topology. But not all elements of such a uniformity will be acceptable for an empirical theory; only those are acceptable which satisfy certain restrictions. Call them *'admissible blurs'*, and symbolize their class by 'A'.

Now structuralism says that the four components listed above are essential constituents of any empirical theory to be taken seriously. Though all of them may be considered as metatheoretical primitive concepts, they are all formally related to the theory's frame M_p. Their relationships (including the relationship between actual and potential models) are as follows:

a) $M \subseteq M_p$

b) $C \subseteq \text{Po}(M_p)$

c) For any $\lambda \in L$, there is an $M_p' \neq M_p$ such that $\lambda \subseteq M_p \times M_p'$.

d) There is a many-one function r so that: $r : M_p \longmapsto M_{pp}$

e) $A \subseteq U \in \mathrm{Po}(M_p \times M_p)$

Consequently, the formal identity of any empirical theory with a certain degree of complexity is given by the array of components M_p, M, C, L, M_{pp}, A. We call this array its '(formal) *core*' and symbolize it by:

$$K := \langle M_p, M, M_{pp}, C, L, A \rangle$$

All components of a theory's core are formal in the sense that, in principle, they may be fully characterized by means of model theory and set theory[3]. However, another basic thesis of structuralism is that the formal core does not exhaust all we have to know in order to know what an empirical theory is and how it works. The gist of an *empirical* theory consists precisely in the fact that its principles are supposed to be applicable to the phenomena which are 'external' to them, in such a way that explanations, predictions, and technological applications are made possible. This 'outside world' is called by structuralists *'the domain of intended applications'*: it is the range of phenomena to which people using the theory *intend* to apply its concepts and laws; it also belongs to the theory's identity because otherwise we would not know what the theory is about.

Structuralism makes three basic epistemological assumptions about the proper way of conceiving this domain of intended applications. First, it is neither 'pure reality' nor 'pure experience' - whatever these expressions may mean. That is, the domain in question does not consist of pre-conceptual 'things-in-themselves' or of sense-data. Scientific theories don't have access to that sort of stuff - if anybody has access to it at all. Rather, the assumption is that the domain of intended applications of a theory is conceptually determined through concepts already available. The real question is whether *all* concepts available or only *some* of them must be employed to describe that domain. In the latter case, the natural assumption is that only those concepts coming from 'outside', viz. the T-non-theoretical ones, will be used. If all the theory's primitive concepts had to be used, we would have to say that the theory's intended applications are conceived of as potential models of the theory; in the second case, they should be interpreted as partial potential models. Symbolize

[3] It is still a matter of debate within structuralism how stringent the claim that the components M_p, M_{pp}, and A of a core are fully characterizable in set-theoretical terms actually is. For the purposes of this introductory chapter, however, it is sufficient to admit that the crucial aspects of these components may be so formalized.

the domain of intended applications by 'I'. Then, the first alternative will be expressed by the formula $I \subseteq M_p$, while the second is expressed by $I \subseteq M_{pp}$. The current hypothesis of structuralism is that, at least in most theories, the most plausible alternative is the latter one.

Second, the intended applications of any given theory don't cover the 'whole universe' - whatever this may be. Intended applications of a theory are multiple and local. They represent 'small' pieces of human experience. There is no such thing as a theory of everything, nor is it reasonable to expect that such a theory will ever show up. (Even the best astrophysical theory won't be able to explain your neighbour's neurotic drive to annoy you, or the current exchange rate between the U.S. dollar and the *deutschmark*.) Each scientific theory has its own domain I; the domains of different theories may coincide, partially overlap, be only loosely related or not related at all. This is all we can say about 'the universe' from a metatheoretical point of view.

Third, to conceive of I as a subclass of M_{pp} (or, alternatively, of M_p) is only a very weak determination of it. It doesn't say much about the specific 'borders' of I within M_{pp} (or M_p). Structuralism claims that, when examining the domains of intended applications of particular theories, we will be able to say much more about their specific nature, but this will hardly be amenable to formalization in terms of model theory and set theory. The reason is that I is a kind of entity strongly depending on pragmatic and historical factors which, by their very nature, are not formalizable (at least not by means of presently available formal tools). It is at this point, at the latest, that structuralism ceases to be a 'purely formalistic' or 'set-theoretical' view of science. This is seen by structuralism itself neither as an absolute virtue nor as an absolute fault. It is rather an unavoidable consequence of the nature of theories and of the tools available to analyze them.

According to structuralism, theories are not sets of statements. But, of course, this is not to deny that it is very important for science to make statements - things that can be true or false, that can be verified, falsified or somehow checked. What structuralism maintains is that theories are not statements but are *used* to make statements - which, of course, have then to be checked. The statements made by means of scientific theories are, intuitively speaking, of the following kind: that a given domain of intended applications may actually be subsumed under the theory's principles (laws, constraints, and links). Let's try to be more precise about this. Let us introduce the symbol '$Cn(K)$' for '(theoretical) content of the theory with the (formal) core K'. If we don't accept the distinction between the T-theoretical and the T-non-theoretical level, or if the theory in question has no use for it, then $Cn(K)$ will be defined as follows:

$$Cn(K) =: \mathrm{Po}(M) \cap C \cap \mathrm{Po}(L)$$

If we do employ the T-theoretical/T-non-theoretical distinction, the definition will be:

$$Cn(K) =: r(Po(M) \cap C \cap \mathrm{Po}(L)).^4$$

In both cases, we may write the so-called central empirical claim of the theory as follows: $I \in Cn(K)$.
This formula expresses a statement 'about the world', and this statement may be checked by means independent of K. It may be true or false. Normally, in any 'really existing' theory, it will be strictly false. But this doesn't make the theory useless. The theory may still be useful, either because there is a subclass $I' \subseteq I$ for which '$I' \in Cn(K)$' is true, or because '$I \in Cn(K)$' is strictly speaking false but *approximately true*. (The 'either-or' is not exclusive here.) What the last proviso may actually mean is a question much discussed in structuralist literature and to which one of the essays in the present volume is devoted.

Since the domain of intended applications Idetermines the identity of a theory as much as its formal core K does, we may define a scientific theory T as a pair $\langle K, I\rangle$, which is used to make the empirical claim '$I \in Cn(K)$'. The official label for such a structure $T = \langle K, I\rangle$ is 'a *theory-element*'. This is the simplest unit which can be regarded as a formal explication of a scientific theory in the intuitive sense.

Some 'real-life' examples of scientific theories can actually be reconstructed as *one* theory-element. However, this is true only for the simplest kinds of theories we encounter in scientific literature. More often, single theories in the intuitive sense have to be viewed as aggregates of several (sometimes a great number of) theory-elements. These aggregates are called *'theory-nets'*. This reflects the fact that most scientific theories have laws of very different degrees of generality within the same conceptual setting. We may say that all axioms of a theory are axiomatic but some are more axiomatic than others. A theory is not a 'democratic' sort of entity. Rather, it is a strongly hierarchical system. Usually, there is a single fundamental law 'on the top' of the hierarchy and a vast array of more special laws (and constraints) with different degrees of specialization. Each special law (usually associated with a corresponding constraint and

[4] In principle, r is defined only at the level of M_p and M_{pp}, which is two set-theoretical levels 'lower' than the one needed for this formula. However, r induces the infinite hierarchy of restriction functions on the higher set-theoretical levels in the standard way. In order to simplify the exposition, we use the same symbol 'r' for all of them.

sometimes a special link) determines a new theory-element. What holds together the whole array of laws in the hierarchy is, first, the common conceptual framework, second, the common distinction between the T-theoretical and the T-non-theoretical layer, and third, the fact that they are all specializations of the same fundamental law. In other words, what holds together all theory-elements composing a theory-net is their common M_p and M_{pp} and the fact that all the laws are successive differentiations of one and the same M_0.

To be more precise, let's introduce the relation of *specialization* between theory-elements defined in the following way. T_j is a specialization of T_i (symbolically: $T_j \sigma T_i$) iff:

(1) $M_p^j = M_p^i$

(2) $M_{pp}^j = M_{pp}^i$

(3) $M^j \subseteq M^i$

(4) $C^j \subseteq C^i$

(5) $L^j \subseteq L^i$

(6) $I^j \subseteq I^i$[5]

Clearly, σ is an ordering relation. A theory-net is a finite set of theory-elements ordered by σ.

A theory-net is the standard structuralist concept of a scientific theory considered synchronically. Let us consider it now diachronically, that is, as a genidentical entity. Then, we have to undertake some substantial modifications of the notion of a theory-net. The reason is that a theory in the diachronic sense is not just a theory-net which keeps existing in the same form through history. Normally, theory-nets experience some more or less drastic changes in the course of scientific development even if those changes don't amount to a scientific revolution or 'paradigm-change'. The theory-nets 'evolve'. The resulting entity is called by structuralists 'a *theory-evolution*'. Essentially, a theory-evolution is a sequence of theory-nets fulfilling two conditions, one for the cores and the other for the domains of intended applications. At the level of cores, we require that each new theory-net in the sequence is such that all its theory-elements are specializations of some theory-elements of the previous net. At the level of intended applications, we require that the domains of the new net have

[5] The relationship between A^j and A^i is more complicated. We leave it out of consideration at present. The reader especially interested in this point may consult *Architectonic*, Ch. VII.2.4.

at least some partial overlapping with the domains of the previous net. By these two conditions, a certain degree of continuity is assured for the theory-evolution.

The relation of specialization is, in a sense, a link between two theories (if we want to call 'theories' the particular theory-elements constituting a theory-net). However, single theory-elements within a theory-net are 'theories' only in a rather Pickwickian sense and the same goes, consequently, for specialization as an 'intertheoretical link'. Genuine intertheoretical relations (like theoretization, reduction, equivalence, approximation, and many others with no particular label) are constituted by links between *different* theory-nets, i.e. between sets of theory-elements differing in their respective classes of potential models, M_p. In these cases, we may plausibly say that we are confronted with genuinely different theories (not only with different laws but also with different conceptual structures) which, however, are interrelated in some interesting way.

Given any theory-net and the links its theory-elements have to other theory-nets, we may assume that a plausible (pragmatic) distinction can be established between 'essential' and 'inessential' links. The first kind of links are those that have to be presupposed if the theory-net in question is to be grasped correctly, if it is 'to work' appropriately in its applications. Now, consider a whole group of distinct theory-nets interconnected by essential links. This is, intuitively, a group of theories 'working together' - being essentially co-ordinated. In a sense, this is a theoretical unit of scientific knowledge. The structuralist name for it is: *'theory-holon'*. It is the most complex unit of science detected by the structuralist program so far. The analysis of these structures leads us to deep epistemological questions like the idea of holism, the possibility of absolute non-theoretical concepts, and the foundationalism/coherentism controversy.

Let's summarize the basic notions introduced so far in the following table.

Structuralism's specific notions and notation

M_p : a class of potential models (the theory's conceptual framework);

M : a class of actual models (the theory's empirical laws);

$\langle M_p, M\rangle$: a model-element (the absolutely necessary portion of a theory);

M_{pp} : a class of partial potential models (the theory's relative non-theoretical basis);

C : a class of constraints (conditions connecting different models of one and the same theory);

L : a class of links (conditions connecting models of different theories);

A : a class of admissible blurs (degrees of approximation admitted between different models);

K : a core, $K =: \langle M_p, M, M_{pp}, C, L, A \rangle$ (the formal-theoretical part of a theory);
I : the domain of intended applications ('pieces of the world' to be explained, predicted or technologically manipulated);
T : a theory-element, $T =: \langle K, I \rangle$ (the smallest unit to be regarded as a theory);
σ : the specialization relation between theory-elements;
N : a theory-net (a set of theory-elements ordered by σ- the 'typical' notion of a theory);
E : a theory-evolution (a theory-net 'moving' through historical time);
H : a theory-holon (a complex of theory-nets tied by 'essential' links).

References

Balzer, W. & Moulines, C.U. & Sneed, J.D., 1987, *An Architectonic for Science*, Dordrecht.
Suppes, P., 1957, *Introduction to Logic*, New York.

Chapter 2: Structuralism As Developed Within the Model-Theoretical Approach in the Philosophy of Science

Werner Diederich

The movement of structuralism is embedded in a broader development which is sometimes referred to as the ***semantic approach*** but will here be described as a 'model-theoretical' one (sec.1). In section 2 structuralism will be compared with van Fraassen's conception, the most successful of the other model-theoretical approaches. Finally, in section 3, I am going to sketch a general frame for the reconstruction of basic features of both, structuralism and van Fraassen's approach.

1. *Historical remarks*

There is a certain family resemblance between a number of philosophies of science, including the ones connected with the names of P. Suppes, J.D. Sneed, B.C. van Fraassen, R. Giere, and F. Suppe. In these philosophies, scientific theories tend to be identified with certain set-theoretical structures; this conception may thus be regarded as 'semantic' (Suppe) or *model-theoretical* (as I prefer to say). While the philosophers mentioned above may be regarded as belonging to a certain movement, which essentially started with P. Suppes, there are some more schools or developments which are similar to this movement, though not directly connected with it: various Polish logicians and philosophers of science, e.g. M. Przelecki – who are, however, indirectly linked to P. Suppes *via* A. Tarski and 'West coast' model theory – and the German physicist G. Ludwig and his followers. However, in this paper I am going to confine myself to the 'Suppesian' philosophies mentioned before.

It all started - if you allow this ahistorical phrase - with Patrick Suppes' method of axiomatizing theories by way of set-theoretical predicates, as laid down in the final chapter of his *Introduction to Logic* (1957). Illustrations for this method were taken e.g. from the mathematical theory of order and group theory and elaborated with the examples of probability theory and classical particle mechanics (resting on earlier axiomatizations by J.C.C. McKinsey, A.C. Sugar, P. Suppes and others). This method of axiomatization thus seems to be indifferent against the difference of mathematics and empirical science; later, it was considered as an extension of the 'Bourbaki Program' in mathematics to theories of empirical science.

Suppes' set-theoretical method for axiomatizing theories was elaborated by Joseph D. Sneed in his *The Logical Structure of Mathematical Physics* which appeared in 1971. Since, by Suppes' method, theories in themselves are not statements but certain set-theoretical structures, one basic task to which Sneed turned was to formulate the ('empirical') *claim* connected with a theory. Sneed's answer differs radically from that of logical-empiricist philosophers of science. A straightforward answer would be that the claim connected with a theory is the claim that the entities to which the theory is to be applied actually bear the structure proposed by the theory. This construal, however, would lead to the well-known 'problem of theoretical terms' as soon as such terms are involved in the description of the entities to which the theory is to apply.

Sneed's way out was to relativize the notion of theoreticity to the respective theory itself. The consequence is that the entities to which a theory may be applied have to be described without the help of the theory's theoretical terms. These terms appear only as variables in the *existential statement* that *some* theoretical components together with the non-theoretical ones of the respective entities bear the prescribed structure (cf. sec.2, below).

Since such an existential statement somewhat resembles a theory's claim as it has been construed by Ramsey, it is sometimes called the 'Ramsey-sentence' of the theory. Certain constraints, i.e., interconnections between various applications of a theory, require a considerable refinement of the Ramsey method. The result is that the claim connected with a number of applications of a theory is just one complex sentence, the 'Ramsey-Sneed-sentence', of that theory with respect to those applications. This consequence drawn by Sneed leads to the impression that his position not only is instrumentalistic (in so far as theoretical terms are only used as variables in a theory's claim), but also *holistic* (cf., however, Gähde's contribution to this volume).

A further major result gained by Sneed was a conception of *theory dynamics* that evidently is much more appropriate to the actual history of science than the naive cumulative picture suggested by logical positivists. In fact, Sneed's theory dynamics set the stage for a reconstruction of Th. Kuhn's and other theories of history of science, a reconstruction subsequently elaborated especially by W. Stegmüller and his followers (cf. my other contribution to this volume).

B. C. van Fraassen started, at about the same time as Sneed, with an 'extension of Beth's semantics of physical theories' (1970) which uses the idea of a *state space*. This was incorporated into the model-theoretical or 'constructive empiricist' view put forward in his *The Scientific Image*

(1980). This view comprises a dichotomy of theoretical (non-observable) and observable entities which is analogous to Sneed's theoretical/non-theoretical dichotomy (cf. sec.2, below). But it commits itself to an antirealist (agnostic) position with respect to theoretical entities and, at the same time, an empiricist position with respect to observable entities. Both have no parallels in structuralism as such; i.e. structuralism itself is epistemologically neutral, although structuralists tend to be empiricists as well.

A third 'Suppesian' line of philosophy of science was developed by F. Suppe since his dissertation of 1967. In his recent *The Semantic Conception of Theories and Scientific Realism* (1989), he advocates a 'quasi-realism', couched in semantic terms somewhat different from van Fraassen's (cf. my review article).

Suppe's book is also valuable for his tracing the 'Origins of the Semantic Conception' in the Prologue and in scattered remarks throughout the book. However, his view on the history of the Semantic Conception seems to be a rather personal one. A main distortion which is of concern here is his exclusion of structuralism from the Semantic Conception; the proposed reason is that 'Sneed's solution utilizes correspondence rules...' (p. 10) or 'retains certain explicit correspondence rules in the form of Ramsey sentences' (p. 19, cf. p. 20). I wonder what idea of correspondence rule might have led Suppe to this interpretation, certainly not one according to which a correspondence rule transports meaning from observational to theoretical terms. Sneed's analysis of theoretical terms is no meaning analysis at all! Also, because Sneed relativizes theoreticity (see above), there is no positivism in his approach as Suppe suggests (p. 10 and 19). Of course, Suppe may coin 'semantic conception' and choose his topics of research however he likes, but the way he characterizes the Semantic Conception leaves no ground for the exclusion of structuralism other than by *fiat*.

Like F. Suppe (with his much read *The Structure of Scientific Theories*, 1974, 2nd ed. 1977), R. Giere contributed a lot to the spread of the Semantic Conception by two textbooks, *Understanding Scientific Reasoning*, 1979, and *Explaining Science: A Cognitive Approach*, 1988. Related to his work are applications of the Semantic Conception by John Beatty and Paul Thompson to biological theories (cf. Suppe (1989), p. 17-19).

2. Structuralism and constructive empiricism

The two most successful semantic or model-theoretical approaches are those by van Fraassen ('constructive empiricism') and Sneed ('structuralism'). On the surface, these two approaches are rather different: while van Fraassen pays almost no attention to the formal elaboration of his view (and quite deliberately so), structuralism is pretty formalized; indeed, in many critics' eyes, it is excessively formalized.

Nevertheless, the two approaches (as other model-theoretical ones as well) share at least the view that theories are basically (sets of) *models*, i.e., certain set-theoretically describable structures. Also shared is the logical form of an *existential statement* that the claim connected with a theory takes. In van Fraassen's case, a theory claims about a given phenomenon that it is (isomorphic to) a substructure of one of the theory's models, and it claims about all relevant phenomena that they fit into one and the same model. According to the structuralist view, the basic form of a theory's claim is that a non-theoretical structure (partial potential model of the theory) is extendible to a full model of the theory and that all 'intended applications' together also fulfil certain constraints.

Some remarks in van Fraassen's work suggest that he thinks of models in the sense of first-order semantics comparable to Przelecki's *The Logic of Empirical Theories* (1969). That would make the use of mathematical structures at least very difficult. I take it, however, that van Fraassen would not principally oppose to a Suppes- and Bourbaki-like characterization of a theory's models (if he were willing to spend some thoughts on formalization at all).

The main difference between constructive empiricism and structuralism is *epistemological*. While structuralism is basically neutral in this respect (although a theory's claim is notoriously called its 'empirical claim'), van Fraassen insists on an absolute notion of observability (of entities, not terms). In structuralism only a *pragmatical* order is suggested in the sense that the typical situation a theory's user is conceived to find him- or herself in is that certain non-theoretically described structures are *given*, and the question is whether the theory applies to them in the way described above. Van Fraassen's concept of *observability* is, of course, problematic. He admits that it is theory-dependent, though not leading to a vicious circle, but only to a 'hermeneutic circle': observability is, in the last end, an anthropological fact, and only as such open to revision according to (psychological etc.) theories.

On the other hand, structuralism does need a more careful analysis of how theories relate to phenomena or '*data*'. Suppes' 'Models of Data' (1962) and other investigations (also by Giere, Suppe, and others) may contribute to a fuller understanding of how intertheoretical connections transport empiricalness.

3. Reconstruction

If we conceive of scientific theories as *genidentical* entities - as we should - we have, first of all, to consider temporally changing (typically: growing) domains of intended applications, say, $A(t)$. (I use '$A(t)$' instead of the structuralist '$I(t)$' in order to allow for an interpretation of van Fraassen's terms in the following way.) If we conceive of $A(t)$, as van Fraassen does, as a set of phenomena, the theory's claim is that all $a \in A(t)$ fit into one and the same model of the theory (as isomorphic to as many 'empirical substructures' of that model, whatever that means). If $A(t)$ is enlarged, say to $A(t') \supset A(t)$, for $t' > t$, one may, in order to fulfil the theory's claim for $A(t')$, have to go over to a *different model* that comprises isomorphic pictures of all $a \in A(t')$. (Such a switch is especially to be expected if one allows for approximate embedding or the like.) The structuralist answer to this kind of situation (and to other situations as well) is the introduction of *constraints*. This device covers van Fraassen's needs (as can be shown easily), but it is more general and is desirable also for independent reasons. It should therefore be preferred to van Fraassen's construal, although it is somewhat cumbersome because it involves classes of models (and hardly helps in the popularization of structuralism).

The following more general consideration leads to the conclusion that we should anyhow go over to the higher level of classes of models. If, like before, a theory is successfully applied to a domain $A(t)$ at time t and intended to be applied to a larger domain $A(t') \supset A(t)$ at time $t' > t$, it does not suffice, as we have seen, to check the additional $a \in A(t') - A(t)$ independently from the already established applications in $A(t)$. The new applications must, in van Fraassen's terms, be isomorphically embedded into one model that also covers $A(t)$. This may be true for a certain model which would not work for a differently enlarged A^* (neither including, nor being included in, $A(t')$) and *vice versa*. Thus all conceivable domains A to which actually (platonically) the theory can successfully be applied, may not be ordered chain-like, and their union may not itself be a domain for successful application. Hence, the 'empirical development' of a theory by expansion of its domain of application may not, in general, be conceived of as the exhaustion of one maximum domain of actual applicability, say A_ω. That a theory is applicable to A may thus not be expressed by the inclusion statement $A \subseteq A_\omega$, but only by a clumsier membership statement $A \in /A$, where $/A$ is the class of all domains to which the theory actually is applicable.

One should not be dogmatic about how exactly the class $/A$ is to be construed out of the theory structure S. (Again, I use 'S' instead of the structuralist 'K' for reasons of neutrality.) The most important ingredient of S is, of course, the set of models, say M, of the theory: $S = < M, ... >$. We have seen that something like constraints have to be introduced, expressed by a set C (of one type higher than M) : $S = < M, C, ... >$. In structuralism, we also have to include some device that indicates the distinction between theoretical and non-theoretical components, say, a function r cancelling the theoretical components in models (and in members of members of C). I take $S = < M, C, r >$ as minimal expression of a theory(-element)'s structure. As I read van Fraassen, this covers also his talk of isomorphic embedding of phenomena: there has to be a model m such that all phenomena fit into $r(m)$. $/A(S)$ would be the class of all such sets of (isomorphic pictures of) phenomena. If van Fraassen's 'one-ness condition' ('one model for all phenomena') is construed as a constraint, as suggested above, the claim a theory makes about a domain A can be expressed, like it is in structuralism, as the claim: $A \in /A(S) := r(PotM \cap C)$. This would be the claim of $T := < S, A >$.

In the minimal characterization of a theory's structure, I deliberately have omitted potential models as something not absolutely necessary. (Also, I am not quite content with the usual explication of this concept, cf. *Balzer & Moulines & Sneed* (1987)). If something like potential models nevertheless is wanted, I would suggest the following construal of the set M_p of potential models of a theory as a non-primitive notion. If α is the set-theoretical predicate defining the structure species of M and τ the type of this structure species, then M_p may be conceived of as the class of structures of type τ.

4. Concluding remarks

What can be won by a careful, formal unification of structuralism and constructive empiricism? Maybe that constructive empiricists get convinced that a certain degree of formalization helps to explicate some basic assumptions of their approach, e.g. the 'oneness condition' (s. sec.3, above). On the other hand, structuralists might learn from an interchange with constructive empiricists to be more flexible in formal matters (e.g. to take into account also state space formulations of theories) and to pay more attention to concrete scientific practice as well as to epistemological questions connected with the use of models.

References

Diederich, W., 1994, 'Review of F. Suppe, The Semantic Conception of Theories and Scientific Realism', *Erkenntnis* 41, 421-426.

van Fraassen, B.C., 1970, 'On the Extension of Beth's Semantics of Physical Theories', *Philosophy of Science* 37, 325-334.
van Fraassen, B.C., 1980, *The Scientific Image*, Princeton.
Giere, R., 1991, *Understanding Scientific Reasoning*, New York, (3rd ed. 1991).
Giere, R., 1988, *Explaining Science: A Cognitive Approach*, Chicago.
Nagel, E. & Suppes, P. & Tarski, A. (eds.), 1962, *Logic, Methodology and Philosophy of Science: Proceedings of the 1960 International Congress*, Stanford.
Przelecki, M., 1969, *The Logic of Empirical Theories*, London.
Sneed, J.D., 1971, *The Logical Structure of Mathematical Physics* Dordrecht, (2nd ed. 1979).
Sneed, J.D., 1989, *The Semantic Conception of Theories and Scientific Realism*, Urbana.
Suppe, F. (ed.), 1974, *The Structure of Scientific Theories*, Urbana, (2nd ed. 1977).
Suppes, P., 1957, *Introduction to Logic*, Princeton.
Suppes, P., 1962, 'Models of Data', In: Nagel, E. & Suppes, P. & Tarski, A. (eds.) (1962), 252-261.

Chapter 3: Scientific Explanation

Thomas Bartelborth

1. Introduction

Before suggesting a philosophical theory it seems a good habit to formulate the requirements and goals this theory has to fulfil. My account of explanation and particularly of scientific explanation is mainly designed to accomplish three aims.[1] As an explication of 'scientific explanation' it should, of course, cover most examples of actual explanations given in science. Furthermore, it should not stand as an isolated theory only confronted with examples but has to be embedded into the wider framework of an epistemology assigning explanations a function in justifications. And, finally, the theory should account for the fact that explanations typically further our (scientific) understanding of why the world behaves as it does.

Every approach to scientific explanation is sensible to examples and counterexamples from science, and a significant part of the philosophical discussion about scientific explanation is devoted to examples – which are not always from science. It is undoubtedly a great weakness of any theory of explanation if it has to reject typical explanations from the social sciences, quantum mechanics or other disciplines as non- explanatory to save the philosophical theory from counterexamples. My intention is to develop a liberal conception of explanation that is broad enough to encompass nearly all cases in which one usually talks of explanation in the sense at issue.[2] In this paper, however, I will not engage in a competition developing fanciful examples and counterexamples to certain approaches to explanation, since this is a point I cannot follow up here extensively. Besides, the structuralist formalism I will use has already proven apt to cope with real-life examples.

Under these considerations, embedding our notion of scientific explanation into a wider epistemological framework becomes more important. It also demands the elaboration of a specific epistemological theory; this was done in Bartelborth (1994). There I proposed a coherence theory of epistemic justification which takes up the conceptions of Lehrer (1990), Bonjour

[1] The following account gives an explication of explanation primarily devoted to scientific explanations because they provide the most definite examples, but it is intended to include common sense explanations too.

[2] Of course 'explain this expression to me' and similar senses of explaining will be excluded in this context.

(1985), and Thagard (1992) and adds diachronic and social aspects to a systematic account of coherence, in which most of the local inferential connections responsible for coherence are explanatory relations. This establishes the epistemological background for a theory of explanation supporting unification approaches like the one I will propose in the following. Though it is here not the place to discuss these epistemological matters, it may ease the understanding and acceptance of my account to have an explanatory coherence theory of justification in mind.

The third goal of the theory of explanation is to suggest a certain conception of scientific understanding. Hempel (1965) already demonstrated that scientific understanding cannot consist in a reduction of the explanandum to something more familiar since many explanations in science explain everyday phenomena, e.g. the shining of the sun, in terms of the much more unfamiliar, e.g. quantum mechanics. A more promising proposal due to Lambert (1988) describes the understanding of a fact E as showing how it fits into a theory T. As a generalization and first approximation I suggest to view scientific understanding as a coherent embedding of a fact into our belief-system. And, of course, this system and the special position where we want to embed the fact must not always be more familiar than the fact. Nevertheless, due to a coherentist epistemology, the whole system is epistemologically more basic than the single fact and our embedding of it furthers our understanding by giving it a place in our model of the world.

As a formal apparatus for analyzing the exact relationship of an explaining theory to the explanandum it is most convenient to use the Structuralist View of theories, since this frame allows for an elucidation, among other things, of phenomena of approximation and theoretical embedding, which are salient features of many scientific explanations. Furthermore, formulating a philosophical theory of explanation in a semantic framework yields an answer to a requirement that has arisen in the course of the debate about scientific explanations, i.e. to specify the contributions of different components of theories to their explanatory force. For instance, the distinction between empirical and theoretical terms in scientific theories, which Hempel regards to be of epistemological importance, plays no role in his classical DN-Schema of explanation. The Structuralist View of the inner structure of theories is much richer and identifies more components of theories and describes how they function in dealing with empirical data. Moreover, characterizing explanations as a certain kind of embedding avoids the so called 'deductive chauvinism' which coerces every explanation in the Procrustean bed of a logical deduction. Many scientific explanations don't have the form of deductive inferences, and in numerous cases we cannot hope to improve them to become valid logical inferences with true premisses. If we believe these cases to be explanations, even though not ideal ones, then we should abjure deductive chauvinism.

2. Approaches to Explanation

In a first step I will give a short evaluation of three prominent strategies in the theory of explanation, with a certain emphasis on my favorite account. A common starting point for nearly all studies of explanation is Hempel's classical DN-Schema characterizing explanation as an argument that has to fulfill some additional requirements. The explanans has to contain an empirical law and the explanandum sentence must be logically deducible from the explanans. For this requirement of Hempel to be effective, one has to specify which sentences are laws, which in spite of all progresses in this area is an open problem in philosophy of science. On the other hand, in science we find many explanations not containing laws. This indicates that the law condition is not as necessary for explaining as Hempel believes.

For Hempel, explanations without laws are not yet complete, but even if he is right, they have in their incomplete formulation a certain explanatory force, which a theory of explanation has to take into account. The DN-scheme, however, cannot deal with these cases, because it permits no degrees of more or less good explanations but only an all or nothing decision. In view of other well-known problems this approach has -such as the asymmetry of explanation or the irrelevant constituents-, we must conclude that Hempel's account is deeply unsatisfactory, and one has to look for alternatives. The DN-scheme can certainly be improved by a suitable pragmatic theory à la van Fraassen (1980) but as Kitcher and Salmon (1987) have proven, the pragmatic aspects of explanations are not sufficient in order to illuminate what makes actual explanations explanatory.

A. The causal approach

An obvious proposal may be found in the causal theories of explanation, according to which mentioning the cause of an event E is an explanation of E. It was elaborated by Salmon (1984 and 1989) and others especially because of its ability to account for the known asymmetry of explanations in the most natural way. But Salmon's theory is confronted with the persistent difficulty to explicate the notion of causality in a way that, on the one hand, it has a definite content and does not boil down to empty talk, but on the other hand remains liberal enough to include most scientific explanations. I doubt whether the causal approach can meet both challenges at once, but I can substantiate this skepticism here only by a few hints. Salmon developed his theory on the background of the special theory of relativity, probably because this conception of causality comes next to our common sense understanding of it. Thus, it is not particularly astonishing that, notwithstanding its statistical formulation, it is inapplicable to explanations in quantum mechanics (cf. van Fraassen 1991, p. 81ff). Since quantum mechanics is not just one theory among others but the best theory we ever had –a view shared by Salmon–, we have a significant restriction in the scope of Salmon's theory. If, on the other side, we give up the common cause principle Salmon employed in his account of causal connections, we have no plausible, substantial idea of statistical causality to offer.

This is not surprising because our understanding of causality already vacillates considerably within the physical sciences. It varies from classical mechanics, which allows for action at a distance and is locally deterministic but globally indeterministic, through the theories of relativity banning action at a distance up to the genuine statistical quantum theories (cf. Earman 1986). These variations involve the risk that we end up with no common properties of causality and, at last, with a deflationary theory of causality à la van Fraassen (1980, p. 124) :

the causal net = whatever structure of relations science describes

After all, even if we can choose a reasonable conception of causality we are still far away from having an explanatory model, since, for example, the 'Big Bang' belongs to the causal prehistory of every event, but we would not be satisfied with explanations of the form: 'all that comes from the Big Bang' (Lipton 1991, p. 34ff). For an explanation, other significant conditions than to be a cause for something have to be fulfilled, going beyond the causal story.

Conversely, many explanations in science don't mention any causes at all, but rather structural relations. Kitcher (1989, p. 426) refers to a 'party trick' in which someone knots a cord around a pair of scissors in such a way that it immediately becomes undone if we draw the cord at the right point but making not the right initial twist one can struggle for hours without getting anywhere. To explain success or failure in this case, we need to know the topological features of the situation and not a story about the causal peculiarities of the cord and our drawing it. Further, non-causal explanations are to be found in explanations by laws of coexistence, as for the pendulum law or Pauli's exclusion principle, and in classificatory explanations (Achinstein 1983, p. 233ff), answering questions like: Why has iron the number 26 in the periodic system? In particular, the account of theoretical explanation of laws, which according to Friedman is a central task for every philosophical theory of explanation, does not fit well into a causal approach —to say the least.

Causal theories of explanation are additionally burdened with some inner difficulties, which seems reason enough to look for another conception. The most promising one, to my mind, is the unification approach which considers the systematization and unification provided by theories as key concepts for their explanatory power. For they do not depend on critical concepts such as 'causation' and have a great potential to distinguish explanations from pseudo-explanations. Two unification theories, Friedman's and Kitcher's, shall be used as guide-lines to this type of metatheory.

B. The unification approach

The first precise proposal towards an elaboration of the concept of unification comes from Michael Friedman (1974). For Friedman, unification of our knowledge primarily consists in a reduction of a multitude of phenomena to just a few laws. His example is the kinetic theory of gases, which can, for example, explain that gases approximatively obey the Boyle-Charles law. If this were the only achievement of the kinetic theory, one might indeed ask why we introduced it at all and are not content with e.g. just the Boyle-Charles law. Requirements of theoretical or ontological parsimony would imply this moderate strategy anyway. Friedman's response is clear: The kinetic theory of gases explains, in addition to the Boyle-Charles law, a whole range of other phenomena like those described by Graham's law of diffusion, and the phenomena of specific heat capacity. Besides, it establishes new connections, which would not have been detected without the kinetic theory, e.g., between the behavior of gases and various mechanical phenomena that we already know. It, therefore, definitely increases the coherence of our belief system. According to Friedman, this is the decisive step for a scientific unification. The theory has reduced a multitude of unexplained and independent phenomena to one. However, Friedman's conception remains troublesome as long as we have no convincing method of counting singular phenomena. For this purpose, Friedman introduced atomic sentences as linguistic counterparts of individual phenomena. Atomic sentences are independently confirmable sentences. Nevertheless Kitcher (1976) showed that Friedman's explication is inadequate since, on his account, only atomic sentences would explain, though in real life explanations we often have to invoke conjunctions of several laws to yield the desired explanatory power. Salmon (1989, p. 94ff) studied this deficit of Friedman's theory in the case of the Newtonian theory of gravitation which, if we follow Salmon, is a conjunction of three independent partial laws for three different ranges of applications: first, for the interaction of heavy bodies, secondly, for the interaction of heavy bodies with light ones, and thirdly for the interaction of light bodies.

To my mind, Salmon's example intuitively is not really convincing as a counterexample to Friedman's idea. It primarily shows that Friedman has to strengthen his conditions of independence of a law and to take structuralist constraints into account. In doing so, he might argue that the three partial laws actually are not independent, because the heavy bodies governed by the first and second law as well as the light bodies governed by the second and third law should be given the same masses in both laws, respectively, which is expressed by an identity constraint in the structuralist sense. How significant this linking of the applications of a theory is already becomes obvious from Salmon's own example. Cavendish was the first who, in 1798, succeeded in measuring the gravitational constant for the case of smaller bodies. Since scientists believed in a universal law of gravitation with a universal constant, applying to heavy bodies in the same way as to light ones, they could determine the mass of the earth by its attractional force on smaller bodies —Cavendish came to 6,6 10^{21} tons which already is very close to today's value of 5,98 10^{21} tons. This constant is further used in the calculation of the masses of other planets etc. All these applications of the gravitational law are not really independent but connected by identity constraints, which are important ingredients of the content of theories. Friedman, I think, could have taken these connections for granted; at any rate, an adequate metatheoretical description of theories should take them explicitly into account to avoid disturbing examples like Salmon's division of Newton's law in three detached 'laws'. In a complete formulation of Newton's original law, we acknowledge a manifest surplus content vis-à-vis its splitted-up version.[3] I will come back to this point in the context of my explication of the organic unity of theories.

In his discussion of Friedman's proposal, Kitcher tries to find his own way to unification, which he had already described in 1976:

"What is much more striking than the relation between these numbers is the fact that Newton's laws of motion are used again and again and that they are always supplemented by laws of the same types, to wit, laws specifying force distributions, mass distributions, initial velocity distributions, etc. Hence the unification achieved by Newtonian theory seems to consist not in the replacement of a large number of independent laws by a smaller number, but in the repeated use of a small number of types of law which relate a large class of apparently diverse phenomena to a few fundamental magnitudes and properties. Each explanation embodies a similar pattern: from the laws governing the fundamental magnitudes and properties together with laws that specify those magnitudes and proper-

[3] Therefore, it meets Watkin's condition of 'organic fertility' discussed below.

ties for a class of systems, we derive the laws that apply to systems of that class." (Kitcher 1976, p. 212)

Kitcher (1981 and 1989) agrees with Hempel to consider explanations as deductions from laws, but, according to him, their explanatory power is correlated to the number of types of arguments they instantiate. That is, the more similar the deductions the more explanatory they are, whereby he measures similarity of arguments by his so-called argument patterns. The aim of Kitcher's unifying account is to generate all the arguments we use in science by as few argument patterns as possible.[4] This is not the place to examine Kitcher's approach in full but I want to mention en passant some difficulties he has to deal with. At first, his examples of unification by argument patterns reveal that we possess clear patterns only if we narrow the range of applications of a theory to simple subclasses. In the case of Newtonian particle mechanics these are one-particle systems, the most elementary systems of Newton's theory. In more general classes, one can only identify 'core patterns' which the argumentations have in common, but they cannot anymore be seen as plain instantiations of the same pattern. Thus, it becomes more difficult to estimate their similarity by argument patterns. Furthermore, Kitcher's commitment to a syntactic, deductivist approach precludes an adequate treatment of statistical explanations and of the approximations we find in quantitative explanations.

But the most serious problem for the pattern approach is the possibility of 'spurious unifications'. We can give a lot of simple schemes for argumentation allowing to derive all or nearly all sentences describing the intended phenomena, for example:

$$\frac{A}{A}$$

These trivial argument patterns are not excluded by the aim of unification alone. We have to add other requirements for the derivation of sentences describing phenomena, namely that it be informative. This is an important demand for all explanations, but Kitcher has not much to say about it. He speaks of the 'stringency' of argument patterns but he can only exclude some simple cases of patterns allowing the derivation of all possible sentences. Stringency is an important idea of Kitcher's account though he has not really substantiated it. The gradual concepts of theoretical and empirical content of theories, as described in the Structuralist

[4] Kitcher follows in (1986) Kant's idea that this procedure is even constitutive for our empirical knowledge.

View, are much more adequate to measure the informativeness of theories. Hence, I will explicate some of Friedman's and Kitcher's ideas on unification within the structuralist framework. The result will be a semantic theory of scientific explanation.

3. Explanation as Embedding

In order to give more content to the slogan that explanations are essentially unifications of our knowledge we should conceive them as embeddings of an explanandum E in a model M. At first, we take 'model' in an informal sense meaning a representation of something. In this sense it includes such mechanical or analogical models as described by Hesse (1963) as well as theoretical models and even the 'simulacra' of Cartwright (1983). This liberal understanding of 'model' can illuminate a corresponding general understanding of 'explanation' embracing common sense explanations as well as scientific explanations. With respect to the latter, I will use structuralism's so-called partial models as explananda and the actual models as embedding models.

A. Unification from a Structuralist Point of View

Some advantages of the structuralist approach are obvious. Models or sets of models need not necessarily be described by laws. They can be stories about a course of events with a narrative character or also 'world' models, e.g. for the world climate, which contain mainly special data and low level generalizations. In this way, the virulent problems of Hempel's deductive-nomological account to give an explication of lawlikeness are avoided. Besides, the embedding approach coincides with the account of scientific understanding presented above, because the embedding of E in a model of an accepted theory demonstrates how E fits coherently into our knowledge. Some further features of the embedding approach –for instance, how it can deal with the asymmetry of explanation– are already laid out in Forge (1980) and (1985), where Forge worked out his 'instance view' of explanation. This can be seen as a precursor of the embedding conception in a structuralist framework. Instead of embedding, Forge requires that E is shown to be an instance of a law or a theory.

Though I think that every explanation can easily be regarded as an embedding, not every instance of embedding is explanatory, of course. The account has to be strengthened by some requirements guaranteeing that the embedding is informative, and the Structuralist View provides the necessary tools to formulate such additional conditions. In a first step, we can observe that the embedding takes place in a sequence of steps.

The initial one is a sort of conceptualization in which we embed the data measured into a partial model. For example, since 1678, Edmond Halley had collected data about comets and noticed similarities between their orbits and regularities in their recurrence, say, every 75 years. On this basis, he assumed that they move on elliptical trajectories, which could be determined more precisely by his data. By this hypothesis, Halley established a basic connection between the orbits of comets and other heavenly bodies and described the motion of comets by a partial model of mechanics. The return of Halleys comet in 1758 provided evidence for his assumption and was a first stage in a deeper understanding of comets.

On a second level, we look for theoretical embedding, i.e. for a model of mechanics with masses and forces explaining the orbits in accordance with Newton's laws of motion. This is the most significant step with respect to scientific explanation. As Friedman puts it:

"A typical, and striking, feature of advanced sciences is the procedure of theoretical explanation: the derivation of the properties of a relatively concrete and observable phenomenon by means of an embedding of that phenomenon into some larger, relatively abstract and unobservable theoretical structure." (Friedman 1981, p. 1)

Kepler's answer to the question why a particular planet was at a certain place at a certain time, viz. that it was moving on an elliptical trajectory and that the places in question were on this trajectory, was a sort of weak explanation. A deeper and causal explanation is provided by Newton's theory of mechanics and, particularly, of gravitation. Friedman and Kitcher established why this is a better explanation. It is because of the unifying power of Newton's theory combined with its stringency and informativeness. But Newton's theory can be informative only by way of its special laws, like the law of gravitation in the case just considered. On the theoretical level, the embedding is realized in different ways. It is realized in theory-nets not only by the basic theory-element but also by the appropriate specializations to which the explananda fit coherently. Additionally, we have to consider the embedding into constraints and the other theory components. In the following, I want to examine how the theory components cooperate with respect to their unifying power.

A first point to mention is that the structuralist form of embedding not into one particular model but into any model out of a given set of models will be unifying only if the set shows some uniformity. Structuralists require that

($\star$) the set of potential models belong to a structure species

in the sense of Bourbaki. The models of a structure species have the same structure and the same base sets: A set of relational structures $M =$

$\{<\mathbf{D},\mathbf{R}>;$ with $<\mathbf{D},\mathbf{R}>=<D_1,\ldots,D_n,R_1,\ldots,R_k>\}$ belongs to a structure species if there exists a complex structure type $\kappa=<k_1,\ldots,k_k>$ such that all R_i belong to type κ_i.

In fact, ($\star$) contributes to unification by theories, since ($\star$) entails that all potential models in a theory-net are homogeneous descriptions of physical systems. In particular, part-whole relations and species-genus relations are identical for all intended applications of the theory. That is a typical feature of normal science. If we speak with Kuhn (1983) of a conceptual net which is subject to some transformations due to theoretical changes, we have to distinguish between revolutionary conceptual revisions resulting in a scientific revolution and small everyday changes —normal evolutions of our concepts. Thagard (1992, 30ff) has proposed to consider shifts in part-whole relations and species-genus relations as a main criterion for conceptual revolutions; this seems intuitively appealing, because such relations decide about the ontological categories and relations a theory assumes. Furthermore, his view is supported by the results of many case studies from the history of science. Therefore, the conceptual embedding into a structure species constitutes a first essential kind of unification for empirical systems.

B. Successive unification by specializations

A severe restriction of the content of a theory leading to a particular increase in the stringency of unification is to be found in the laws of a theory restricting the set of its models. The embedding of certain partial models into the models of a theory corresponds to Friedman's requirement to subsume phenomena under laws. The more intended applications can be embedded by a theory the greater will be its unifying power. However, the embedding by theory-nets takes place not in one big step but in a succession of steps corresponding to several specializations of the net, each one only concerning a certain subclass of the intended applications of the whole theory. All intended applications of Newtonian mechanics, for example, obey to Newton's second law of motion —more or less exactly —, but only a section of it can be subsumed under Hooke's law or under the law of gravitation etc. First of all, the partial models are embedded in the basic theory-element. But in most cases that hardly proves to be a really stringent embedding, as case studies on different theories have shown: In many cases, the basic theory-element considered alone has an empty empirical content. A substantial empirical assertion will be supplied primarily by subsumption of partial models under the special laws which an advanced theory contains. A theory-net can be extended even into the area of technical applications of a theory. For example, Hooke's law can be specialized to certain types of material determining the spring constant or the elasticity of a deformed solid in a specific manner. By the process of successive embedding into the theory-elements of a net we gain more and more insights about the explanandum.

Any theory should yield a systematization of a lot of phenomena of different types but, at the same time, also provide valuable insights on these phenomena. These two requirements are in a certain tension to each other, so that we have to weigh them up. The conception of theory-nets permits to meet both requirements in different ways on the various levels of the net. Towards the basic element we observe an increasing systematization, i.e. more phenomena can be embedded into single theory-elements, and in the other direction the net shows increasingly more information about specific phenomena. The conceptual structure of the potential models and the basic axioms of the theory-net constitute a general frame for all the information supplemented by the constraints and links of the net. All information and systematizations contained in a net enter into its empirical assertion; this will be the focus of our attention in the rest of this essay. For this purpose, I am now going to specify the contributions of the other theory components to the unification effects of a theory.

C. Unification by consistency requirements

In addition to the laws of a theory, which are essential to the stringency of its theoretical unification, the innertheoretical cross connections between potential models, i.e. the constraints, are important factors of systematization by theories. They bind together several potential models which exhibit coherent assignments of values to certain magnitudes. Therefore, constraints have a unifying effect on the values of a theory's concepts while the conditions determining potential models uniformize the conceptual structures in the net. Often, we find identity or invariance constraints requiring, for example, that the mass of a particle should be equal if the same particle appears in different models. This constraint guarantees that 'mass' denotes a genuine and objective property of particles and indicates that we refer to the same concept in different models. Gähde (1989) has examined the import of constraints for the empirical content of theory-nets more closely and has provided case studies from the history of science which demonstrate their significance in actual scientific practice. Constraints assure that theories are not just disjunctions of singular laws, in which case they would disintegrate into independent subsystems and could not be regarded as the organic units they are supposed to be.[5]

[5]This point will be discussed in more detail in connection with the empirical claim of theory-nets.

This point can be illustrated by means of the conception of local models employed in the Structuralist View. Local models only represent single applications of the theory, i.e., in general, they have a limited space-time area and include only some aspects of the world. For example, a mechanical model may represent only a few particles and their masses, distances and some forces on them but not their colors and shapes, their meaning to human beings etc. Because of this feature, local models are able to describe particular experimental results and particular observations. They are the smallest units theories speak about, and structuralists prefer them to the global models used, e.g., by van Fraassen in his state-space approach, since they allow a more detailed metatheoretical account of the interaction between theories and specific applications. The effect of constraints can also be described as follows: the local models fulfilling a constraint can be embedded collectively into a global model of the theory.[6] This means that they imply a further step of unification by embedding.

Links play a unifying role between different theory-elements. They establish that certain magnitudes do not only bear the same names but in fact are the same in different theories or different specializations of the same theory-net. The term 'pressure', e.g., refers to the same magnitude in thermodynamics and in hydromechanics, and we have to describe this fact in an adequate reconstruction of the respective theories and their relation to each other since it has a significant effect on their empirical contents. Furthermore, links can bind concepts with different names, as, e.g., temperature and the kinetic energies of the molecular particles of a gas. Theories interconnected by links result in greater holistic complexes – structuralists sometimes call them 'theory-holons' – unifying scientific knowledge on a still higher level and so assuring that our theories can be embedded into a coherent picture of the world.

D. Unification of phenomena

Any talk of unifying theories has to mention as well the subsumption of phenomena under theories, which was Friedman's main topic. The degree of unification directly depends on the number of phenomena and perhaps on the number of intended applications that can be embedded. But how can we divide the set I of intended applications into subsets representing types of applications or phenomena?[7] For the intended applications

[6] For an example of how the distinction between local and global models works, see Bartelborth (1993), where this distinction is examined for the case of the general theory of relativity.

[7] I am mainly following Bacon and others in preferring instantial variety for inductive support, while Carnap gave more importance to instantial multiplicity (cf. e.g. Cohen 1989), but it is of course possible to extend or modify the approach to include the

of Newtonian mechanics we get a natural division of the following form: systems of planets and other heavenly bodies, pendulums, systems of colliding particles,... or finer divisions in a similar manner. Much more trying is the question of how to state precise criteria for the determination of types of applications, which is connected with the question of which sets represent natural kinds and which only arbitrary collections of things.

A difficult task, which I will not tackle in this paper, is to give necessary and sufficient conditions for membership in a natural kind; however, one can provide reasonable sufficient conditions in certain special cases for belonging to separate natural kinds on the basis of an existing theory-net $N = (T_i)_{i \in J}$. The potential models x and y from T_0 are of different kinds with respect to a theory-net N, if there exists a theory-element T_i in N with: $x \in I(T_i)$ and $y \notin I(T_i)$.

No matter how we get a division in natural kinds, at least we have an intuitive partition of the intended applications of each theory-element T_j in most actual examples from science: $I(T_i) = \cup_j P_{i,j}$

where $P_{i,j}$ represent the phenomena considered. With the coarsest partition $\Pi(N)$ including all the partitions $P_{i,j}$ of the net we obtain the set of all phenomena the net is supposed to explain. Let's say N unifies a phenomena set $\mathbf{P} \subseteq \Pi(N)$, if all the phenomena $P_j \in \mathbf{P}$ can be embedded completely and collectively into the theoretical content of the net.[8]

E. Empirical claim and organic unity

In order to be unificatory and thereby explanatory, theories not only have to be more than mere conjunctions of singular statements but also more than just arbitrary conjunctions of laws. The laws must be formulated in a common language and have a common range of applications, which should be interconnected in such a way that it does not admit decomposition into entirely independent sets of applications. Actually, accepted theories exhibit a certain organic unity that arbitrary sets of propositions lack. Though many philosophers of science have seen this point, only few have proposed an explication of the concept '(organic) unity' of theories.

One interesting but meanwhile abandoned suggestion was Friedman's, already discussed above. A similar, more promising one, stems from Watkins (1984, p. 203ff) who talks about the 'organic fertility' of theories. According to Watkins, theories should have a synergistic surplus content in addition to the conjunction of their parts. A theory T does not meet his requirement of organic fertility if there is a division of $T's$ axioms into two sets T' and T'' such that:

second form of induction.

[8] How the theoretical content of a theory-net can be specified will be discussed below.

$$(\star) \qquad CT(T) = CT(T') \cup CT(T'')$$

where '$CT(T)$' denotes the content of T, i.e. the set of singular propositions derivable from T.[9] In the case of $(\star), T$ is only a conjunction of two distinct theories and we have no reason not to work with T' and T'' separately. Watkin's account, however, as every purely syntactical approach, leads to some problems concerning the possibility of a reformulation of theories which he wants to solve by means of five rules for natural formulations of axiom systems (Watkins 1984, p. 208f). But he is not sure to have succeeded in this strategy.

A more natural way to express an organic unity requirement is offered by the Structuralist View. Ulrich Gähde has already done important work on the decompositions of theories and their empirical claims; I will rely on it in the following. To obtain a precise formulation of organic unity and other conditions of unification, we first have to determine the empirical claim made by a theory. In the case of theory-elements there is a simple standard form, but for theory-nets we need a new auxiliary concept. For a theory-element $T =< M_p, M, M_{pp}, C, L, I, A >$ with potential models, actual models, partial models, a General Constraint, a General Link, the intended applications, and the admitted blurredness, respectively, we obtain as its theoretical content:

$$Cn_{th}(T) := Po(M) \cap C \cap Po(L)$$

and as its projection, the empirical content, on the T-nontheoretical level: $Cn = r[Cn_{th}(T)]$, with the restriction function r.
Then, we may formulate the empirical claim of T :

$$I \in Cn(T).$$

To state the empirical claim for a whole theory-net $N = (T_i)_{i \in J}$ is much more complicated because we have to take into account the conditions of consistency between different specializations in the net. The first author to have provided an adequate formulation of it is Ulrich Gähde (1989, 66ff). In the following, I will employ his proposal with a slightly different notation making the claim easier to read, and more apt for the following analysis. For this purpose, I introduce the class of embedding functions $\mathbf{EMB}(T) := \{e : M_{pp} \to M_p;$ with $r \circ e = id_{M_{pp}}\}$, i.e. functions respecting the bundle structure of the fibre bundle (M_p, r, M_{pp}). Thus the embedding functions assign to every partial model x a potential model y such that $r(y) = x$. Now we reformulate the empirical claim in a form which can be

[9] This is comparable to the empirical content in the structuralist view, but be aware of the differences between a syntactic and a semantic approach.

extended naturally to theory-nets:

Empirical Claim of a Theory-Element T:
$\exists e \in \mathbf{EMB}(T)$ with: $e(I) \in Cn_{th}$.

The empirical claim of a theory T is true if we can find an embedding or theoretical extension for all intended applications simultaneously that lies in the theoretical content of T. For two theory-elements T_0 and T_1 Gähde (1989, 66ff) formulates the following empirical claim:

$$\exists X_0 \exists X_1 [r(X_0) = I_0 \wedge X_0 \in Cn_{th}(T_0) \wedge r(X_1) = I_1 \wedge X_1 \in Cn_{th}(T_1) \wedge X_1 \subseteq X_0]$$

It is necessary to explicitly mention the two extension sets X_0 and X_1 to guarantee that models contained in both sets I_0 and I_1 be extended to the same actual models of T. This condition is a consistency requirement for the extensions. By means of the embedding functions we get the reformulation:

$$\exists e \in \mathbf{EMB}(N)[e(I_0) \in Cn_{th}(T_0) \wedge e(I_1) \in Cn_{th}(T_1)].$$

In the case of a general theory-net $N = (T_i)_{i \in J}$ we find:

(EC) Empirical Claim of a Theory-Net N:
$$\exists e \in \mathbf{EMB}(T_0) \forall j \in J[e(I_j) \in Cn_{th}(T_j)].$$

In (EC) we are looking for an embedding function e that simultaneously assigns to every set of intended applications I_j of a specialization T_j an element of the theoretical content of T_j. It shows how every specialization step of the basic theory-element T_0 strengthens the empirical claim by supplying a new requirement on the embedding function. Therefore, the function e represents all instances of embeddings in N on all levels of specialization.

As a last step we can take approximations into account:

(AEC) Approximative Empirical Claim of N :
$$\exists e \in \mathbf{EMB}(T_0) \forall j \in J[e(I_j) \sim\in\sim CN_{th}(T_j)],$$

with '$\sim\in\sim$' as a sign for approximative membership in the sense of the class of admitted blurs A.

These are the means by which we can give the requirement of organic unity of a theory $T =< K, I >$ a precise form à la Watkins. If we represent the empirical claim that the core K of the theory is applicable to the intended applications I by: $\mathbf{CLAIM}(K, I)$, we then want a theory T to have as few decompositions of the form

$$\exists K'K''I'I'' with : \mathbf{CLAIM}(K, I) \leftrightarrow \mathbf{CLAIM}(K', I') \wedge \mathbf{CLAIM}(K'', I'')$$

as possible. A special type of such decompositions, namely:

$\exists I'I'' with : \mathbf{CLAIM}(K, I) \leftrightarrow \mathbf{CLAIM}(K, I') \vee \mathbf{CLAIM}(K, I'')$

was studied by Gähde (1989, 130ff). He obtained the following result: If the set I of intended applications can be decomposed into two constraint independent sets X and Y with $I = X \cup Y$ then the empirical claim is correspondingly decomposable. The concept of constraint independence is defined as follows:

$I, Y \subseteq M_{pp}(T)$ disjunct and not empty are called constraint independent iff: $\forall e \in \mathbf{EMB}(T) : [e(X) \in C(T) \wedge e(Y) \in C(T)] \leftrightarrow e(X) \cup e(Y) \in C(T)$.

This is the situation Salmon had in mind in his above-mentioned counter-example against Friedman's theory; but actually the example doesn't hold in the case of Newton's theory of gravitation because the different sets of bodies are connected at least by the identity constraint for the mass function. Thus, the consistency of a subset of applications cannot guarantee the coherence of their union. Now we could easily prove Gähde's theorem:

If there exists a disjunct decomposition of I in constraint independent sets X and Y, then it holds:

$$(\star) \quad \mathbf{CLAIM}(K, I) \leftrightarrow \mathbf{CLAIM}(K, X) \wedge \mathbf{CLAIM}(K, Y)$$

And this result can be conferred to theory-nets N where a similar theorem holds. The independence property is now given by:

$X, Y \subseteq M_{pp}(T_0)$ disjunct and not empty are called constraint independent with respect to the net $N = (T_i)_{i \in J}$ iff: $\forall e \in \mathbf{EMB}(T) \forall i \in J$: $[e(X) \in C(T_i) \wedge e(Y) \in C(T_i)] \leftrightarrow e(X) \cup e(Y) \in C(T_i)$.

The condition of constraint independence exhibits in an intuitive way how theories may be decomposable into independent subtheories thereby gradually loosing their unifying power. In such a case, the composed theory may just be a conjunction of two separate parts. Only constraints connecting different areas of application of a theory in a coherent way can make sure that there is one global model for all of them. So, decomposable theories only require the existence of global models for certain subsets of I and therefore have a smaller unifying effect.

F. Dimensions of Unification

Now we are in a position to examine the dimensions of unification that make out a good explanation. Unifying theories – if we first think of a single theory-element T – should be able to embed as many phenomena as possible into models as informative as possible, and they should unify the phenomena not by binding them together in form of a conjunction but in systematizing them in an organic theory. The parameter for the informativeness of a theory is its theoretical content:

1) $Cn_{th}(T) := Po(M) \cap C \cap Po(L)$

The smaller this set, the more informative the embedding in T since a small $Cn_{th(T)}$ excludes more situations than a large one. The parameter for the systematization force of T is the number of phenomena or the range of phenomena which can be embedded simultaneously:

2) $\mathbf{RP}(T) := \{Q \subseteq \prod(T) : \cup Q \in Cn_{th}(T)\}$

And the third parameter is a measure for the organic unity of T. It yields the set of all conjunctive decompositions of T.

3) $\mathbf{CD}(T) := \{Q : Q \text{ is a conjunctive decomposition of } T\}$

By a conjunctive decomposition of T I mean a set of empirically non-empty theory-elements $(T_i)_{i \in D}$ (with no superfluous elements) such that:

$$\mathbf{CLAIM}(K, I) \leftrightarrow \wedge_{i \in D} \mathbf{CLAIM}(K_i, I_i)$$

We can also specify what is meant by an empirically non-empty theory-element: $\exists Q \subseteq M_{pp}(T)$ with: $Q \notin Cn_{th}(T)$
It is a theory that does not admit a trivial embedding, i.e. one in which every set of possible applications can be embedded into the theoretical content of the theory.

Now we have the three most important parameters for the unifying power of a theory-element T. They allow for a comparison between two theories T and T' with respect to their informativeness and systematization effect, assuming the two theories to have a common non-theoretical level of description for the systems they are applied to, at least with respect to the set-theoretical typifications of their concepts. For example, it is possible to compare relativistic theories with pre-relativistic ones, since both can be formulated within a general framework of space-time with four-dimensional magnitudes, as was done in Havas (1964) or Bartelborth (1987). On the whole, T' yields a stronger embedding than T – and has more explanatory power –, if the following three conditions are met:

a) $Cn_{th}(T') \subseteq Cn_{th}(T)$ (T' has more Theoretical Content than T)

b) $\mathbf{RP}(T) \subseteq \mathbf{RP}(T')$ (T' unifies a greater Range of Phenomena than T)

c) $\mathbf{CD}(T') \subseteq \mathbf{CD}(T)$
(T' admits fewer Conjunctive Decompositions than T)

Of course, this is only a partial ordering since T may be better than T' with respect to the first two parameters while T' is better with respect to the third. For such cases, I dont't know any general rule deciding which theory to prefer, but, nevertheless, the three parameters seem to be reasonable clues for discussing which theory is more unificatory.

This is only a first sketch of the way how to estimate the quality of an embedding. Many reasonable improvements suggest themselves. We could, for instance, try to correlate the 'distance' between types of intended applications with the unifying power of T or compare such types by means of the probability of their occurrence with respect to our background knowledge etc. By means of the proposed parameters one could evaluate particular embedding functions in an obvious way; but I will now go on by transferring the present results to theory-nets and not follow the other suggestions any further in this essay.

In theory-nets there is no simple theoretical and empirical content for the whole net but a content for each specialization separately. Therefore, the formal expenditure to establish parameters for the unifying power in theory-nets is greater than for theory-elements. For a theory-net $N = (T_i)_{i \in J}$ with the index set $J := \{1, \ldots, n\}$ we have to introduce a content-function **CF** to specify its theoretical content:

1) $\mathbf{CF} : [Po(M_p)]^n \rightarrow Po(\mathbf{EMB}(N))$,

which is defined for $< Q_1, \ldots, Q_n > \in [Po(M_p)]^n$ by:

$$\mathbf{CF}(< Q_1, \ldots, Q_n >) := \{e \in \mathbf{EMB}(N); \forall i \in J : e(\cup Q_i) \in Cn_{th}(T_i)\}.$$

CF assigns to every n-tuple of possible intended applications $< Q_1, \ldots, Q_n >$ the set of all functions embedding the sets of Q_i into the theoretical content of the corresponding theory-element T_i. This property characterizes **CF** as an adequate parameter for the strength of the whole net. If **CF** assigns very large sets to n-tuples $< Q_1, \ldots, Q_n >$ of types of intended applications then it is less informative to embed them than in cases where **CF** only allows for one or a few embedding functions. The worst case for a theory is

$$\forall < Q_1, \ldots, Q_n > \ \mathbf{CF}(< Q_1, \ldots, Q_n >) = \mathbf{EMB}(N).$$

In this case, every specialization of the net has the trivial theoretical content $M_p(T)$. To provide informative explanations, even $\mathbf{CF}(< I_1, \ldots, I_n >)$ should not be equal to the set of all embedding functions of the net. The empirical claim of N is now specified by:

$$\exists e \in \mathbf{CF}(< I_1, \ldots, I_n >) \text{ or simply } \mathbf{CF}(< I_1, \ldots, I_n >) \neq \emptyset.$$

Furthermore, we have to determine, in analogy to the range of phenomena

RP(T), a range **RP**(N) for theory-nets. It has to be an n-tuple of sets of phenomena since we have embeddings of phenomena on all levels of specialization within the net.

2) $$\mathbf{RP}(N) := \{< Q_1, \ldots, Q_n > \in (Po[\prod(N)])^n; \quad \mathbf{CF}(< Q_1, \ldots, Q_n >) \neq \emptyset\}$$

In **RP**(N) we find all n-tuples of sets of phenomena that are simultaneously embeddable in N. The third parameter should be a measure for the unity of N. As in the case of theory-elements, it consists of the set of all possible conjunctive decompositions of N.

3) $$\mathbf{CD}(N) := \{F = \{N_1, \ldots, N_k\}; \quad \mathbf{CLAIM}(N) \leftrightarrow \mathbf{CLAIM}(N_1) \wedge \ldots \wedge \mathbf{CLAIM}(N_k)\}$$

The comparison of two nets N and N' runs very similar to the simpler case of theory-elements and involves similar presuppositions in addition to the assumption that the two nets have the same structure of specializations. This is surely a restrictive condition but nets that are distinct only at a few terminal places can be compared by allowing slight technical modifications of the nets as repetitions of theory-elements so that we may obtain similar net structures without changes in the content of the nets. Nets completely differing in structure can, at least in principle, be translated into complex single theory-elements, as Gähde (1989, 125ff) has proven, and so can be compared in this way. For nets with identical structures we find that N' has greater (or equal) unifying power than N if the following three conditions are met:

a) $\forall < Q_1, \ldots, Q_n > \in [Po(M_p)]^n$ it is:
$\mathbf{CF}'(< Q_1, \ldots, Q_n >) \subseteq \mathbf{CF}(< Q_1, \ldots, Q_n >)$ (Content Function)

b) $\mathbf{RP}(N) \subseteq \mathbf{RP}(N')$ (Range of Phenomena)

c) $\mathbf{CD}(N') \subseteq \mathbf{CD}(N)$ (Conjunctive Decompositions).

As in the case of theory-elements, a) - c) provide a helpful guideline for theory comparison even in situations that don't allow for a plain ordering.

By this, we have found a promising research program towards a metatheory of explanation via unification that has more to say on non-trivial unifications than its predecessors and can avoid some of their weaknesses. It does not rely on any particular conception of laws and is not committed to an unrealistic deductive chauvinism since a set of models which can be determined without laws and embeddings into a model represent weaker relations than logical deductions. Furthermore, the embedding approach can determine the precise point where unifications become spurious in Kitcher's sense, namely in the case of theories with trivial empirical content, and it can even measure the non-triviality of unifications by a natural parameter. It can take approximations into account, which are a typical phenomenon of nearly all quantitative explanations. Relativistic theories produce better explanations than Newtonian ones because they need smaller sets of admitted blurs, which increases the information content. This is an additional argument for the claim that explanation admits gradations and that theories of explanation should be able to give precise measurements or grounded estimations of explanatory power.

On the other hand, the idea of explanation as embedding is general enough to include common sense explanations and can easily be extended to incorporate explanations of whole theories or reductions. This small segment of the problem-solving capabilities of the semantic embedding account of explanation may be sufficient to justify the view that it is the most promising approach today.

References

Achinstein, P., 1983, *The Nature of Explanation*, Oxford.

Asquith, P.D. & Nickles, T. (eds.), 1983, *Philosophy of Science Association 1982*, vol. 2, East Lansing.

Balzer, W. & Moulines, C.U. & Sneed,J.D., 1987, *An Architectonic for Science*, Dordrecht.

Bartelborth, T., 1988, *Eine logische Rekonstruktion der klassischen Elektrodynamik*, Frankfurt a.M.

Bartelborth, T., 1993, 'Hierarchy versus Holismus. A Structuralist View on General Relativity', *Erkenntnis* 39, 383-412.

Bartelborth, T., 1994, *Wissenschaftliche Erklärungen und menschliche Erkenntnis. Eine diachronische Kohärenztheorie der epistemischen Rechtfertigung*, Habilitationsschrift (unpublished), Berlin.

BonJour, L., 1985, *The Structure of Empirical Knowledge*, Harvard.

Butts, R. (ed.), 1986, *Kant's Philosophy of Physical Science*, Dordrecht.

Cartwright, N., 1983, *How the Laws of Physics Lie*, Oxford.

Cohen, L.J., 1989, *An Introduction to the Philosophy of Induction and Probability*, Oxford.

Earman, J., 1986, *A Primer on Determinism*, Dordrecht.

Forge, J., 1980, 'The Structure of Physical Explanation', *Philosophy of Science* 47, 203-226.

Forge, J., 1985, 'Theoretical Explanation in Physical Science', *Erkenntnis* 23, 269-294.

Friedman, M., 1974, 'Explanation and Scientific Understanding', *The Journal of Philosophy* 71, 1-19.

Friedman, M., 1981, 'Theoretical Explanation', in: Healey (1981).
Gähde, U., 1989, *Theorie und Hypothese*, Habilitationsschrift (unpublished), Bielefeld.
Havas, P., 1964, 'Four Dimensional Formulation of Newtonian Mechanics and their Relation to the Special Theory of Relativity', *Reviews of Modern Physics*.
Healey, R.A., 1981, *Reduction, Time, and Reality*, Cambridge.
Hempel, C.G., 1965, *Aspects of Scientific Explanation*, New York.
Hesse, M.B., 1963, *Models and Analogies in Science*, London.
Kitcher, P., 1976, 'Explanation, Conjunction and Unification', *Journal of Philosophy* 73, 207-212.
Kitcher, P., 1981, 'Explanatory Unification', *Philosophy of Science* 48, 507-531.
Kitcher, P., 1986, 'Projecting the Order of Nature', in: Butts (1986).
Kitcher, P., 1989, 'Explanatory Unification and the Causal Structure of the World', in: Kitcher,P. & Salmon, W.C. (1989), p. 410- 505.
Kitcher, P. & Salmon, W.C., 1987, 'Van Fraassen on Explanation', *Journal of Philosophy* 84, 315-330.
Kitcher, P. & Salmon, W.C. (eds.), 1989, *Scientific Explanation*, Minnesota Studies in the Philosophy of Science.
Kuhn, T.S., 1983, 'Commensurability, Comparability, Communicability', in: Asquith, P.D. & Nickles, T. (1983).
Lambert, K., 1988, 'Prolegomenon zu einer Theorie des wissenschaftlichen Verstehens', in: Schurz G. (1988).
Lehrer, K., 1990, *Theory of Knowledge*, London.
Lipton, P., 1991, *Inference to the Best Explanation*, London.
Salmon, W.C., 1984, *Scientific Explanation and the Causal Structure of the World*, Princeton.
Salmon, W.C., 1989, 'Four Decades of Scientific Explanation', in: Kitcher, P. & Salmon, W.C. (1989).
Schurz G. (ed.), 1988, *Erklären und Verstehen in der Wissenschaft*, Oldenbourg.
Thagard, P., 1992, *Conceptual Revolutions*, Princeton.
van Fraassen, B.C., 1980, *The Scientific Image*, Oxford.
van Fraassen, B.C., 1991, *Quantum Mechanics*, Oxford.
Watkins, J., 1984, *Science and Scepticism*, Princeton.

Chapter 4: Structuralism and the Interrogative Model of Inquiry

Matti Sintonen

1. Ask What Your Theory Can Do For You!

Structuralism is often identified as an ontological stance challenging the received wisdom of the received view of theories. The received view is that theories are sets of sentences, and the received wisdom is that sets of sentences are ideally perspicuous when axiomatized within a formal language. Looking from this angle structuralism is part of the anti-linguistic turn in the philosophy of science. Its New World cousin, the semantic view, is another participant in the rebellion, but similar voices have been heard also from outside this loose alliance.

It is, of course, occasionally important to pause over ontological issues. But from the point of view of salient philosophical applications, what theories can do for a scientist is at least as important as what they are. It suits the spirit of structuralism to ask what the scene looks like if we put the methodological horse before the ontological cart. One important insight of structuralism is that we ought to ask, not just what a scientific theory is, but also what it is good for.

So what are theories good for? The original structuralist proposal of theories as pairs of conceptual cores and sets of intended applications already contains the answer. Theories are *tools* which scientists can use in the process of defining, refining, searching and assessing answers to empirical and theoretical questions. Actually, I shall argue that the most sweeping philosophical gain of structuralism is that, according to it, theories also produce or give rise to questions. The questions can be broad or specific, and, logically speaking, of a variety of types, from yes-no - questions to wh-questions and even to explanation-seeking why-questions. Note that construing theories as tools does not commit one to an instrumentalistic view of the aims of inquiry.

My aim here is to examine some of the unexplored possibilities emerging from the consilience of structuralism and this interrogative view of inquiry. But first, why bother with questions? The short answer is that questions are needed in philosophy of science because scientific problems emerge into consciousness as various types of questions. Moreover, questions and answers are the currency that scientists and scholars are accustomed to deal in. Any problem or research program, whether in basic or applied science (Sintonen (1990c)), can be portrayed as a quest for suitable answers to a set of questions. The somewhat longer answer is that an explicitly interrogative construal of structuralism gives new impetus to the study of pragmatic aspects of inquiry: the increasingly keen cooperation between historical, philosophical and social studies of science has made clear that science is a form of social enterprise in which the primordial language games of questioning and answering are practised. This is not just the banal observation that rhetorics is important. Rather, there are relatively precise tools which can be used to study the conditions under which a question is legitimate or sound, and an answer acceptable or conclusive, for an inquirer or a scientific community, and within a certain context.

The view also backs up structuralist diachronics and theory dynamics. Rather than as finished products, theories ought to be viewed as evolving research programmes which originate around some fundamental but often inarticulate principal questions. These are then made more precise during the process of inquiry, and answered with the help of a series of what later will be called operational questions and answers to these. Quite a few of the problem- or puzzle-solving models in philosophy of science can be fruitfully recast in the interrogative guise (cf. Kuhn (1970a), Laudan (1977), Nickles (1980)). It is somewhat surprising that the explicit question-answer idea has not been developed in greater detail, for it brings back to the metatheoretic picture an aspect which, by and large, has been monopolized by historical studies of science.

I shall first prepare ground by taking a brief look at the idea of putting questions to Nature (section II), at erotetic logic (III), and at its applications to inquiry and knowledge-seeking (IV). Section V outlines how the interrogative and structuralist views might meet, section VI addresses some difficulties in the erotetic programme, arguing that erotetic logic must be married with a structured notion of a theory. This insight is put to further use in section VII dedicated to the heuristics and logic of discovery, and in section VIII, where I'll return to the interrogative notion of explanation, the home base of the entire approach, and argue that the emerging structuralist-interrogative idea is needed to make sense of the explanatory commitments of a scientist or a scientific community.

There are many obstacles to be overcome before the question- answer perspective and the structuralist view can merge, such as the insistence of standard erotetic logics to make reference to some specific language in which the questions are phrased. Consequently, although some of the applications rest on a relatively secure basis, others are more akin to proposals for a more detailed study.

2. Putting Questions to Nature

The interrogative idea – that knowledge acquisition in general and scientific inquiry more specifically is a question-answer process – is an early metatheoretic insight. Whether a literal truth or a metaphor, it informed both Aristotle's views on explanation and Bacon's attack on the Aristotelian tradition. According to Bacon (1620) and Kant (1787), the task of an inquirer was to force (even to torture, as Bacon put it) Nature to give conclusive answers to questions of the inquirer's choice. Thus, in the interrogative tradition, unsolicited observations and experiments can be looked upon as yes-no- questions or wh-questions put to Nature, and explanations can be viewed as answers to explanation-seeking why- questions (See Gale (1978) and Sintonen (1984a)).

The notion that inquiry is putting questions to Nature was dear to such fathers of modern natural science as Newton (cf. the Queries in his Optics) and Darwin, who expressly viewed his projects in the form of strategically planned series of questions. The notion survived in the thought of many philosophers and scholars from Collingwood (1939, 1940) to Gadamer (1960), but its serious comeback in modern analytic philosophy of science can be dated to Carl Hempel and Paul Oppenheim's classic paper (1948) and to R.G. Braithwaites monograph (1953) on scientific explanation. Both works resisted the extreme positivist notion that explanations belong to metaphysics and not to science proper, and both adopted the interrogative view in which explanations are answers to why-questions.

Despite this, there was no detailed attempt to make use of idea. It is a striking fact that the first sentence of Hempel's and Oppenheim's (1948, p. 245) classic paper reads: 'To explain the phenomena in the world of our experience, to answer the question 'why?' rather than only the question 'what' is one of the foremost objectives of empirical science.' Yet, the interrogative analysis did no real work, for the explanatory relationship was conceived as a two-placed relation between sets of sentences, where these sets were not construed as questions and answers. Similarly, Braithwaite (1953, p. 319) embraced but did not refine the idea when he wrote that 'Any proper answer to a 'Why?' question may be said to be an explanation of a sort'.

I have suggested two reasons for this lack of progress on the interrogative front (Sintonen 1989). One was the lack of adequate logical tools to deal with questions and answers, and the other one was the fear of subjectivism and psychologism. The first shortcoming was obvious: there was no erotetic logic to deal with questions and answers (nor a logic for belief change which could have explicated growth of knowledge in terms of epistemic states, quite independently of the questions and answers which might give linguistic expression to those states). Without erotetic and epistemic logic at hand, the intuitively appealing way of questions must have appeared formally intractable.

The fear of subjectivism and psychologism was equally obvious. Precision and intersubjective control are the *sine qua non* of science. If different inquïrers are not just interested in different questions (which would be acceptable), but are also entitled, legitimately, to choose different answers even to one and the same question, fundamental methodological values are in jeopardy. On such a view, what counts as an acceptable answer would be an anthropomorphic affair. Michael Scriven (1975, p. 4) decided to bite the positivist bullett and submitted that the notion of explanation *does* owe its meaning to subjective and psychological factors. Scriven's (1959a, 1959b, 1962) vehement attack on Hempel's and Oppenheim's (1948) model generalized the parallel criticisms of Norwood Russel Hanson (1958) and Stephen Toulmin (1961). The attack challenged the entire logical-empiricist idea of rational reconstructions by logical tools.

3. Erotetic and Epistemic Logic

The emergence of erotetic and epistemic logics marked a beginning of a new era. I am forced to cut a very long (and occasionally messy) story short, and to make a number of simplifying assumptions. There are numerous erotetic explications around, from linguistically oriented speech act theoretic models to logical models of problem-solving and question-answering (cf. Achinstein (1983), Sintonen (1984a), Tuomela (1980)). Crucial for a choice among them is the suitability of an erotetic logic as a basis of a comprehensive interrogative account. The highly influential imperative-assertoric version of Belnap and Steel (1976) followed Hamblin's (1958) dictum, according to which there is an extremely tight relationship between questions and answers. According to Harrah (1963: 450) the standard situation is one in which the questioner knows the set of potential answers, and Belnap (1963) wrote that the analysis starts from the (two-person) situation in which both the querier and the prospective answerer know in advance what would count as an answer. Here the latter's task amounts to choosing the right one (or ones). This is unduly restrictive. As Thomas Nickles (1980, 1981) and other friends of discovery observed, such a logic is useless in the study of scientific problem-solving. Similarly, the theory of explanation needs a less restrictive relationship between questions and answers.

Although I shall start with the semi-formal notation of Belnap and Steel, further developing is along the lines of Hintikka (1974, 1975, 1976) and Åquist (1969, 1971) in which epistemic notions play a crucial role in the elucidation of what counts as a satisfactory or conclusive answer. There have been prior attempts to construct a comprehensive interrogative model of inquiry, such as Bromberger's (1965, 1966 and especially 1971) and Gale (1978; see also the useful bibliography in the latter). Yet Hintikka's interrogative model (I-model for short) combines heuristic potential with a precise epistemic and erotetic logic, and consequently offers perhaps the most promising comprehensive model.

The important notions in erotetic logic are those of a question, its desideratum, its presupposition, and its direct answers. I shall not try to define these notions formally, but intuitively, a question is a request, addressed to a prospective respondent, to cite an answer which brings about the desideratum of the question, i.e., the state of affairs in which the questioner knows a true and conclusive answer to her question. Direct answers, of varying logical types and complexity, are propositions or singular terms which are intended to bring about the desideratum. The (logical) presuppositions of a question are obtained via a syntactic operation (to be explained in short). Semantically, a presupposition is a proposition whose truth is necessary for the question to have a (true or false) direct answer.

A bit more precisely, we distinguish between propositional questions and wh-questions. In a simple propositional yes-no - question of the form 'Is p the case?' the questioner desires to know whether a proposition is true. Thus the question sentence (1) has disjunction (2) as its presupposition, and its direct answers include its two disjuncts (3) and (4):

(1) Is particle p at location s?
(2) Particle p is at location s or particle p is not at location s.
(3) Particle p is at location s.
(4) Particle p is not at location s.

Wh-questions, such as who-, where-, and when-questions differ from these in that they receive singular terms which specify individual persons, spatial locations, instances of time, etc., as answers. The semi-logical forms of the wh-questions 'Where did Wittgenstein teach school?' and 'Where is particle p?' have the form (?x)p(x), here (5) and (6), respectively,

(5) (?x) (Wittgenstein taught school in x)?
(6) (?x) (Particle p is at location x)?

where x ranges over places or locations (and can have more structure than the semi-logical form expresses, see below). The presuppositions of (5) and (6) are the propositions (7) and (8), i.e., that Wittgenstein did teach school somewhere, and that particle p is somewhere (has some spatial location or other):

(7) (Ex) (Wittgenstein taught school in x)?

(8) (Ex) (Particle p is at location x)?

Presuppositions of questions may have presuppositions of their own, such as the presupposition of (7) that there was someone called Wittgenstein, or that physical particles are the sort of entities that occupy space. Now questions (5) and 6) have (direct) answers only if (7) and (8) are true. There is also a simple way to tell what a candidate answer to a wh-question is: it is simply a substitution instance of the (syntactically defined) matrix of the question, i.e., the open sentence following the quantifier. There is also a simple relationship between the logical forms of the questions and their presuppositions, precisely because the question has a true direct answer only if its presupposition is true. If the presupposition of the question is false, the question only admits of true corrective replies; thus if (7) were false, no substitution instance of its matrix would yield a true answer, and the proper response would be 'But Wittgenstein never taught school'.

The relationship between a wh-question and its (direct) answers can now illustrate what a pragmatically relativized satisfactory or conclusive answer to a wh-question is. An answer to (5) or (6) is conclusive for an inquirer, in a context, if it causes (brings about) the desideratum, and if the inquirer can say 'I know where Wittgenstein taught school (or where particle p is)'. A direct answer, even a true or correct one, may fail in this, if it does not meet the questioner's pragmatically determined conditions vis-à-vis type, uniqueness, and identifiability. A name or a definite description need not be satisfactory, if the inquirer wants to know not just a name or description but also wants to be in a position to locate it on her map. Answerhood proves to be important, for the innocent question (6), in the language-in-use of a putative scientist, can have a pragmatic rider which requires that direct answers specify the location by using a particular descriptive vocabulary and frame of reference.

Such pragmatic requirements on answerhood could equally well be called pragmatic presuppositions of questions (to mark the difference with their logical presuppositions), and they prove valuable in the elucidation of pragmatic (contextual, perceptual, idiosyncratic or communal) *scientific commitments* (See Sintonen (1984a) and (1985)). Although the borderline between semantics and pragmatics is notoriously messy, Hintikka's model makes a virtue out of the potential vice: there is a simple (though sometimes pragmatically speaking indeterminate) condition as to when a direct answer satisfies the questioner: the answer, together with the inquirer's background knowledge, must suffice to entail the desideratum, here 'I know where Wittgenstein taught school'. A good litmus test of pragmatic conclusiveness is that the answer does not give rise to important additional questions concerning, say, the location of the place[1]. Finally, there are the explanation-seeking why- and how-questions which occupy a halfway house between wh-questions and yes-no-questions (and can have the surface form of a what-question). They introduce variables over reasons, causes and manners of operation, expressed in semi-logical notation as '(?f)p(f)?'. Thus the questions 'Why do magnets lose their magnetism when heated?' and 'How do leucocytes kill bacteria?'can be expressed by either (9) or (10) and (11):

(9) (?f) Magnets lose their magnetism when heated for the reason f.
(10) (?f) Magnets lose their magnetism when heated because f.
(11) (?f) Leucocytes kill bacteria in the manner f?

The (logical) presuppositions are obtained by dropping the quantifiers (See e.g. Achinstein 1983, Tuomela 1980 and Sintonen 1984a).

4. *Knowledge-Seeking Games*

Epistemic logic helps to elucidate the relationship between a question and its answers, but this static notion does not capture the process in which questions and answers are sought. In Hintikka's I-model the latter aspect is incorporated by looking upon inquiry as a game (against Nature) where an Inquirer strives for a cognitive objective by subjecting Nature to an array of questions, and by trying to achieve the objective by help of Her answers and the Inquirer's own background knowledge. One important type of objective is that of finding an answer to a big initial question, called in the I-model the *principal question*, by subjecting nature to an array of small *operational questions*. Another type of objective is that of trying to establish a predetermined conclusion B by a like interrogative procedure. This latter type of Shakespearean *Fragestellung* – B or not B? - can, however, be regarded as a variant of the former, for such questions can be represented as yes-no-questions.

Apart from the coarse division to types of questions, questioning games can also be graded with respect to the degree of complexity of the answers allowed. The extremes are games in which all answers confine to atomic propositions and their negations, and games in which formulas of any given quantificational complexity are allowed. The dominant view in

[1] It does not mean, though, that there are, in science, final answers in Popper's (1972) sense. In fact, one important desideratum in theory choice is fruitfulness in Kuhn's (1972b) sense, and one explicate for this is that the theory gives rise to well-defined further questions. But suggesting further big questions is different from leaving gaps in an otherwise satisfactory answer.

philosophy of science has adopted, usually without argument, the atomistic assumption that all answers coming from Nature are atomic. However, questions in experimental contexts characteristically receive so-called AE-answers expressing functional dependencies (See Hintikka (1988)). As Hintikka (1985) has argued, there is also a continuum of mediating types of inquiry having the following property: the richer the background theory, the simpler can the questions (and answers) be.

Now, despite variations on a common theme, all knowledge- seeking games (I-submodel) share the crucial feature that the steps in the knowledge-seeking process can be divided into deductive and interrogative moves[2]. Deductive moves represent deductions from premises the inquirer already possesses, whereas interrogative moves, answers to questions, provide new information. It is not required that the inquirer is aware of the accompanying mental episodes or their verbal expressions. Thus premeditated experiments (and thought experiments) as well as unsolicited observations count as answers to questions put to Nature. A process of querying and answering can be codified in terms of the semantical tableaux employed by Beth and Hintikka, with explicit rules governing deductive rules, and further rules governing admissible interrogative moves (cf. Hintikka (1981a), (1981b)). Again, the theme allows for variation, but in general items cannot be rushed from the right hand column to the left hand one in a tableau. It is also assumed that before an inquirer can raise a question its presupposition must occur in the left column. An important idea is the requirement that any given game is tied to a particular model M and its language, so that nature's answers are presumed to be true in the model M. In the simple game in which the aim is to prove a predetermined conclusion C the inquirer has succeeded when she or he has closed the tableau. What the inquirer attempts to do, then, is to prove the conclusion by forcing nature to give unambiguous answers to her or his questions, answers which the inquirer can then avail in the interrogative derivation of the chosen conclusion.

5. The Nascent Structuralist-Interrogative Idea

After this stage setting, let us see how the questions-answers idea fits structuralism. There are no formal innovations in the following attempt

[2]The account given here is in many ways simplified, and based on Hintikka's early views in Hintikka (1976), (1981a) , (1981b) and (1985). The more up-to-date model developed by Hintikka and his group is based on what is called independence-friendly logic (IF-logic). IF-logic gives a more unifying account of epistemic and erotetic logic, and makes it possible to explicate, in a more satisfactory way, the nature of knowledge-seeking in which the desiderata of questions are phrased in promissory terms. For the more recent developments, see e.g. Maunu (1993).

to pave the way from the interrogative model to the structuralist view. I shall simply follow the terminology of *Architectonic*, and indicate if deviations are made. Assume then that a theory-element T is an ordered pair $< K(T), I(T) >$ in which the theory-core $K(T)$ is a quintuple $< M_{pp}(T), M_p(T), M(T), GC(T), GL(T) >$. Here $M_{pp}(T)$ stands for the potential partial models of the theory, those structures of which it makes sense to ask whether they can be enriched with theoretical functions so as satisfy the laws M of the theory, $M(T)$ for the laws of the theory, and $M_p(T)$ for potential models. $GC(T)$ and $GL(T)$ the global constraint and the global link of T, understood in the usual way. The range of intended applications $I(T)$ is delineated intensionally through some paradigmatic exemplars, and theory claims are construed in the standard way. This mode of determining $I(T)$ means, in question-theoretic parlance, that the set of questions which fall in the domain of responsibility of $K(T)$ is (somewhat) fuzzy: theories have some degree of autodetermination, and since an object or configuration of objects must exhibit similarity or analogy with an exemplar, there is no God's eye point of view as to exactly what a theory is forced to deal with. Yet, fuzziness is not anarchy, and already this much suffices to remedy a difficulty in some problem-solving models[3].

It is now easy to see what a structuralist-interrogative idea looks like, in rough outline. Claims and questions are two types of speech acts which differ with respect to their propositional attitudes but not necessarily in contents. They can be vague or precise, global or local. Thus, where an inquirer can claim that the set $I(T)$ belongs to the content of T, she can also raise the question whether this is the case. Similarly, just as the inquirer can pick out a single element from the set $I(T)$ and claim that it can be enriched to become a model of the theory, she can raise the question whether this is the case. In fact, depending on her scholarly commitments (and mood), her mental life might include also other propositional attitudes, for she or he might suspect that this is the case, take it to be highly likely that it is the case – or fear that it is not the case. But if we pertain to the questioning mood, we have here a yes-no -question concerning a structure in the set of intended applications, raised in terms of the non-theoretical language of the theory-core.

Before proceeding to more details, we must deal with the fundamental cleavage between the statement and structuralist views, if only briefly. Since practically all approaches to erotetic logic require a suitable formal language in which the logical forms are couched, we must find a way of

[3]There is also another reason for fuzziness, viz., that questions are characteristically raised in the language-in-use of the scientist or scientific community. This aspect of fuzziness will occupy us more towards the end of the paper.

incorporating language in some way. The traditional choice is to use a specific language (say, first-order predicate logic) coupled with suitable quantifiers and concepts used in erotetic and epistemic logic. Alternatively, one might stay within model-theoretic bounds but leave the nature of the formal language open and resort to 'abstract logic' (See Pearce and Rantala (1983) , Rantala (1980), Sintonen (1984a)). This is because requiring theories and questions to be couched in, say, first-order language may lead to some empirical theories and questions being beyond formal reach.

The non-statement line of thought speaks against incorporating languages into theory-elements. Already Carnap (1956: 241) observed that the language of science is 'mainly a natural language' augmented with a few explicit conventions introduced to govern the use of some special symbols. Structuralists and semanticists do not wish to deny that language is important but they do maintain that its role has been exaggerated[4]. Consequently, in the structuralist notion language is not baked into an empirical theory. Yet, although the structuralist framework does not provide for formation rules which in the model-theoretic framework tell how individual, predicate and function constants can be pooled to form terms and sentences, language is not simply ignored. In fact, a species Σ of structures (in the structuralist sense) determines a non-logical vocabulary which can be employed to express the content of the claims – and questions – made by help of the theory (for details, see *Architectonic*, p. 17-18).

Since my focus is on the applications of the interrogative idea and not on its formal foundations, and since it is possible to assign to the elements of a core their linguistic counterparts (see Niiniluoto (1980), Tuomela (1978)), the choice hinges on matters of expository and heuristic fruitfulness. Now, one such consideration concerns the fact that there are several relevant languages in any given context. There is what I have already called the scientists language-in-use, the medium used in informal communication. But there are also the various languages possibly associated with the various species of structures. To steer around the choice, I shall follow Carnap and the structuralist and not incorporate language into theories. Should the need arise to make the infe-

[4] See e.g. Van Fraassen (1970: 326), where van Fraassen explicitly says that there are natural interrelations between the syntactic-axiomatic and the semantic orientations: 'These interrelations, and the interesting borderline techniques provided by Carnap's method of state-descriptions and Hintikka's method of model sets, would make implausible any claim of philosophical superiority for either approach'. Recent developments also outside physical sciences have highlighted the advantages of not baking languages into theories. Thus the semantic theorists have shown that it is possible to stay, within metatheoretic explications, very close to the scientists' own formulations. I have discussed the nature of biological theories, especially evolutionary theory, in Sintonen (1990b), (1991) and (1993).

rences within the I-model more precise, one can always phrase the Σ - types in a linguistic form. Thus, although empirical claims have the form $I \in Cn(K)$, for purposes of the I-model this is simply shorthand for the equivalent linguistic claim. The same goes for questions. But to go to the substance, take classical particle mechanics as an example. Its basic theory element $T(CPM)$ contains the core $K(CPM)$ and set $I(CPM)$ of intended applications, (see *Architectonic*, III.3.), where $K(CPM) = \langle M_p(CPM), M(CPM), M_{pp}(CPM), GC(CPM), M_p(CPM)\rangle$ and $I(CPM) \subseteq M_{pp}(CPM)$ is such that

(1) $I_o \subseteq I(CPM)$ where I_o contains the solar system, pendulum, projectile, harmonic oscillator U

(2) all members of $I(CPM)$ are sufficiently similar to those of I_o.

The definition of a potential model $M_p(CPM)$ is important, for $M_p(CPM)$ specifies the descriptive apparatus :

$M_p(CPM)$: x is a potential classical particle mechanics ($x \in M_p(CPM)$) iff there exist $P, T, S, c_l, c_2, s, m, f$ such that

(1) $x =< P, T, S, \mathbb{N}, \mathbb{R}, c_l, c_2, s, m, f >$

(2) P is a finite, non-empty set and S, T are sets

(3) $c_1 : T \rightarrow \mathbb{R}$ and $c_2 : S \rightarrow \mathbb{R}^3$ are bijective

(4) $s : P \times T \rightarrow S$ and $c_2 o s_p o c_1 = r_p$ is smooth for all $p \in P$

(5) $m : P \rightarrow \mathbb{R}^+$

(6) $f : P \times T \times \mathbb{N} \rightarrow \mathbb{R}^3$

Now in the definition of $M_p(CPM)$ the symbols P, T, S (doing the work of terms in a model-theoretic presentation) stand for the base sets of particles, time intervals and spatial regions, $\mathbb{N}$ and $\mathbb{R}$ for the auxiliary mathematical sets of natural and real numbers. The function symbols s stands for the position function which assigns the individuals from base set P their locations at any given time. m is the symbol for the (real-valued) mass function and f for the force function such that m assigns each particle in P its mass, and f assigns particles (from P) and time-intervals (from T) the total force acting on the particles at each time[5]. The functions c_1, and

[5] Technically, since there can be various types of forces acting on a particle at any given time, function f should be split into component functions, indexed by natural numbers. This explains the form of (6) $f : P \times T \times \mathbf{N} \longrightarrow \mathbf{R}^3 : f(p,t,i) = a$ then reads: the i-th component force acting on p at t. See *Architectonic*, p. 31).

c_2 map time instants to real numbers and space-points to real 3-vectors, thus giving coordinations of time and space. It is clear from this that the functions in (3)-(6) in the definition for $M_p(CPM)$ provide not just a (non-logical) vocabulary but also a 'grammar' which makes it possible to express some claims and questions, and not others[6]. Thus, since the position function c_2 is a real-valued function from the sets of particles and time-intervals to space points, and since the mass function does not have an argument place for time, the 'grammar' of CPM allows for the question (12) but not for (13),

(12) $(?x)$ Particle p is at location x at time t?
(13) $(?x)$ The mass of particle p at time t is x?

Question (13) makes no CPM-sense, although there might be theories in which it does make sense.

Now we already had question (6),i.e. $(?x)$ (Particle p is at location x)? above, a question phrased in the languge-in-use. How are the (6) and (12) questions related? My suggestion is that there usually (perhaps always) is more to a question than meets the naked and untutored ear, in that the semantics of natural language, or the language-in-use at large, does not determine, in unambiguous terms, what vocabulary the answers must employ. A question sentence may leave out, e.g. for reasons of pragmatic convenience, pertinent information, but that information can be filled in. Thus a holder of CPM is pragmatically tuned to CPM -answers, and is prone to regard (12) as a CPM -reformulation and refinement of a question in which time is not explicitly mentioned. By the same token, since space has the classical three-dimensional structure, a full answer should specify spatial coordinates in accordance with function c_2. Therefore answers to the CPM -holders question and to its elliptic form (6) should have the form 'Particle p is at location $< s_1, s_2, s_3 >$ at time t_1'.

The notion of a presupposition of a question was syntactically characterized, but presuppositions also serve the purpose of providing a necessary condition for a question to have a true direct answer. Since $M_p(CPM)$ specifies what vocabulary must be employed to describe a range of intended applications, and hence what **kinds** of questions are legitimate, it

[6]This account parallels an early suggestion by Scott Kleiner (1970), suggesting that the extra-logical symbols with appropriate physical interpretations (in Kleiner's terms, the vocabulary) and the 'proper syntax' which governs the symbols literally divides claims made by help of CPM into those that are legitimate and those that are illegitimate. The main difference between our accounts is that whereas Kleiner wishes to provide to questions and answers their logical forms, I have chosen not to do so. Thus, the structure on answers imposed by the laws M comes through a pragmatic commitment to employ the components $M_p(CPM), M(CPM), M_{pp}(CPM), GC(CPM), M_p(CPM)$ of CPM . Yet, the laws themselves are not formalized.

also constrains admissible answers. This gives an explication of the notion that M_p's are systems *of which it makes sense to ask whether they are models of the theory*. The ascent from M_{pp}'s to M_p's describes a process in which a question in the language-in-use of a scientific commitment gets more content. But of course the laws of the theory M accomplish more by specifying physically (or otherwise nomologically) possible systems. Thus

$M(CPM)$: x is a classical particle mechanics ($x \in M(CPM)$) iff there exist $P, T, S, s, m, f, c_1, c_2$ such that
(1) $x = (P, T, S, N, \mathbb{R}, c_1, c_2, s, m, f)$
(2) (2)$x \in M_p(CPM)$
(3) for all $p \in P$ and $a \in \mathbb{R} : m(p)D^2r_p(a) = \sum f(p, c_1^{-1}(a), i)$.

The example looks trivial because CPM-grammar is a settled issue. But one only needs to take a look at theory formation and model building in biology and the social sciences to appreciate that constructing a vocabulary for the description of a phenomenon is perhaps the most important task. Just consider population genetics, where the aim could be that of building dynamical mathematical models which predict and explain changes of gene frequencies in populations as a function of time. Actual populations are extremely complex, and consequently the task of finding (or perhaps more appropriately, constructing) a set of variables, parameters and laws sufficient to describe the evolution of a population is extremely difficult. Whether these fields are just immature or plainly different is an open issue.[7]

6. Bromberger's Programme

The simplistic characterization of $T(CPM)$ does not of course spell out all that goes into describing CPM -systems, but it suffices to show that theories are devices which make it possible to raise and answer yes-no-questions, wh-questions, and, finally, why-questions of various generality. It also helps to overcome some difficulties in the interrogative program. I shall illustrate these difficulties by returning to the developments in the 1960's and 70's (see end of Section II), to what I call the Bromberger Programme (BP, for short). BP, formulated in Bromberger (1965), (1966)

[7]Lewontin (1984: 6-7)) nicely summarizes the problem: '... the problem of theory building is a constant interaction between constructing laws and finding an appropriate set of descriptive state variables such that laws can be constructed.... We do not really know what a sufficient description of a community is because we do not know what the laws of transformation are like, nor can we construct those laws until we have chosen a set of state variables.....there is a process of trial and synthesis going on in community ecology, in which both state descriptions and laws are being fitted together.'

and (1971) and summed up in (1992), addressed both the logic and the methodology of question-answer analysis in the philosophy of science. The guiding intuition was that one should aim at stating the 'principles which govern the acceptability of any alleged contribution to science' (Bromberger (1971: 49)). This general aim covers not just the developing of an erotetic logic for the various types of questions, but also the characterization of what the question-answer relationship is, of how scientific questions arise, and of how answers are sought (by search algorithms). The ambitious programme was not carried to completion, but it did provide useful criteria of adequacy. The first criterion of adequacy was an erotetic logic sufficiently rich to characterize the various types of questions that might arise in scientific contexts. In BP this was first-order logic augmented with erotetic operators (although Bromberger admitted that it was not rich enough, as it stood). I shall no longer dwell on this, but hasten to an additional insight. BP accepted the standard erotetic view that if a presupposition of a question is false, the question does not arise.

However, a question may fail to arise for two intuitively speaking different reasons (cf. Belnap (1969)). Bromberger, and later van Fraassen (1980) groped for a way to distinguish between questions which fail to arise simply because the inquirer has made a factual mistake, and questions which are, by the lights of an inquirer or community of inquirers, not of the type admitting direct answers. The latter type of failure is more radical. For example, a question raised in the CPM -community which presupposes that the mass of a particle depends on time or location is at odds with the prevailing view.

Van Fraassen addressed the difficulty through the notion of relevance built into his theory of explanation. I am forced to confine myself to the bare essentials here. The idea was that a why-question arises only if its presupposition P_k (van Fraassen's topic) is true, only if the propositions in its contrast class $X = \{P_1, P_2, ...P_k\}$ are all false, and only if there is one proposition A that bears a certain relevance relation R to the ordered pair $< P_k, X >$. To spell out the idea by help of an example, to ask why Peter got the disease called paresis is to presuppose that Peter did get paresis (P_k), that others in the relevant relevance class did not (that, let us say, Paul and Mary did not, although they could have), and that there is a cause (expressed by A) which gives a true answer to the question. Van Fraassen's analysis is a refinement of the Belnap-Steel account briefly outlined in section II. However, it was criticised by Kitcher and Salmon for a failure to impose any restrictions on the relevance relation R between the putative answer A and the (logical) presupposition P_k (see Salmon (1989, 141)). The only requirement is that the answers 'rely on scientific theories and experimentation (and not on old wives' tales)'. Even this is in jeopardy, for the relevance relation might be based on straightforwardly extrascientific interests (cf. Sintonen (1989)). And of course, if anything can be relevant, nothing really is. Van Fraassen seems to embrace the impossibly woolly view that explanation is a purely antropomorphic pleasure (see van Fraassen (1980, 87)).

I think that van Fraassen's account is, nevertheless, a step towards fulfilling BP's criteria of adequacy. Although Bromberger's (1971) proposal represented some progress in the attempt to develop formal conditions under which scientific questions are 'legitimate', it is unlikely that erotetic logic itself helps much. The reason is simply that conditions of legitimacy are largely pragmatic, as van Fraassen's analysis acknowledges.

However, it does not follow that nothing of interest can be said of the pragmatic factors needed in the theory of explanation, or indeed of the process in which the relevant scientific vocabulary is fixed. My suggestion is that the weak constraints imposed by the logic of questions must be supplemented by pragmatic constraints arising from the group commitments of theory-holders. The reason why legitimacy appears to be opaque is that it is a pragmatic notion, and can only be explicated by help of a structured and dynamic notion of a theory. This goes not just for erotetic logic but also for the further desiderata of developing what Bromberger (1971) called 'a theory of theories' which could explicate how sets of questions are identified and how answers are derived.

To see the potential import of structured background knowledge, consider van Fraassen's problem of assessing answers, once a salient question is fixed. Take a specific question (in which the presupposition, relevance relation, and contrast class are fixed). Then there still is, according to van Fraassen, the problem of determining the portion of background knowledge which is relevant when the virtues of rival answers are measured. 'The evaluation', van Fraassen (1980, 147) writes,' uses only that part of the background information which constitues the general theory about these phenomena, plus other 'auxiliary' facts which are known but which do not imply the fact to be explained.' However, no one has had much to say about the crucial metatheoretic problem of delineating the relevant portion. And van Fraassen concluded that this must be an additional contextual factor.

Although van Fraassen is, I think, right in insisting on the importance of contextual factors, the further claim that explanation is a purely anthropomorphic pleasure (or that pragmatic equals psychological, as Scriven writes), does not follow. What is needed, however, is some way to represent the way questions may acquire or lose legitimacy in theory-evolution.

Scientific progress – increase in question-answering power – is not just adding new true answers to old questions, but rephrasing and refining old questions through conceptual enrichment. And for describing them one needs conceptual equipment to represent not just the static relationship between a question and its answer, but the dynamic way of representing changes in the descriptive apparatus, and for telling what questions are considered as legitimate[8].

The dynamic viewpoint is this. Questions and answers can be phrased using different vocabularies. A question can concern (its presuppositions may refer to) a system which is described in the language-in-use of a scientific community, and only later be described in the language (indeed, languages) of an established theory-element. Now, diachronically speaking, there can be explanation-seeking why-questions, expressed in the informal language of the scientific community, before the $M_p(T)$ and $M(T)$ of the theory-element T have been articulated. And once $M_p(T)$ is available, one can raise the further question if structures within the set $I(T)$, now enriched with theoretical functions, are also models.

The transition from the 'pretheoretical' language to the theoretical one is of course a crucial step, but the boderline is difficult to draw in individual cases. But it suffices for our informal purposes to note that a question can stay 'in the air' until it is first refined, and then answered (more or less conclusively), by holders of a theory-element. This answering amounts to embedding the non-theoretically described system into a theory-element. One way in which this process could be characterized, then, is that the inquirer's initially vague why- question gets more content during the process of inquiry.

7. The Process of Inquiry: Heuristics and the Logic of Discovery

Let us next turn to the further desiderata in BP. How are questions generated and answers sought? All interrogative construals start with the assumption that questions arise from a structured form of curiosity. Knowledge comes from a combination of knowledge and ignorance. Bromberger (1971: 51) acknowledged this by writing that ignorance is 'not one big undifferentiated glop, one huge unstructured nothingness'. Although a scientist may be in a 'predicament' with respect to a proposition describing a phenomenon, her background knowledge imposes constraints on admissible answers.

[8] This gives one reason why truth as a goal of scientific inquiry is problematic: before one can assess whether an answer is true, one must assess its legitimacy. This no doubt was the view of the early analytic philosophers and positivists when they insisted that questions of meaning are prior to questions of truth.

The best way to come to terms with this is to assume, just as we did, that questions can have more structure than their natural- language expressions suggest, i.e., that questions arise from conceptually deep problem-situations (See Nickles (1981)). And one way to make this notion more precise is to assume that questions inherit this structure from a rich background theory which both refines them and gives heuristic guidance in the search for answers. It is here that the resources of the structuralist-interrogative view come to full view. The metaphor is that a set of intended applications $I(T)$ provides food for thought and the theory-core $K(T)$ provides the cutlery. There are several details in this metaphor worth spelling out. First, a theory specifies how questions arise, and thus paves the way to one desideratum in BP. Secondly, the pragmatized notions of a Kuhn-theory or of paradigm governed theory evolution throw throw light on the heuristics and logic of discovery. Third, in the process of specialization (and especially, theoretization), questions receive new content. Fourth, we can show, albeit informally, that the so-called problem-solving models have been on the wrong track in focusing on questions. For, paradoxically, often imperceptible movement in background knowledge, and especially improvements in mathematical tools, provides the impetus for inquiry. And finally, we can see, when the metaphor is turned into a literal truth, that there is a way in which the conclusiveness conditions on answers can be applied to questions that, on the face of it, seem beyond reach.

How do questions arise? Collingwood (1939, p. 30-31) wrote, in a perceptive remark, that Bacon's *Novum Organum* and Descartes *Discourse on Method* turned a new page in the understanding of inquiry. They both expressed the 'principle that a body of kowledge consists not of 'propositions', 'statements', 'judgments' or whatever name logicians used in order to denote assertive acts of thought'. Rather, knowledge consists of propositions 'together with the questions they are meant to answer; and that a logic in which the answers are attended to and the questions neglected is a false logic'. We can see that structuralism not only falls in the interrogative camp, but that it is in marked contrast with the statement view in that it incorporates the set of intended applications within the theory. Consequently, theory-elements literally feed in empirical questions concerning structures, and theory-nets contain open gaps of the form: what special laws (and theoretizations) do we need to say more about such and such structures? These are nothing but theoretical questions.

However, single theory-elements do not suffice to describe all aspects of question-answer dynamics. Theories do not hatch as finished products sufficient to deal with all forthcoming applications. Rather, they are conceived of as gappy structures which must be nurtured until they turn into powerful theories. Theory-elements in science characteristically conspire to form theory-nets N, sequences of theory-elements $T_1, T_2, ...T_n$ connected with one another by the specialization relation, and theory-holons H which may contain theory-elements from different nets. A theory-net in turn has one or more basic theory-elements $B(N)$ and a number of specialized theory-elements $< K_i, I_i > \in N$, introduced to make more specific claims about some more limited classes of applications ($I_i \subseteq I_o$). The basic core of the theory-net may then give rise to several branches of specializations, and the result may be a hierarchic tree-structure.

To make the connection to the interrogative model of inquiry still more explicit, we can follow the early suggestion of Balzer and Sneed (1977) and define an explicitly pragmatized notion of a Kuhn-theory $< K_o, I_p, N_p, M >$, where K_o is the fundamental core, I_p its set of paradigmatic exemplars, and N_p an associated paradigmatic theory-net, and M a function which lays down restrictions on net-expansions (it thus formally explicates promising lines). Such a notion, or its genealogical descendants (cf. *Architectonic*) allows us to study the paradigm-governed development of a theory over time. When a Kuhn-theory is proposed, only some of the intended applications in I_p have been shown to be models of the theory (in Newtonian mechanics these consisted almost entirely of the gravitation phenomena). It is a task for later generations to refine and expand the theory-net to cover the remaining envisioned but so far unexamined or unsuccesfully examined applications. A theory-evolution represents such a historical development: it is a finite sequence of $< K_0, I_p >$-based theory-nets $N_1, N_2,...$ such that each $N_i + 1$ contains at least one theory-element obtained by specialization from an element in the historically preceding theory-net N_i.

Reference to specialized theory-elements discovered by a certain time (during a certain historical period associated with a theory-evolution) is a welcome addition to the interrogative perspective which assumes as undefined primitives inquirers (and their communities), and pragmatic presuppositions for questions. Consider, then, a scientific community, such as the CPM -holders during a certain historical period. The extant theory-net and its elements present a host of more or less well-defined questions. In the rough order of well-definedness, there are predictive singular yes-no -questions concerning systems on which the community decides to focus during the period. There are also clarifying, elucidating and classifying wh-questions, requiring refinement of terms or values or classification of phenomena in classes of intended applications as answers. There are important wh-questions concerning values for constants, and why-questions requiring explanatory answers: whenever a theory-element yields wrong predictions for a $y \in I_i$, these predictions give rise to a why-question of form (14): (14) (?f) (The system y behaved in a manner O because f)?, where there may or may not be a well-defined contrast class.[9]

Now one important type of big initial question is a what-question of the form: what special laws are needed to capture, within the standards imposed by a Kuhn-theory T, the known facts and regularities within a so-far unexamined or recalcitrant intended application? The interrogative idea leads to a picture in which an inquirer or a scientific community attempts to find an answer to such a question. Questions therefore indicate gaps in the theory-net.

To return to the initial image of theories as tools, it is clear that the basic theory-element, with all its components, is the basic tool, but of course all already confirmed special laws are included in the kit. First of all, there are the vocabularies and the grammar imposed by the basic theory-element. Moreover, there are exemplars and the values embodied in the exemplary solutions to previous questions which both restrict answers and provide heuristic guidance: if there is a pattern of answer already used succesfully elsewhere, a good strategy is to attempt to establish an analogy. A good example of the grip of analogical models would be the role of the inverse square laws in the development of *CPM* (see Sintonen (1985)). Finally, there are the constraints and links, in force during the period, which narrow down possible and plausible interpretations for generalizations and laws.

Inquiry, then, proceeds by attempting to refine questions by finding suitable vocabularies to express answers, and by exploring possible laws expressed by means of these vocabularies. The I-model differs from the traditional hypothetico-deductive notion of theory-construction in many ways, but one important departure is that auxiliary singular (and general) facts can and must be surveyed during an interrogative derivation. When joined with the idea that search for specializations and theoretizations involves strategic thinking, the model refines the familiar notion that asking a good question at the right time is more effective than performing countless but aimless deductions. It is also clear that portraying inquiry as such a game explicates some important aspects which are hard to formulate linguistically: as the historicist critics of the received view have emphasized, scientists not just follow rule books but also must make practical choices. (See Stegmüller (1979), p. 34-35.) There is a tacit dimension shot through with non-propositional know-how. One important scientific skill is that of

[9] For more flesh around this abstract description, see the development of (Newtonian) *CPM* in Moulines (1979), and Architectonic.

recognizing analogies and similarities, and another one that of organising one's questions in an efficient way, with respect to money, time and other resources (See Sintonen (1993)).

There is a chicken-and-the-egg problem in problem-solving models which can now be addressed, if only briefly. Thomas Nickles (1980) has argued that a viable problem- solving model starts with the assumption that (conceptually deep) problems and not theories are the cenral units of analysis. I have argued (Sintonen(1985)) that this is a false contract: questions can only be identified on the basis of theories, and the structure they have is inherited from theories. But the point can be put more strongly than this, for often developments in background theories are decisive, in the two ways described: they enable new types of vocabularies and grammars for the expression of empirical questions, and they provide tools for answering them. This phenomenon has not been clearly articulated in the literature, but it follows from the I-model. The conditions for a conclusive answer to a who-question includes the requirement that the questioner be able to identify the person. Analogously, the conclusion for a principal why-question may require not just that the inquirer knows that there is a function of a certain type; she may actually have to know, to identify, what that function is. The generalization of this observation is that the development of mathematics, constructing well-behaved theories for base-sets, is crucial for progress in science. Just witness the vistas opened by the emergence of differential and integral calculus, or Hamiltonian equations, or practically any mathematical theory.

Let us take, next, the metatheoretic problem of the logic of discovery (thus named, though never seen in daylight). There is a sense in which the structuralist-interrogative proposal, however informally cashed out here, comes closer to the very idea than views advanced before. The standard objection to discovery programmes has been that the generation of new ideas and specific hypotheses ultimately is a matter of guess and luck, for there can be no logical way of having ideas (for a review, see Nickles (1980)). With the rise of the hypothetico-deductive notion of theory formation, these issues were demoted to the psychology of creative thinking and to the history of science. Altough there has been a reappraisal, even the friends of discovery seem to think that the issue is not the logic but the heuristics of discovery.

We can agree that discovery is too important to leave it to psychologists and historians. But we can do better than this. It is often impossible to divide the process of inquiry into the two stages or even contexts of discovery and justification. We can now see why. First, one must specify what exactly is discovered. As Kuhn (1957) has shown, it is not clear what it is that Copernicus was the first to discover. Newtonian mechanics or *CPM* is an equally tough case, for althoug there was a baptizing period of the basic idea, it took some two hundred years to mature. And during this process of development discovery and justification episodes alternated. This is of course a built-in feature of the sketched structuralist-interrogative model. The I-model takes it for granted that there are justificatory moves in the midst of the pursuit of a full-fledged theory, and even during the generation of a basic idea. The friends of discovery are right in the demand to modify the discovery/justification bifurcation of recent decades, for historical case studies show that it is untenable. But it does not follow that there is no logic to discovery.

8. Why-Questions and Explanatory Commitments

I indicated, in the opening section, that the interrogative view of inquiry has had its comeback in the theory of explanation. However, although Hempel and Oppenheim (1948) commenced with the observation that explaining phenomena amounts to answering why-questions, the perspective was not endorsed in the analysis. I also suggested two reasons for this, viz., the lack of an erotetic logic, and the fear of subjectivism. In what follows, I shall suggest that we can salvage the rational core of the initial intuition. A satisfactory view of explanation emerges hand in hand with a satisfactory theory notion. This is a continuation of the line already taken: erotetic logic must join hands with a rich theory notion, one which indicates how structured background knowledge constrains potential answers.

To spell out this suggestion, we need to take a closer look at explanation-seeking questions. The most serious attempt to salvage the covering law notion within an interrogative yet objective setting was in *BP*. Although it made some progress, there is a special difficulty about explanation-seeking why- and how- questions (including 'What is the reason for q?' -questions) which *BP* addressed, but did not overcome. For many why- questions are opaque to the inquirer: their desiderata are formulated in promissory or vague terms. In terms of the logic of questions the difficulty is this. Propositional yes-no -questions (and which-questions) and wh-questions (see section II above) are transparent to the inquirer, in that the addressee is typically offered a list of alternatives (as from a tray of hors d'oeuvres, as Belnap and Steel (1976, 17) put it). And even if there is no finite list, a question should at least indicate in rather precise terms what ***kind*** of entity would count as a direct answer.

Now why-questions which can be translated into which-questions do fit this trouble-shooting pattern, but there are why-questions where the difficulty is the unavailability of clear potential answers. These are the cases where there is a sound and partially pragmatically motivated question for which there either is no candidate theory-element and hence no descriptive vocabulary, or for which there is a descriptive vocabulary but no candidate laws, and finally, questions for which there are candidate laws, but not one which fits the target phenomena with tolerable accuracy. It is to BP's credit that it singled out these why-questions. The above account refines the BP notions of p- and b-predicaments (Bromberger (1962)). In such predicaments, the inquirer, for Bromberger the rational ignoramus, does not even know what kinds of facts to look for.

Belnap and Steel specifically excluded such philosophical or scientific puzzles, precisely because they are erotetically stubborn. For an encompassing I-model the difficulty is serious, for the questions-to-nature idea seems to turn into a powerless metaphor. If answerhood cannot be settled by help of an observation or experiment, Nature cannot be forced to give unambiguous answers. There is no direct way in which Nature can be forced to answer why-questions. To put it in terms of the insight of the tradition, where Nature is not offered a well- defined choice, She is uncooperative (or does not even understand the questions).

I have suggested that the main reason why Hempel's and Oppenheim's suggestive interrogative opening – that explanations are answers to why-questions – did not turn into a satisfactory analysis was simply that the semantics of why-questions is not as well-behaved as that of wh-questions (see Sintonen (1989)). To take the simple example, consider Ignaz Semmelweiss's

(?p) There are childbed fevers because p?

When aired, this question was an obviously sound and legitimate natural-language question. However, although it carried the pragmatic type-tag of a causal query, its semantics gave no clues whatever to plausible and acceptable potential answers. In fact, in search of a true and conclusive answer, Semmelweiss ran through a series of potential answers which cited a number of possible causal agents from overcrowded wards and different climates to careless handling and the fear induced in mothers by priests visiting the sick and the deceased.

Hempel (1966) used this example to illustrate the impossibility of the narrow empiricist notion of inquiry; inquiry does not start with fact hunting, but rather with a precise problem or a question. But what is a precise question? Where do precise questions come from? Hempel had the answer on his lips, but neither he nor Bromberger managed to give it an adequate formulation. The reason is of course that the entire logical-empiricist *Fra-*

gestellung in the theory of explanation is tied to the logical-empiricist view of theories. And from the point of view of scientific explanation (as well as other metatheoretic applications), the crucial difference does not lie in the ontology - whether theories are set-theoretic or linguistic entities - but in the sundering apart of theories and intended applications.

The way out of this involves two steps. First, the notion of answerhood must be widened to cover also why-questions. Secondly, to avoid trivializing answerhood, answers must be required to conform to pragmatic restrictions imposed by background knowedge. The first step comes from the notion of answerhood in Hintikka's I-model, the second from the structuralist notion of a theory. The notion of a conclusive answer in the I-model amounts to a direct response which is not just of a correct syntactic and semantic type but also suffices to bring about the desideratum of the inquirer. As I indicated above, a correct direct answer is conclusive for an inquirer in a context only if it implies, together with what the inquirer background knowledge is (and with what can be inferred in the situation), the truth of the desideratum. Although logically precise, the notion gives pragmatic leeway. For different inquirers may have different desiderata in raising a question, and may also have different epistemic backgrounds. It is therefore possible that different inquirers pick out different answers, although these answers may come from the same complete story (see Railton (1989)).

The second step involves acknowledging that this is as far as erotetic logic can take us, and that further help can only be obtained from theories sufficiently strong to narrow down admissible answers. In a paradigm-covered evolution of a theory-net, the basic theory-elements, neighbouring elements, as well as links serve to pinpoint lacunae of various sizes and shapes. What is missing both from Hempel's and from the original BP is the acknowledgment that background knowledge is not all that amorphous, at least not within the mature sciences. Normal scientific problem-solving simply is a process of answering well-defined questions. However, problems are determinate to various degrees: what is needed for a viable problem-solving model is an account in which problems are made more precise and thereby amenable to solution by help of a theory. The description of inquiry as a process in which new pragmatic presuppositions are designed for constraining admissible answers is based, in part, on this intuition.

Moreover, there is a qualitive change in the types of questions, for the rich theory notion chops the unmanageable initial why- questions into more manageable wh-questions and 'small' yes-no-questions. A query which starts as a loose why-question, formulated in the language-in-use of the scientific community, and concerning an area of experience, turns (in the 'paradigmatic' stage in which there is a conceptual core $K(T)$ with its vocabularies) into a what question: what special laws are needed to govern this particular set of applications $I_0 \subseteq I(T)$? What further types of theoretizations could be added to the existing elements to yield more stringent questions, and thereby claims? It is a further task to find yes-no- and which-questions portraying specific alternatives and still narrowing down admissible substitution instances.

There are still further advantages in the structuralist theory-notion, concerning explanation and the aims of science. The account given above makes no specific commitment as to the rationale of explanation. To follow the terminology of Wesley Salmon (1984, 1989), there seem to be three basic intuitions concerning explanation. According to the epistemic view (this is Stegmüller's (1983) informational view) the function of scientific explanations is to make (singular) events nomically expected. The modal view insists that explanations must somehow show that an event had to happen, and the ontic view has it that explanations specify causal mechanisms. This classification is not completely successful, for, e.g., the recent revival of the unificationist notion does not fit in easily (see e.g. Friedman (1974), Kitcher (1989), Sintonen (1983) and (1984), Thagard (1978).

Salmon regards the erotetic view as a variant of the epistemic conception, but this is artificial: all three can be given an erotetic formulation, distinguished in terms of the types of questions, and the answers allowed (see Sintonen (1984), (1989)). I shall not go into the details of the model, but rather point out that something like the structuralist notion is needed to make sense of both the informational and the unificationist views, the two most plausible accounts.

Note, to begin with, that the structuralist notion covers both explanations of laws (or generalizations) and singular facts. Establishing an intertheory relation or link between a basic core and its specializations (and theoretizations) counts as theoretical explanation. Similarly, a deduction (or, more typically induction) on the level of a theory-element counts as a singular explanation (provided some further restrictions are heeded). But do we need to choose between the epistemic and the unificationist view? Is there a unitary notion of explanation? I think these questions must remain open, for the time being. The erotetic view as such is non-committal, because it allows both answers which increase the expectedness of a phenomenon, and answers which unify the inquirer's total view. Despite Salmon's recent turning away from the epistemic approach to the ontic one, an answer which merely increases nomic expectedness may suffice.

Yet there is a grain of truth in the criticisms of the epistemic notion which points towards a conceptual link between the epistemic and the unificationist one. For suppose we agree that the rationale of a scientific explanation is to increases one's ability to foresee phenomena. Such increase could then be represented by Peter Gärdenfors's (1980) sophisticated pragmatic model of explanation. The model enriches the logical tools of previous formal models by an account of belief states and belief changes. On that model, explanations are assessed from the point of view of the epistemic states of a rational ignoramus whose *belief function* assigns to sentences in her language, describing singular events, their degrees of belief. Assuming, then, that a sentence S is unexpected, i.e., has a low belief value for the inquirer in a context, and assuming that the event described nevertheless materializes, the inquirer has a problem which in our erotetics is translated into the why-question '$(?p)S(p)?$'. Now what would be a good or conclusive answer? Gärdenfors does not insist on an answer which brings high value, but on one that raises the value a little bit at any rate. The more the better. This increase must be relative to the initial epistemic state, otherwise we would land up with the paradoxes that troubled the deductive-nomological model[10].

The proposal gives a precise measure for the explanatory relief administered by an answer. It suggests the plausible advice of aiming at the answer which, when added to the initial belief state, maximizes the belief value of S. It excels in explicative virtues, and it is a decisive step towards a viable epistemic notion. However, one question remains partially open. Although Gärdenfors's model is dynamic in that it depicts the path of the rational ignoramus from the innocent knowledge state (in which there is no serious thought that S materializes) to the intermediate state (where S could no longer be denied), to the happy end state, there is no explanation of why she would want to proceed to the end state. Assuming, namely, that making things expected (here, raising belief value), is all there is to explanation, why should she bother searching for an answer? For once she learns that S, its belief value reaches the maximum and couldn't possibly be raised. Furthermore, in statistical explanations of an S, the belief value of S is *higher* in the problem state than in the allegedly harmonious end state because probability laws only make S probable to the specified degree. Therefore, if high belief value is the desideratum, one should be happy in the problem state in which one knows that S, but does not know why.

There are also further reasons to think that the epistemic notion alone is insufficient. We live amidst a plethora of sentences whose belief values

[10] See Stegmüller (1983) for a lucid account of the paradoxes. See also Sintonen (1984a) and (1989) .

are high - but which can nevertheless be explained. On the other hand, there are events which are highly improbable but do not worry us a bit. These commonplaces give rise to a general difficulty. The epistemic conception proceeds on the premise that explanation-seeking questions arise because there are phenomena which are either improbable or even impossible, given available knowledge. However, there are weaker forms of problem situations than improbabilities and inconsistencies. In fact, mere compatibility between a theory and a description of a phenomenon can be a problem if there is a 'premium' for solving it (Laudan (1977)), or if the phenomenon falls in the domain of responsibility of a theory (Nickles (1981)). The difficulty of the epistemic notion is that, to put it in Aristotle's terms, it is unable to differentiate between knowledge that something is the case and reasoned knowledge. Gärdenfors's explication takes a step towards a solution by providing the rational ingnoramus with a dynamic pespective. Yet, there is nothing to tie these epistemic states together, nor motivation for the inquirer to worry over isolated beliefs or weaker types of epistemic disintegrity. So, what makes the inquirer responsible for his commitments in the problematic epistemic state? When is there a premium for solving a problem?

According to the traditional view, the range of applications falls outside the theory. Theories are taken to have universal scope, or else the applications must be separately specified. Since the intended applications are inessential for the theory's identity, phenomena which do not contradict (or which are not improbable vis-à-vis) a theory do not emerge as problems. In contrast to this, the structuralist notion of a theory-net specifies, albeit not extensionally, its domain of intended applications. Moreover, since a theory, or rather, a holder of a theory, typically makes the very strong claim that all members of a set of intended applications are models of the theory, there is a constant discrepancy between what a theory claims to be good for, and the results harvested by a given time. Unfinished business is therefore the rule in empirical sciences, and since a theory-net both specifies a set of open questions and some tools for answering them, even unrelated phenomena can give rise to pressing questions. Thus an inquirer can be in the claiming and querying moods at the same time. During the course of evolution of a Kuhn-theory, the communal bond to the basic theory element (or theory-elements, if there are more than one) remains strong, although claims and queries concerning specializations and theoretizations come and go.

References

Achinstein, P., 1983, *The Nature of Explanation*, New York and Oxford:

Oxford University Press.

Åqvist, L., 1965, *A New Approach to the Logical Theory of Interrogatives*. Uppsala.

Åqvist, L., 1971, 'Revised Foundations for Imperative, Epistemic and Interrogative Logic', *Theoria* 37, 33-73.

Bacon, F., 1902, *Novum Organum, or True Suggestions for the Interpretation of Nature*. Ed. by Joseph Devey, American Home Library, New York, (First published in 1620).

Balzer, W. and Sneed, J.D., 1977/78, 'Generalized Net Structures of Empirical Theories', *Studia Logica* 36 and 37, 195 - 211 and 167-194.

Balzer, W., Moulines, C.U., and Sneed, J.D., 1987, *An Architechtonic for Science*, Dordrecht.

Belnap, N.D., 1969, 'Questions: Their Presuppositions, and How They Can Fail to Arise', in K. Lambert (ed.) *The Logical Way of Doing Things*, New Haven and London.

Belnap, N.D. and Steel, T.B., Jr., 1976, *The Logic of Questions and Answers*, New Haven and London.

Braithwaite, R.B., 1953, *Scientifc Explanation. A Study of the Function of Theory, Probability and Law in Science*. Cambridge University Press, Cambridge.

Bromberger, S., 1965, 'An Approach to Explanation', In: Butler, R.J. (ed.)., *Analytical Philosophy, 2nd Series*, Oxford: 72-103.

Bromberger, S., 1966, 'Why Questions', In Colodny, R. (ed.), *Mind and Cosmos*, Pittsburgh, 86-111.

Bromberger, S., 1992, *On What We Don't Know When We Don't Know Why*.

Carnap, R., 1956, *Meaning and Necessity*, Chicago, (2nd edition).

Collingwood, R.G., 1939, *An Autobiography*, London.

Collingwood, R.G., 1940, *An Essay in Metaphysics*, Oxford.

Fraassen, B.C. van., 1970, 'On the extension of Beth's Semantics of Physical Theories'. *Philosophy of Science*, 325-338.

Fraassen, B.C. van., 1980, *The Scientific Image*, Oxford: Clarendon Press.

Friedman, M., 1974, 'Explanation and Scientific Understanding', *The Journal of Philosophy* LVVI, No. 1., 5-19.

Gadamer, H.-G., 1960, *Wahrheit und Methode, Grundzüge einer philosophischen Hermeneutik*, Tübingen.

Gale, S., 1978 'A Prolegomenon to an Interrogative Theory of Scientific Inquiry', In Hiz, H. (ed.), *Questions*, Dordrecht, 319-345.

Gardenfors, P., 1980, 'A Pragmatic Approach to Explanation', *Philosophy of Science* 47, 405-423.

Hamblin, C.L., 1958, 'Questions', *Australasian Journal of Philosophy* 36, 159-168.

Hanson, N.R., 1958, *Patterns of Discovery*, Cambridge.

Harrah, D., 1963, *Communication. A Logical Model.* Cambridge, Mass.

Hempel, C., 1965, *Aspects of Scientific Explanation and Other Essays in the Philosophy of Science.* New York.

Hempel, C. and Oppenheim, P., 1965, 'Studies in the Logic of Explanation', Reprinted in Hempel, 245-295, (Originally published in 1948).

Hempel, C., 1966, *Philosophy of Natural Science*, Englewood Cliffs, New Jersey.

Hintikka, J., 1976, *The Semantics of Questions and the Questions of Semantics*, Acta Philosophica Fennica, Vol. 28, No 4. North Holland.

Hintikka, J., 1981a, 'On the Logic of an Interrogative Model of Scientific Inquiry', *Synthese* 47, 69-83.

Hintikka, J., 1981b, 'The Logic of Information-Seeking Dialogues: A Model', In W. Becker and K. Essler (eds.), *Konzepte der Dialektik*, Frankfurt-am-Main, 212-231.

Hintikka, J., 1984, 'The Logic of Science as a Model-Oriented Logic'. In Asquith, P.D. and Kitcher P. (eds.), *PSA 1984* vol 1, East Lansing, Michigan, 177-185.

Hintikka, J., 1985, 'A Spectrum of Logics of Questioning', *Philosophica* 35, 135-150.

Hintikka, J., 1987, 'The Fallacy of Fallacies', *Argumentation* I, 211-238.

Hintikka, J., 1988, 'What Is the Logic of Experimental Inquiry?', *Synthese* 74, No. 1, 173-90.

Hintikka, J., 1989, 'The Role of Logic in Argumentation', *The Monist* 72, No. 1.: 3-24.

Kant, I., 1933c, *Critique of Pure Reason*, translated by Norman Kemp Smith, The Macmillan Press, London and Basingstoke, (Published in 1787).

Kitcher, P., 1989, 'Explanatory Unification and the Causal Structure of the World', In *Minnesota Studies in the Philosophy of Science*, Vol. XIII. Ed. by P. Kitcher and W.C. Salmon. Minnapolis: University of Minnesota Press, 410-505.

Kuhn, T., 1970, *The Structure of Scientific Revolutions*, Chicago (2nd edition).

Kuhn, T., 1977, 'Objectivity, Value Judgement, and Theory Choice', In T. Kuhn, *The Essential Tension: Selected Studies in Scientific Tradition and Change*, Chicago.

Laudan, L., 1977, *Progress and Its Problems. Towards a Theory of Scientific Growth*, London and Henley,

Maunu, A., 1993, 'Independence-Friendly Epistemic Logic and its Uses', In O. Koistinen and J. Räikkä (eds.), *Good Reason*, Turku.

Moulines, C.U., 1979, 'Theory-Nets and the Evolution of Theories: The Example of Newtonian Mechanics', *Synthese* 41, 417 - 439.

Nickles, T. (ed.), 1980, *Scientific Discovery, Logic, and Rationality*, Boston Studies in the Philosophy of Science, Vol 56, Dordrecht-Holland/Boston-U.S.A./London-England.

Nickles, T., 1981, 'What is a Problem that We May Solve It?', *Synthese* 47, 85-118.

Niiniluoto, I., 1980, 'The Growth of Theories, Comments on the Structuralist Approach', *Proceedings of the Second International Congress for History and Philosophy of Science, Pisa, 1978*. Dordrecht, 3-47.

Pearce, D. and Rantala, V., 1983, 'New Foundations for Metatheory', *Synthese* 56, 1-26.

Popper, K., 1972, *Objective Knowledge*, Oxford.

Railton, P., 1989, 'Explanation and Metaphysical Controversy', In *Minnesota Studies in the Philosophy of Science*, Vol. XIII. Ed. by P. Kitcher and W.C. Salmon, Minnapolis, 220-252.

Rantala, V., 1980, 'On the Logical Basis of the Structuralist Philosophy of Science', *Erkenntnis* 15, 269-286.

Salmon, W., 1984, *Scientific Explanation and the Causal Structure of the World*, Princeton, New Jersey.

Salmon, W., 1989, 'Four Decades of Scientific Explanation', In *Minnesota Studies in the Philosophy of Science*, Vol. XIII. Ed. by P. Kitcher and W.C. Salmon. Minnapolis, 3-219.

Scriven, M., 1959a, Explanation and Prediction in Evolutionary Theory, *Science* 130, p 477-482.

Scriven, M., 1959b, Truisms as the Grounds for Historical Explanations, In Gardiner, P. (ed.), *Theories of History*, New York, 443-475.

Scriven, M., 1962, 'Explanations, Predictions, and Laws', In H. Feigl and G. Maxwell (eds.), *Minnesota Studies in the Philosophy of Science*, Vol 111, University of Minnesota Press, Minneapolis, 170-230.

Scriven, M., 1975, 'Causation as Explanation', *Nous* 9, 3-15.

Sintonen, M., 1989, *The Pragmatics of Scientific Explanation*, Acta Philosophica Fennica, Vol. 37. Helsinki.

Sintonen, M., 1984, 'On the Logic of Why-Questions', *PSA 1984, Vol. One*. Edited by P.D. Asquith and P. Kitcher. East Lansing, Michigan, 168-176.

Sintonen, M., 1985, 'Separating Problems from their Backgrounds: A Question-Theoretic Proposal', *Communication and Cognition* 18, 25-49.

Sintonen, M., 1989, 'Explanation: In Search for the Rationale', *In Minnesota Studies in the Philosophy of Science*, Vol. XIII. Ed. by P. Kitcher and W.C. Salmon. Minnapolis, 253-282.
Sintonen, M., 1990, 'Taming a Regulative Principle: From Kant to Schlick', In Rescher N. (ed.), *Aesthetic Factors in Natural Science*, Lanham, Maryland, 47-56.
Sintonen, M., 1990, 'How to Put Questions to Nature', In Knowles, D. Ed., *Explanation and Its Limits*, Cambridge, 267-284.
Sintonen, M., 1990c, 'Basic and Applied Research: Can the Distinction Still Be Drawn?', *Science Studies*, 23-31.
Sintonen, M., 1991, 'How Explanatory Theories Face Reality', *Synthese* 89.
Sintonen, M., 1993, 'In Search of Explanations: From Why-Questions to Shakespearean Questions', *Philosophica* 51, 1-27.
Stegmüller, W., 1979, *The Structuralist View of Theories, A Possible Analogue of the Bourbaki Programme in Physical Science.* Berlin, Heidelberg and New York.
Stegmüller, W., 1983, *Erklärung, Begründung, Kausalität*, Zweite, verbesserte und erweiterte Auflage, Berlin, Heidelberg, New York.
Thagard, P., 1978, 'The Best Explanation: Criteria for Theory Choice', *The Journal of Philosophy* LXXV, 76-92.
Toulmin, S., 1963, *Foresight and Understanding.* New York.
Tuomela, R., 1978, 'On the Structuralist Approach to the Dynamics of Theories', *Synthese* 39, 211-231.
Tuomela, R., 1980, 'Explaining Explaining', *Erkenntnis* 15, 211- 243.

Chapter 5: Pragmatic and Diachronic Aspects of Structuralism

Werner Diederich

In the following pages I am going to report efforts, made by structuralists, to reconstruct Kuhn's ideas of developmental patterns of science (sec.1). These explications have been made with the help of certain pragmatic notions (sec.2) which allow to conceive of 'theory dynamics' in a precise way (sec.3). Finally the issues of incommensurability and realism will be addressed to (sec.4).

1. Reconstructing Kuhn's conception of scientific development

In the early seventies the philosophers of science community only hesitatingly received structuralism as put forward by Joseph D. Sneed in his *The Logical Structure of Mathematical Physics* (1971). The one notable exception is Wolfgang Stegmüller's enthusiastic response in his *Theorienstrukturen und Theoriendynamik* (1973, engl. translation 1976), and this book was then mainly received for its 'reconstruction' of Thomas Kuhn's theory of how science develops in alternating phases of continuity and discontinuity. Stegmüller confessed how difficult it has been for him, as a philosopher in the tradition of logical empiricism, to accept Kuhn's 'irrationalist' image of the history of science and how he felt relieved when he got to know Sneed's concept of a scientific theory and learnt how this concept may be used to describe in a precise way the patterns of scientific development as found by Kuhn.

Many philosophers of science were rather surprised when in 1975 Th. Kuhn in a way endorsed the structuralist reconstruction of his ideas. This happened on a symposium, held at the International Congress for Logic, Methodology, and Philosophy of Science in Ontario/Canada, with Sneed, Stegmüller, and Kuhn (see *Erkenntnis* 1976). This surprise is telling for a then (and even now) widespread misunderstanding of the method of formal explication which had been excessively used by logical empiricists, but by itself in no way commits its user to this sort of philosophy. Maybe this misunderstanding had been encouraged, at least for German readers, by the fact that Stegmüller's 1973 book had appeared as the second part of his *Theorie und Erfahrung* (1970), a mostly conventional account of the 'received view', as if Kuhn's idea could somehow be incorporated into and reconciled with logical empiricism. It should be stressed, however, that Stegmüller made quite clear that for him the old view's applicability was confined to certain synchronic or a-temporal structures.

As is well known, Kuhn distinguishes between 'normal' and 'revolutionary' science. Both kinds of science have characteritic features. The overall pattern of scientific development is considered to consist of alternating phases of these two kinds. Let us begin with *normal science*. It is cumulative and reigned by a 'paradigm', i.e. basically a certain theoretical structure successfully applied to a number of problems by methodological standards which need not be explicit but are internalized by the adherents of the paradigm, i.e. the members of the 'scientific community' connected with the paradigm. (I do not go into the details of Kuhn's well-known picture of normal science.) The most typical aim of normal science is the transfer of paradigmatic solutions to further cases and to eventually confirm that all the theory's 'intended applications' (a structuralist term) actually are accessible to the paradigmatic methods and types of solution.

The claim that the paradigm's theoretical structure is extendable to further cases is, of course, subject to possible failure: the scientific community may not be able to actually apply the theory to certain cases and thus may be urged (by resistance of 'reality') to remove these cases from the domain of intended applications. In this sense there may be no continually strengthened claim connected with the paradigm's theory and thus an inductivist perspective of science is challenged. But Popperian falsificationism is questioned, too. The failure to apply the theory in a certain situation does not falsify the respective claim of the theory (since it may be applied to this situation later on), let alone the theory itself; strictly speaking, the latter is not even possibly subject to falsification, since it is, in itself, no statement at all, but a certain structure (and a heuristic to generate claims about phenomena of a certain domain). This is, very roughly, the idea of the 'non-statement-view' of theories, stressed especially by Stegmüller in the early times of structuralism. What comes closest to the falsification of a theory in Popper's sense is the abandonment of a whole paradigm during a 'scientific revolution' (see below). But as long as a scientific discipline develops normally there is a certain 'immunity against falsification': a trait of science stressed by Kuhn and underestimated by Popper (and Feyerabend).

It is not my purpose here to argue for or against Kuhn, but to report, and, to a certain degree, re-evaluate the structuralist reconstruction of his ideas. As far as normal science is concerned my impression is that, all in all, structuralism fairly represents Kuhn's picture of scientific development, if one puts less stress on the sociological and psychological connotations in

Kuhn's conceptions. And if scientific theories roughly are the way structuralists describe them (as I tend to think), the reconstruction of Kuhn makes a strong case for Kuhn against Popper and other critics. (Some of the concepts used for this reconstruction are discussed in sections 2 and 3, below.)

Most philosophers of science, however, considered Kuhn's idea of normal science less spectacular than his concept of *scientific revolution*. The occurence of such breaks in the history of science runs counter to the widespread belief in a continuous and cumulative development of science. Logical empiricism had not really delt with questions of theory dynamics, but certainly most of its adherents shared the conviction that science steadily progresses. (By way of irony, Kuhn's epoch-making book appeared as a volume of the 'International Encyclopedia of Unified Science', founded by Neurath and other empiricists.)

As has been repeatedly observed, there are scientific revolutions where the pre-revolutionary paradigm is completely abandoned, and others where the older paradigm is somehow incorporated into the new one. The revolution in chemistry connected with Lavoisier is often regarded as a revolution of the first sort, while the Einsteinian revolution 'keeps' the Newtonian solutions to a vast number of problems where velocities are much smaller than the velocity of light.

Sneed and Stegmüller focus on this latter kind of revolution and ask: how can the relation between pre- and post-revolutionary paradigm precisely be described? Again they concentrate on the theoretical structures comprised in the respective paradigms. They propose that after a successful revolution (of the second type) the older theory can be *reduced* to the newer one. Because of the phenomena of incommensurability etc. (cf. sec. 4, below), a reduction relation that works as here desired cannot be of the Nagel type (i.e. basically consisting in logical derivation of the old theory from the new one, after defining the old theory's concepts through the new theory's ones). The reduction relation proposed by structuralism works via a certain correlation between the so-called partial potential models of the two theories, i.e. mainly between the two theories' respective intended applications. This correlation is required to 'transfer' the old theory's successful applications into successful applications of the new one. No definitions or derivations need to be involved in this correlation (cf. again sec. 4, below).

There are, however, in my eyes, severe problems connected with this construal of 'theory dislodgement'. Firstly, there is the problem that the formal conditions put forward for the reduction relation seem far too weak in the sense that, taken as defining 'reduction', would allow for lots of unintended 'reduction relations'. (This kind of problem seems typical not only for structuralism in general, but for other sorts of formal explication as well; it tends to surface whenever formal conditions are put forward not (only) as necessary, but (also) as sufficient conditions. I will not go into this here.)

More severe is the fact that the aspect of scientific revolutions focussed on by structuralism is but one in a number of characteristics, maybe even only a minor one (in some historians' eyes). The re-orientation of scientific communities with all its sociological and psychological factors may legitimately be considered as far more important for Kuhn's views. That does not, of course, devaluate the deep insights into intertheory relations won by the structuralist concept of reduction; but primarily it is this side effect of the reconstruction of Kuhn, not so much this reconstruction itself, which remains interesting.

In 1977 Paul Feyerabend - by then known as a radical critic of (analytic) philosophy of science - published an appraisal of structuralism in *The British Journal for the Philosophy of Science*, in which he delivered a detailed critique, not just a harsh rejection on general grounds, of the 'Sneedification' of Kuhn. Stegmüller responded with a booklet in 1979 (cf. Diederich (1982)). After this exchange there evidently has been only little further work on the structuralist reconstruction of Kuhn. However, some of the pragmatic and diachronic concepts developed in this course have been further refined and elaborated, and some episodes of the history of science have been reconstructed in a fairly detailed way (see sections 2 and 3 below).

2. Pragmatics

In the first accounts of structuralism by Sneed and Stegmüller (1971 and 1973, respectively) the key pragmatic concept was 'to have (or hold) a theory' or '*über eine Theorie verfügen*', respectively. Holding a theory, in this sense, roughly amounts to the corroborated belief that a certain expansion of the theory's core actually applies to the theory's 'intended applications'. Later on, after the notion of 'expanded theory-core' had been replaced by the concepts of 'theory-element' and '(specialization)net of theory-elements' (cf. Balzer & Sneed (1977/78)), C. Ulises Moulines introduced the notion of a *pragmatically enriched theory-element* (Moulines (1979)). Such a theory-element was conceived as comprising, besides a theory core K and a domain I of intended applications, a *scientific community, SC*, and an *historical period, h*. Still later on, this concept was modified to that of a *diachronic (idealized) theory-element*, $\langle K, I, G\rangle$, where G is a scientific generation, i.e. the set of those members of a scientific community who are active within a certain historical period (cf. *Balzer & Moulines & Sneed* (1987), ch. V, esp. p. 216).

More or less all these concepts serve the same purpose, namely to express propositional attitudes of persons with respect to theories. Since I do not here intend to go into more technical considerations, let me use the following concept as the basic one: a ***pragmatic theory-element*** is a pair $< K, SC >$, where K is a theory-element core and SC a scientific community. This kind of minimal (semi-) formalization closely resembles the one given by Moulines in a 1991 paper (Moulines (1991a)). Note that in this formalization the domain of intended applications, I, is not specified. The reason why there is no need for such a specification is the following. The set I of intended applications of a theory-element has always been understood as a pragmatic notion (despite the formal requirement that I is a subclass of the set M_{pp} of partial potential models). If, however, SC is introduced as a component in a (pragmatic) theory element, I may be defined via SC and a propositional attitude (cf. Moulines (1991 a)): I is the set of those partial potential models, for which SC intends to show that they belong to the content of the core (i.e. that the core K can, in fact, be successfully applied to them).

Defining I via SC, by the way, corresponds to a 'sociological' reading of Kuhn: not a common belief in (the applicability of) a certain theory/paradigm defines a scientific community, but *vice versa*. At least at the time when Th. Kuhn wrote his 1969 postscript to *The Structure of Scientific Revolutions* (1962) he seemed to prefer, methodologically, this order of the two key concepts. Indeed, the strongest version of Kuhn's sociologism allows to consider K itself as dependent on SC. I would like, however, to leave the question of primacy of one of these two components open and stick to the neutral construal of a theory-element as $< K, SC >$.

3. Diachronics

The key diachronic concept is that of a '***theory-evolution***' (cf. *Architectonic* (1987), V.2, p. 216 ff). In a simplified way, this notion may be described as based on both the pragmatic notions considered in sec. 2 above and the '***specialization***' relation between theory-element cores (defined by inclusion of the respective classes of models, etc. - see also Ch. I of this volume). By this relation, (pragmatic) theory-elements may be tied together to form '(pragmatic) theory-nets'. Each member of such a net is understood to refer to the same scientific community. (I.e., in the minimal formalization of sec. 2, above, a pragmatic theory-net is a specialization net of theory-element cores, each of which is attached to the same scientific community.)

Temporal growth of a pragmatic theory-net is construed as a sequence of such nets where every net 'contains' the 'earlier' ones in the sense that all its 'knots' (i.e. theory-elements) are linked to knots of earlier nets by way of specialization. Such a 'growing' theory-net is called a *theory-evolution*. This notion is still very 'theoretical': an inventive (or fanciful) scientific community may easily produce rapidly growing theory-nets as mere paper-work. To capture real (scientific) life examples of theory-evolution constraints (in a non-technical sense) were put on the net-members' domains of intended applications with respect to *confirmation*: the domains of confirmed instances should grow ('progressive theory-evolution') and eventually comprise those instances which, at earlier stages, were only assumed as cases to which the respective core is, actually, applicable ('perfect theory-evolution'). If we introduce, as a further primitive pragmatic notion that of a *paradigmatically* acknowledged set of intended applications, I, we may also define, in a fairly straightforward way, what it means for a theory-evolution to be *Kuhnian*.

A suspicious reader might have gotten the impression that all these pragmatic and diachronic notions are somewhat sterile. In fact, there are only very few theorems which have been derived from these definitions. However, the main spin-off should not be seen in an elaborated metatheory of the development of science, but in concrete reconstructions which become possible by applying the structuralistically sharpened tools to actual scientific episodes. Such reconstructions have successfully been carried out by Moulines for classical partial mechanics and simple equilibrium thermodynamics (cf. *Architectonic* (1987), ch. V, sections 3 and 4).

4. Incommensurability and Realism

In a way, the classical problems of incommensurability (Feyerabend, Kuhn) have been circumvented by structuralism: meaning variance is no topic in an extensionalist, model- theoretical approach. More specifically: since the claim connected with a theory is about certain *partial* potential models of the theory, any successive theory that replaces only the 'theoretical superstructure' is about the same kind of entities and may thus be duductively connected with the previous one (contrary to Feyerabend's incommensurability). Sometimes, however, this type of comparison does not quite work, because – like with classical and (special) relativistic particle mechanics – the partial potential models (i.e., in the example, kinematic systems) are to be taken *modulo* different transformation groups. But again, in such cases the underlying theories – the respective geo-chronometries – may be checked against each other (cf. Sneed (1979), Balzer (1979)).

In my eyes, the ontological commitment of a theory does not extend beyond its non-theoretical level; the theoretical components merely are mathematical devices to construe the models into which the non-theoretical structures should fit. If, however, a theory forms the basis of / is presupposed by a 'higher' one (linked to the former by 'theoretization'), its theoretical structure is used to describe the objects of application of the higher theory and thereby looses its ontological innocence. If the theoretical parts of a theory are said to explain the non-theoretical parts, a structuralist should not subscribe to an 'inference to the best explanation' (in the sense of '... best explanans') but only to an inference to the best explanandum: one does not have to believe in theoretical entities before they are referred to by a higher theory.

References

Balzer, W., 1979, 'Ursprung und Rolle von Invarianzen in der klassischen Kinematik', In: Diederich, W. (ed.) (1979), 115-148.

Balzer, W. & Moulines, C.U. & Sneed, J.D., 1987, *An Architectonic for Science*, Dordrecht.

Balzer, W. & Sneed, J.D., 1977/78, 'Generalized Net Structures of Empirical Theories', *Studia Logica* 36, 195-211, 37, 167-194.

Diederich, W., 1982, Stegmüller, 'On the Structuralist Approach in the Philosophy of Science', *Erkenntnis* 17, 377-397.

Diederich, W. (ed.), 1979, *Zur Begründung physikalischer Geo- und Chronometrien*, Bielefeld: Schriftenreihe des Universitätsschwerpunkts 'Mathematisierung der Einzelwissenschaften'.

Feyerabend, P.K., 1979, 'Changing Patterns of Reconstruction', *The British Journal for the Philosophy of Science* 28, 351-369.

Kuhn, T.S., 1962, *The Structure of Scientific Revolutions* (Int. Enc. of Unified Science, vol. 2, No. 2), Chicago, 2nd ed., enlarged, with a Postscript-1969, 1970.

Kuhn, T.S., 1976, 'Theory-Change as Structure-Change', *Erkenntnis* 10, 179-199.

Moulines, C.U., 1979, 'Theory-Nets and the Evolution of Theories: The Example of Newtonian Mechanics', *Synthese* 41, 417-439.

Moulines, C.U., 1991a, 'Pragmatics in the Structuralist View of Science', In: Schurz, G. & Dorn, G.J.W. (eds.) (1991), 313-326.

Moulines, C.U., 1991b, 'Pragmatisch-diachronische Aspekte der Wissenschaftstheorie', *Untersuchungen zur Logik und zur Methodologie* (K.-M.-Univ. Leipzig) 8, 1-21.

Niiniluoto, I. & Tuomela, R. (eds.), 1979, *The Logic and Epistemology of Scientific Change* (Acta Philosophica Fennica 30), Amsterdam.
Schurz, G. & Dorn, G.J.W. (eds.), 1991, *Advances in Scientific Philosophy*, Amsterdam.
Sneed, D.J., 1971, *The Logical Structure of Mathematical Physics*, Dordrecht, 2nd ed. 1979.
Sneed, D.J., 1976, 'Philosophical Problems in the Empirical Science of Science: A Formal Approach', *Erkenntnis* 10, 115-146.
Sneed, D.J., 1979, 'Theoretization and Invariance Principles', In: Niiniluoto, I. & Tuomela, R. (eds.) (1979), 130-178.
Stegmüller, W., 1970, *Theorie und Erfahrung* (Erster Teilband), Berlin.
Stegmüller, W., 1973, *Theorienstrukturen und Theoriendynamik* (Theorie und Erfahrung, Zweiter Teilband), Berlin, (Engl. transl. 1976).
Stegmüller, W., 1976, 'Accidental ('Non-Substantial') Theory Change and Theory Dislodgement: To What Extent Logic Can Contribute to a Better Understanding of Certain Phenomena in the Dynamics of Theories', *Erkenntnis* 10, 147-178.
Stegmüller, W., 1979, *The Structuralist View of Theories*, Berlin.

Chapter 6: Truth Approximation by the Hypothetico-Deductive Method

Theo A.F. Kuipers[1]

1. Introduction

According to the leading expositions of the method of hypothesis or hypothetico-deductive (HD-)method by Hempel (1966) and Popper (1934/59) the central aim of the HD-method is to find out whether a theory is (at least observationally) true or not.

No doubt, HD-testing is a matter of deriving test implications from the theory, and they have to be tested on their turn. The standard interpretation of the test results is the subject of this paper. Assuming a number of idealizations, the test results of the HD-method are supposed to lead either to a falsification or a confirmation of the theory. Since Popper's emphasis on the idea that general empirical theories cannot be verified but only falsified, it has even become a rule to assume that the establishment of a conclusive counterexample is the ultimate goal of testing a certain theory. It is important to note that the term 'confirmation' and Popper's favourite term 'corroboration' both have the connotation that the theory has not yet been falsified. Whereas confirmation/corroboration asks for further tests of the same theory, conclusive falsification asks for a new theory to be tested. Hence the HD-method was interpreted as aiming at falsification and subsequent rejection of the theory.

Due to Kuhn and Lakatos, who followed Duhem and Quine, it became clear that in practice there were lots of complications and so possibly good reasons for avoiding the conclusion of falsification and for refusing the dismissal of a theory after its falsification could be afforded. Hence, according to Lakatos and others, who, unlike Kuhn, were not willing to give up the idea of methodology, the HD-method was to be replaced by some more sophisticated method.

In this paper it will be argued that the apparent devaluation of the HD-method of testing was premature because, due to the one-sided truth-value perspective, its logic never has been well understood. An elementary analysis in *Section 2* will show that it essentially is a method for evaluating a theory in terms of its instantial failures and its explanatory successes,

[1] The author would like to thank Wolfgang Balzer for his stimulating remarks.

i.e., counterexamples as well as explained facts. That is, falsification and explanation constitute the complementary aims of the method.

From this it follows that, even under ideal conditions, a theory, whether or not it already has been conclusively falsified, should not be dismissed, or at least not as long as there is no alternative theory having a better test record. This basically instrumentalist view of the HD-method immediately suggests how to compare two theories in the light of the available evidence, i.e., the test results. This naturally leads to the idea of comparative HD-testing and of the rule of success that prescribes choosing the best theory after a sufficient comparative testing process.

In *Section 3* it will be shown that the rule of success, and hence the HD-method, is efficient for approaching the (strongest relevant) truth within a given context in the following sense. One can explain a difference in success by the comparative hypothesis that the more succesful theory is closer to the truth than the less successful one. Moreover, the opposite possibility, i.e., that the less successful theory be closer to the truth than the more successful one, is excluded. The claims are based on the structuralist explication of greater truthlikeness (see Kuipers (1992)).

In *Section 4* we will compare the merits of the instrumentalist methodology with realist and empiricist methodologies. The standard methodological and sociological interpretations of non-falsificationist practice are put into question, in view of the fact that there is a straightforward justification of the instrumentalist methodology in terms of truth approximation. Moreover, so-called inference to the best explanation, according to which the best explanation may be accepted (for the time being) *as true*, is shown to be in need of important corrections. Our analysis naturally suggests the rule: inference to the best theory *as the closest to the (strongest relevant) truth*.

In *Section 5* our picture of theory evaluation is compared with some important methodological points raised by Popper, Lakatos, and Nowak. We will first evaluate Popper's and Lakatos' emphasis on novel facts from the truth approximation perspective. Then it will be shown that Lakatos' relativization of the notion of a crucial experiment directly follows from that perspective. Finally, it will be argued that the present account of theory evaluation can be seen as an explication, including corrections, and justification of the methodology of Lakatos, but that it also leaves room for the idealization and concretization methodology developed by Nowak and others. Both methodologies are special cases of the instrumentalist methodology, and hence they are functional for truth approximation.

The final *Section 6* concludes, among other things, with a proposal suggested by the main lessons contained in this paper, namely, to replace the term *Context of Justification* by Context *of Evaluation.*

2. The HD-Method and its Instrumentalistic Use

Every statement raises two questions, namely, an instantial and an explanatory one: to what extent is it true, and what does it explain? It is intuitively clear that there is a tension between these two aspects of a statement. The stronger it is, the more it will be able to explain, but the more likely will it be false. Conversely, the weaker a statement is, the more likely is that it will be true, but than the less it will be able to explain. As we will show, the HD-method is a systematic method to answer both questions in detail, at least for the time being, in an effective and efficient way. The analysis will directly suggest the notion of a momentary evaluation report of a theory, consisting of explanatory successes and instantial failures. In turn, this immediately suggests a plausible way of comparing theories and the instrumentalist rule of theory choice.

A statement or a conjunction of statements is called a theory (or hypothesis, but we will use the term 'hypothesis' for rather more specific purposes) in so far as we are interested in an answer to the two questions it raises and as long as they have not been answered. A theory may or may not use so-called 'theoretical terms', besides so-called 'observation terms'. Of course, what is observational is not to be understood in some absolute, theory-free sense, but highly depends on the level of theoretical sophistication.

The HD-argument

Testing a theory not only presupposes that there is an intensionally specified domain of intended applications of the theory, but also that this domain is essentially fixed. It is important to note that the latter assumption is not standard within the structuralist approach. With this qualification in mind, we indicate the set of intended applications standardly by I. In this section we will generally use statement terminology together with structuralist notations, anticipating specific, but not always standard, structuralist reinterpretations of the statement terminology, which will be presented in the next section.

A theory M is tested by deriving from it General Test Implications (GTI's), each of which is to be tested.

Let us first circumscribe what we mean by a GTI. A GTI deals, like the theory itself, with I, and is general due to the fact that I is intensionally specified, and hence not related to a specific (object or) system and place and time. In at least one of these respects I is supposed to be general, but not necessarily in the universal sense. That is, I may be restricted within certain boundaries, e.g. to all systems of a certain type, to all places in a region, to all times in an interval. Moreover, within these boundaries it may concern actual cases or it may concern all 'empirically possible cases'. If a GTI is (assumed to be) true, scientists use to speak about a general fact; more specifically, in the first case about an accidental general fact, and in the second case about a lawlike or nomic general fact, or simply a law. Note that philosophers usually hesitate to talk about general facts, but we prefer to follow the usage of scientists.

A GTI is assumed to be testable in principle as well as in practice. It is testable *in principle* if it is formulated in observation terms. To be also testable *in practice* several specific conditions, depending on the context, will have to be satisfied.

A GTI is an implication or conditional statement in the sense that it claims that to all cases in the domain I satisfying certain initial conditions C a certain other individual fact F applies. The conditionally claimed fact can be, like the initial conditions, of (simple or compound) deterministic or statistical nature. In sum, a GTI is formally of the form:

$$G: \textit{ for all } x \textit{ in } I \; [\textit{ if } C(x) \textit{ then } F(x)\;)]$$

Finally, a GTI is an implication in another respect, namely it is assumed to be an implication of M, i.e., to follow from M by some logico-mathematical derivation. *Scheme 1* represents this *HD-argument*, in which LMC indicates the suggested Logico- Mathematical Claim and MP the Modus Ponens. Of course, it is also assumed that M is necessary for the derivation, i.e., G is not a logico-mathematically provable truth.

$$M$$

$$\text{LMC}: \; \textit{if } M \text{ then } G$$

$$\text{-----------------------MP}$$

$$\text{GTI}: G$$

Scheme 1

Note that the HD-argument in the form above neglects auxiliary (empirical or semantic) hypotheses. Hence, we presuppose that G, and therefore I, C and F, are formulated in the (observational) vocabulary of M.

One of the most well-known examples of a theory with several general test implications, and with rather disparate initial conditions, is Newton's theory of gravity, yielding, among other things, corrected versions of the laws of Galilei and Kepler as GTI's.

Falsification or explanation

Now we start to concentrate on testing GTI's. As a rule, we will just speak of individual and general facts, always assuming that these facts have been convincingly established. Of course, we may nevertheless be mistaken in this, but for the time being we assume that we are not.

Successive testing of a particular GTI will lead to two mutually exclusive results. The one possibility is that sooner or later we come across a falsifying instance or counterexample, i.e., some x_0 in I such that $C(x_0)$ and not-$F(x_0)$, where the latter conjunction may be called a falsifying combined fact.

Assuming that LMC is correct, a counterexample of G is, of course, also a counterexample of M, for not only not-G can be derived from the falsifying combined fact, but also not-M by Modus (Tollendo) Tollens. In other words, we obtain a *falsification* of M by a falsified GTI. The relevant counterexample will be recorded as an *instantial failure* of (G and) M.

The alternative possibility is that, despite variations, all our attempts to falsify G fail, i.e., lead to the predicted results. The conclusion associated with repeated success of G is, of course, that G is established as true, i.e., as a general (reproducible) fact.

Now, the acceptance of G as true is usually called also a confirmation or corroboration of M. This terminology is most inconvenient in case M has already been falsified via another general test implication. Moreover, it does not do justice to the fact that G was derived from M, which means that it (its truth) can be explained by M. Hence, it is more plausible to speak of *explanation* and to call G a general fact explained by M and to be recorded as an *explanatory success* of M.

It may well be that certain GTI's of M have already been tested long before M was taken into consideration. Therefore, the corresponding counterexamples and explained general facts are of course to be included in the test record of M.

Hence, at each moment t, the test record or, better, the *evaluation* report of M (see below) consists of two components:

> *the set* $IF(M,t)$ of instantial failures, i.e., the established counterexamples of M, each associated with a certain GTI of M,
> *the set* $ES(M,t)$ *of explanatory successes*, i.e., the established GTI's of M, that is general facts explained by M.

Note, first, that the two components are concerned with matters of different nature. A counterexample represents a system at a certain place and time, satisfying certain conditions. An explanatory success is a general statement considered to be true. Later on we will see how both can be represented set-theoretically, i.e., as a single structure, and as a possibly infinite set of structures, respectively, both connected with a certain claim.

Note also that we already speak of an explanation of G by M when G is accepted as true and M implies G. Hence, we do not assume that M is supposed to be true. Although it is frequently assumed that a statement can only explain a fact if it is true, in the context of theory evaluation it is useful and customary to talk about explanation independent of its truth. In this way a theory can have instantial failures, and hence have been falsified, and still have explanatory successes. In other words, it enables us to treat the instantial and the explanatory evaluation of the theory as separate questions. The talk of 'potential explanation' would only be possible as long as the theory has not been falsified, and furthermore it is misleading, because it obscures the fact that the derivability of a general fact is a merit irrespective of the fact that the theory is true or false.

Note, in addition, that the explanation of a fact is assumed to be just the standard minimum in another respect as well: it means the derivation of the relevant description of that fact from the theory and perhaps some other assumptions. Hence, we do not assume any further restrictions e.g., causal explanation. We leave the flagpole-problem aside. Note, by the way, that this problem does not even arise when explaining general facts. It only arises when explaining individual facts. Finally, our notion of explanation is of standard, derivative form. We will leave more problematic non-derivative types of explanation aside. Given that the first aspect is liberal and the second a matter of cautiousness our ('object-level') use of the notion of explanation will be acceptable not only for realists, but also for empiricists and instrumentalists.

The derivation of a general test implication from a theory essentially is a prediction of a certain general fact, and if this predicted general fact turns out to be true, this general fact is, according to the above standards, explained by the theory: this is the famous symmetry between explanation and prediction (restricted to general facts). The only difference between explanation and prediction lies in the question whether the relevant fact was established before or after the derivation. This allows us to interchange explanation and prediction, just depending on whether the supposed fact has, or has not yet been established as such. (For discussions about the notion of explanation, see e.g. Schurz (1988)).

Instrumentalist theory comparison: the rule of success

The analysis of the HD-method presented so far has hopefully made explicit what most readers roughly used to think about the testing of a single theory. However, the interest of spelling out the details will become clearer when we start to focus on comparative theory evaluation.

According to falsificationist methodologies a conclusively falsified theory becomes essentially useless, and so the rules for theory selection essentially are restricted to not-yet-falsified theories. For example, according to all probabilistic methodologies, including the Bayesian, falsified theories are generally thrown on the scrapheap of 'theories with zero posterior probability', and theory selection applies only to 'theories with nonzero probability'. Realist methodology too prescribes to refrain from falsified theories, for its supposed rule for theory selection, viz. inference to the best explanation (IBE), is assumed to imply that the best explanation is true, which is openly incompatible with accepting that it is false. Even within Lakatos' approach it is not altogether clear whether there is already enough room for falsifying instances under naive conditions. Although (constructive) empiricist methodology is usually not presented as falsificationist, it behaves essentially the same way: a falsified theory crosses the fundamental border of observational adequacy, for which scientists are supposed to strive. Of course, within every approach it is possible to contest that we are dealing with a genuine falsification, for all kinds of complications are possible.

Our point is that idealized methodology should already leave room for continuing interest in a falsified theory. First, it is perfectly possible that the theory contains an important core of truth, in which case other general test implications will pass their tests, leading to the establishment of new general facts and corresponding explanatory successes. Second, even new tests leading to new falsifications are very useful, because they provide new, relevant individual facts and corresponding instantial failures. Hence, at least as long as there has not been invented a better theory, it remains useful to continue to test the old theory, to get a better understanding of its explanatory strengths and instantial weaknesses.

The described test attitude is roughly representative for theory and practice of an instrumentalist methodology. As the term 'testing' is laden with the connotation of 'establishing the truth value', it makes indeed sense to speak of the testing of general test implications and other testable generalizations, but to speak of the testing of theories is less adequate from the instrumentalist perspective. As for theories, the term 'evaluation' is much more appropriate.

We will now first study instrumentalist methodology in more detail. After analysing its relation to truth approximation in the next section we will compare its merits with those of the falsificationist methodology, in particular in its realist as well as improper empiricist versions.

A main question for the instrumentalist is of course what makes a new theory better. Our intuitive answer will be that the new theory has at least to save the established explanatory strengths of the old one and not to add new instantial weaknesses on the basis of the former tests. The following definition is the obvious formal interpretation of this idea.

> Theory $M^\star$ is (at time t) *at least as successful* as theory M iff the set of instantial failures of $M^\star$ at t forms a subset of that of M and the set of explanatory successes of M at t forms a subset of that of $M^\star$.

> Theory $M^\star$ is (at time t) *more successful* (or better) than theory M iff $M^\star$ is at t at least as successful as M, and in at least one case the relevant subset is a proper subset.

It is plausible that we can also define in these terms the notion of 'the most successful theory thus far among the available alternatives' or, simply, 'the best (available) theory'.

It should be stressed that the diagnosis that $M^\star$ is more successful than M does not guarantee that this remains the case. For the diagnosis is based on thus far established facts, and new evidence may change the comparative judgment. But it is easy to see that it cannot be turned the other way around, i.e., that M becomes more successful than $M^\star$, for whatever happens, M either always has additional counterexamples or $M^\star$ has additional explanatory successes.

It should be conceded that frequently it will not be possible to establish the comparative claim, let alone the claim that one theory is more successful than all its available alternatives. In other words, these definitions do not guarantee a constant linear ordering, but only an evidence-dependent partial ordering of the relevant theories. Of course, one should interpret this as a challenge for refinements, e.g. by introducing different maximally successful theories or by a quantitative approach.

Be this as it may, we are now in a position to introduce our explication of the core of instrumentalist methodology by the following rules. The first one will be selfevident: as long as there is no best theory, one should continue the HD-evaluation of all available theories in order to explore the domain further. We will concentrate on the second rule, applicable in the case that one theory is more successful than another, and hence in the case that one theory is the best.

Suppose theory $M^\star$ is at t more successful than theory M. Consider the following comparative success hypothesis:

CSH: $M^\star$ (is and) will remain more successful than M.

CSH is an interesting hypothesis, even if $M^\star$ is already falsified. Given that $M^\star$ is known to be more successful than M at t, CSH consists at t of two 'weak' components, an instantial and an explanatory one:

ICSH: all instantial failures of $M^\star$ are instantial failures of M.

ECSH: all explanatory successes of M are explanatory successes of $M^\star$.

where 'all' is to be read as 'all past and future'.

These two components are testable generalizations. Following ICSH, we may derive a GTI from $M^\star$ that does not follow from M, and test it. If we get a counterexample of this GTI, and hence of $M^\star$, it may or may not be also a counterexample of M. If it is not, we have a falsification of ICSH. Alternatively, following ECSH, we may derive a GTI from M which cannot be derived from $M^\star$, and test it. If it becomes accepted this means falsification of ECSH. Of course, in both cases, the opposite test result is not only in favour of the corresponding comparative subhypothesis, and hence of CSH, but it increases the registrated success difference. In the following we will call (the combination of) these two ways of testing of CSH *comparative HD-testing*. Applied to theories not yet falsified, this relates to one aspect of Popper's idea of severe testing: severe testing as comparative testing.

The rule of theory selection we have already alluded to is the following:

Instrumentalist rule of success (IRS) If CSH has passed 'sufficient tests', then choose, for the time being, the more successful theory.

Of course, what 'sufficient tests' are, is a matter of degree, and hence of dispute. IRS assumes a kind of inductive generalization to conclude, for the time being, that CSH is true. If CSH is (assumed to be) true, the relevance of further comparative HD-testing of $M^\star$ against M vanishes. Choosing the more successful theory, i.e., $M^\star$, then means, whether or not it has already been falsified, that one can start to concentrate on the independent HD-evaluation of that theory, and continue with the attempt to invent new interesting competitors, that is, competitors that are at least as successful.

Later we will generalize the comparative hypothesis and the rule to the case that there is a best theory, and clarify the relation of the generalized IRS to the rule called inference to the best explanation (IBE), to which it apparently resembles. However, it is important to note already now that IRS should not be interpreted as claiming the chosen theory to be true.

At this point it is sufficient to conclude that our analysis of the logic of HD-evaluation of theories naturally suggests the definition of 'more successful', the comparative success hypothesis, the comparative HD-testing of it, and the instrumentalist rule of success (IRS). HD-evaluation of theories and HD-testing of the comparative success hypothesis provide the right ingredients for applying the definitions first of 'more successful' and later of IRS, respectively. In short, the HD-method is functional for IRS. In the following section we will show that, under certain conditions and in a certain sense, IRS and hence the HD-method are functional for approaching the (strongest relevant) truth.

3. Truth Approximation by the Rule of Success

It is wellknown that since the end of the seventies a couple of alternatives to Popper's definition of (greater) truthlikeness have been proposed (Tichy and Oddie, Kuipers, Miller, Niiniluoto, Brink and Heidema, Schurz and Weingartner. Cf. Kuipers (1987a), where most of them are presented and discussed). Whatever their relative problems and merits, they don't have the famous drawback of Popper's definition, independently discovered by Miller and Tichý (see Kuipers 1982) or (1987b) for the proof of this claim for my own approach). On the other hand, almost of them exhibit the feature of language dependence, introduced by David Miller and still conceived by him (Miller (1990)) as a serious problem for any explication of truthlikeness. However, the latter claim is a matter of serious debate (see e.g. Niiniluoto (1987), Mormann (1988)). That is, at least up to now, almost definitions of truthlikeness are essentially language dependent, but the question is how this fact should be evaluated. Is it dramatic, unavoidable or even desirable? Whatever the conclusion is, the same conclusion then has to be drawn for the idea that one theory is more successful than another, for it follows immediately from analogs to Miller's argument that this notion, if explicated along similar lines, is equally language dependent. Hence, language dependence is not only specific of greater truthlikeness, but also concerns the explication of other comparative notions commonly used by philosophers and scientists of all sorts.

It will not be surprising that our own explication of truthlikeness serves our present purposes. Our explication of the idea that one theory is closer to the truth than another (Kuipers (1982), (1984), (1987b), (1992)) starts from a realist version of the structuralist approach of theories. That is, theoretical terms are intended to refer, and theories are intended to make true observational and theoretical claims. More specifically, theories are assumed to be concerned with what is empirically possible within a set of conceptual possibilities, which seems to be the core of theorydirected natural science. 'The truth' is supposed to refer to the strongest truth in the context, called the strongest relevant truth. From this explication it naturally follows that, in general, any theory will be more successful than another if it is closer to the truth. Of course, to test the comparative hypothesis that a new theory is closer to the truth than an old one will only be possible on the basis of their respective success. In this section we will only present what is directly essential for the purpose of the present article. For further conceptual analysis and for illuminating pictures the reader is referred in particular to Kuipers (1992).

Naive truth approximation

The basic idea is the following: given a fixed domain D of natural phenomena (states/-situations/systems) and a set M_p of potential models or *conceptual possibilities* designed to characterize them, there is a unique, time-independent subset I of all *empirical possibilities*, which is the great unknown and hence the target of theory-directed research in the domain. Hence, we assume a special interpretation of the set of intended applications I. (See Kuipers (1994) for different ways of interpreting the set of intended applications).

Some remarks are in order. 1) M_p is the *conceptual frame* of a research program for D, and is, technically speaking, a set of structures of a certain similarity type. 2) Obviously, I is M_p-dependent (*conceptually relative*), but nevertheless also reality-dependent (*realism*): 'I is D seen through M_p' captures both aspects nicely. 3) However, I does not represent 'the actual world' (no descriptive realism), but the set of empirical possibilities, as the basic target of our theorizing (modal realism): it is D seen through M_p from the perspective of what is empirically possible. This set of assumptions may be called conceptually relative modal realism, or simply *constructive realism.*

In practice, the determination of D, M_p (hence I) and M is a matter of interaction. Above, and in what follows, we abstract from this interaction by assuming that D, M_p and hence I are fixed. The only things that are supposed to change are the theories to be considered. This is rather an idealization. In particular, it will be difficult to get fundamental flexibility of the determination of D into a more realistic picture. However, as far as the fixation of M_p is concerned, the idealization may be less dramatic than it may seem at first sight. To begin with, we will take stratified theories into consideration. Moreover, otherwise refined versions, based on an underlying notion of structure-likeness, are suggested by Kuipers (1992). In particular, although there may be a fundamental incommensurability between different conceptual frames, such versions suggest the possibility of practical commensurability, paving the way for an extension of the analysis in this paper to theories couched in different conceptual frames. Coming back to the possibility of changing D, it is now easy to see that the basic problem here is that the target changes, not due to conceptual changes but to apparent changes in the natural systems we are interested in. In Kuipers & Vos & Sie (1992) we have shown that there is a partial analogy between truth approximation and design research. In the latter context, the target, i.e., the desired product, may well change. It is plausible to study the extrapolation of the present analysis to the case of a changing domain by further exploring the indicated analogy.

Given some fixed M_p and the indicated intensional characterization of a fixed subset I, the goal of theorizing is to obtain an extensional characterization of it, and theories precisely aim at such characterizations. More formally, a theory M is reinterpreted as a subset M of M_p together with the claim that $M = I$, i.e., the claim that M is a (perfect) characterization of the unknown I. Note that, according to our definition, we define the claim of a theory much stronger than the standard claim formulated within the structuralist approach: I is a subset of M. We will reserve that weaker claim for hypotheses. The consequences of a theory M are of course the consequences of its claim. We will call a theory true when its claim is true and false when its claim is false, with the immediate implication that there is just one true theory, viz. theory I. For this reason I may indeed be called *the truth*.

The following definition is crucial for the comparative notion of truthlikeness of theories.

> Theory M^* is *at least as close to the truth as* M iff
> $I - M^*$ is a subset of $I - M$, and
> $M^* - I$ is a subset of $M - I$.

The first clause is, for obvious reasons, called the descriptive or *instantial clause*; it says that all empirical possibilities wrongly excluded by M^* are also excluded by M. The second clause is called the *explanatory clause*, for reasons which will become clear later on. It states that all possibilities wrongly admitted by M^* are also admitted by M.

The definition of 'at least as close to the truth' can be transformed into a definition of 'closer to the truth' as follows:

> Theory M^* is closer to the truth than M iff
> $I - M^*$ is a subset of $I - M$, and

$M^\star - I$ is a subset of $M - I$ and
at least one of the subsets is a proper subset.

The above notion of truthlikeness is called naive, because similarity between structures is not taken into account. In Kuipers (1992) we have defined a refined notion of truthlikeness, based on an underlying notion of structure-likeness. However, in the present paper naive truthlikeness suffices to our purposes. Moreover, the possibility of an extension of the main point of the present section follows directly from the 1992-article.

Truth approximation by the HD-method and IRS

The claim that one theory is closer to the truth than another, in the above sense, is a perfect example of an empirically testable comparative hypothesis. As we will see, it predicts, and hence explains, that the first will always be at least as successful as the second.

The strong truth appoximation hypothesis that $M^\star$ is closer to the truth than M trivially implies the following weak truth approximation hypothesis WTAH, the hypothesis of our primary interest:

WTAH: theory $M^\star$ is at least as close to the truth as theory M.

In terms of a comparative application of the HD-method the main theorem to prove is that WTAH has both weak components of CSH as general test implications:

Main theorem: WTAH implies, assuming that all instantial failures and explanatory successes are proper, and have mutually been checked adequately:
ICSH: all instantial failures of $M^\star$ are instantial failures of M
ECSH: all explanatory successes of M are explanatory successes of $M^\star$

The conditions of proper failures and successes and adequate mutual check will be made clear in the course of the proof.

The following proof of the main theorem is an adaptation and combination of related proofs in Kuipers (1984/1987b/1989). We start with the instantial comparative success implication ICSH. Recall that the sets of instantial failures of theories M and $M^\star$ in the light of the available evidence at time t were indicated by $\mathrm{IF}(M,t)$ and $\mathrm{IF}(M^\star,t)$, respectively.

The assumption that the instantial failures of a theory are proper means that they are conceptual possibilities that are empirical possibilities, excluded by the theory. This amounts to the assumption that there have not been made mistakes when applying the instantial (falsificationary) side of the HD-method. Formally, the proper instantial failures assumption is equivalent to the assumption that IF(M,t) is a subset of $I - M$ and IF($M^\star, t$) of $I - M^\star$. Note that this assumption implies that the instantial failures are couched in terms of M_p, i.e., the language of the theories under evaluation. The assumption of adequate mutual check of instantial failures means that x is in IF($M^\star, t$) if x is in IF(M, t) and x is not in $M^\star$, and, vice versa, that x is in IF(M, t) if x is in IF($M^\star, t$) and x is not in M.

Assume now WTAH, particularly that $M^\star$ is instantially at least as close to I as M, i.e., $I - M^\star$ is a subset of $I - M$. What we have to prove is that IF($M^\star, t$) is a subset of IF(M, t). Let x be a member of IF($M^\star, t$). The proper instantial failure assumption implies that x belongs to $I - M^\star$. Following WTAH it then also belongs to $I - M$. The adequate mutual check assumption implies what we wanted to prove, viz. that x has also been registrated as a member of IF(M, t).

The proof of the explanatory comparative success implication ECSH needs a number of preparations. A (noncomparative) *hypothesis* H is interpreted as a weak version of a theory: H is a subset of M_p, and the claim is that I is a subset of $H, I \subseteq H$. Again, hypothesis H is true/false if the claim is true/false, respectively. Note that this definition corresponds to the standard structuralist definition of a theory. Note also that there is more than one true hypothesis. Note, finally, that hypothesis I is the strongest true hypothesis.

The set of hypotheses Q(M) following from theory M can be represented by the set of subsets of M_p which include M, for the associated hypothesis-claims follow from the theory-claim. Now it is not difficult to prove the 'bridge-theorem' that the explanatory clause ($M^\star - I \subseteq M - I$) is equivalent to the condition that all true hypotheses following from, and hence explained by, theory M, also follow from – hence are explained by – theory $M^\star$. This explains the name of the clause.

Formally this is expressed by:

$$(\star) \qquad Q(M) \cap Q(I) \subseteq Q(M^\star) \cap Q(I)$$

To prove this, start from (*) and note first that $Q(M) \cap Q(I) = Q(M \cup I)$. Hence ($\star$) is equivalent to the claim: $Q(M \cup I)$ is a subset of $Q(M^\star \cup I)$. On the level of sets this is equivalent to the claim: $M^\star \cup I$ is a subset of $M \cup I$. On its turn, this is trivially equivalent to the claim that $M^\star - I$ is a subset of $M - I$ (QED).

Now we can turn to the proof of the explanatory comparative implication ECSH. Recall that ES(M, t) and ES($M^\star, t$) indicate the sets of explanatory successes at time t of M and $M^\star$, respectively.

The assumption that an explanatory success L of a theory M, i.e., a member of ES(M, t), is proper means that L is a subset of M_p such that it includes M as well as I as subsets, for the claim of a true hypothesis explained by a theory has to be derivable from the theory as well as from I itself. Similarly, a member L of ES($M^\star, t$) must include $M^\star$ as well as I as subsets. The proper explanatory successes assumption amounts to the assumption that there have not been made mistakes when applying the explanatory side of the HD-method. The assumption of adequate mutual check of explanatory successes means that L is in ES($M^\star, t$) if L is in ES(M, t) and L includes $M^\star$, and, vice versa, L is in ES(M, t) if L is in ES($M^\star, t$) and L includes M.

Assume now WTAH, particularly that $M^\star$ is explanatorily at least as close to I as M, i.e., $M^\star - I$ is a subset of $M - I$, of which we demonstrated already that it is equivalent to the assumption that $Q(M) \cap Q(I)$ is a subset of $Q(M^\star) \cap Q(I)$. What we have to prove is that ES(M, t) is a subset of ES($M^\star, t$). Let L be in ES(M, t). The proper explanatory successes assumption implies that M as well as I are subsets of L, and hence that L is a member of $Q(M) \cap Q(I)$. Following WTAH, L then is a member of $Q(M^\star) \cap Q(I)$, i.e., an hypothesis following from I and $M^\star$. The adequate mutual check assumption implies what we want to prove, viz. that L has been recorded as a member of ES($M^\star, t$).

So far for the proof of the main theorem. It immediately follows that IRS (and hence the HD-method) is functional for approaching the truth in the following sense. If theory $M^\star$ is more successful than theory M, $M^\star$ may still be closer to the truth than M, which would explain that $M^\star$ is at least as successful as M, and M cannot be closer to the truth than $M^\star$, for otherwise M could not be less successful than $M^\star$.

The stratified case

Up to now we did not assume a distinction between an observational and a theoretical(-cum-observational) level. However, it is not difficult to extrapolate (the proof of) the main theorem for the stratified case along the explanatory line (Kuipers (1989)), thererby leading to the conclusion that the fact that one theory is on the observational level explanatorily more successful than another can be explained by the hypothesis that the first theory is on the theoretical level explanatorily closer to the truth than the second. Due to the many-one character of the projection of the set of theoretical structures onto the set of observational structures, an analogous result cannot be proved straightforwardly for the instantial side. But a further investigation of this side will turn out to be very instructive.

Let us start with some general means and assumptions. Let M_p and the subset I_t of M_p indicate on the theoretical level the sets of conceptual and empirical possibilities, respectively. M_{pp} and I_0. refer to the corresponding possibilities on the observational level. Let π indicate the projection function from M_p onto M_{pp}. Let πM, for a subset M of M_p, indicate the set of projections of all the members of M. On semantical grounds we may assume that πI_t is a subset of I_0 (called I-projection), for if something is empirically possible on the theoretical level, it will remain so when we skip the theoretical components. However, I_0 ne just contains some degenerate cases of M_p-structures, e.g. structures with empty domain and relational sets for these non-referring components.

For the explanatory line we only have to prove that the explanatory clause of truthlikeness on the theoretical level, implies that on the observational level, for we already know that the latter implies being at least as successful on that same level. Hence, to be proved is: if $M^\star - I_t$ is a subset of $M - I_t$ then $\pi M^\star - I_0$ is a subset of $\pi M - I_0$. The antecedent is equivalent t_0 : $M^\star - M \cup I_t$ is empty; this in turn implies that $\pi M^\star - \pi M \cup \pi I_t$ is empty; from I-projection we get that $\pi M^\star - \pi M \cup I_0$ is empty, which is an equivalent version of what was to be proved.

A similar proof is impossible for the instantial line. Before we start to study the instantial comparative claim in detail, we introduce a plausible claim associated with the theoretical frame, to which we already alluded:

> Reference-claim: I_0 is a subset of πI_t,
> and hence, due to I-projection, $I_0 = \pi I_t$.

The label of this claim is derived from the assumption that the substantial (non-vacuous) extension of structures transforms empirical possibilities systematically into empirical impossibilities, with the consequence that πI_t is a proper (perhaps even empty) subset of I_0, if and only if one or more of the added components (of non-logico-mathematical nature) are – essentially, not just occasionally – nonreferring components.

Suppose now that $M^\star$ is instantially at least as close to I_t as M, that is, I_t - $M^\star$ is a subset of $I_t - M$, or, equivalently, $M \cap I_t - M^\star$ is empty. Suppose further that this is not the case on the observational level, that is, there are counterexamples to the claim that $M^\star$ is instantially at least as close to I_0 as M, that is, $I_0 - \pi M^\star$ is a subset of $I_0 - \pi M$. Such counterexamples can be established by realizing the relevant empirical (observational) possibilities. Hence, let c belong to $\pi M \cap I_0 - \pi\ M^\star$, that is, the empirical possibility c is an instantial failure of πM^*, but not for πM. Now, from I-projection it follows, that c either belongs to πI_t or to $I_0 - \pi I_t$. Which of these alternatives is the case, we cannot decide without further assumption. But if c belongs to $I_0 - \pi I_t$ it is a counterexample to the reference-claim of the frame. On the other hand, if c belongs to πI_t it is not difficult to derive that there have to be x in $M - M^* \cup I_t$ and z in $I_t - M \cup M^*$ such that $\pi(x) = \pi(z) = c$, in which case it is a counterexample to the following general claim:

for all x in $M - I_t$ and z in $I_t - M$: if $\pi(x) = \pi(z)$
then there is y in $M \cap I_t$ such that $\pi(x) = \pi(y) = \pi(z)$.

This claim is for trivial reasons equivalent to the claim:

for all x in M and z in I_t: if $\pi(x) = \pi(z)$
then there is y in $M \cap I_t$ such that $\pi(x) = \pi(y) = \pi(z)$

or, simply, $\pi M \cup \pi I_t \subseteq \pi(M \cap I_t)$

and, hence, together with the mathematical fact that the converse inclusion is always the case, we get the:

Relative adequacy claim (for M) : $\pi(M \cap I_t) = \pi M \cap \pi I_t$.

The label is related to the fact that the claim guarantees that (relative to πI_t) appropriate observational models of the theory can be extended to (of course, relative to I_t) appropriate theoretical models. It should be realized that the antecedent in the first two versions of the nonmathematical part of the claim requires perfect correspondence between an observational model and the relevant projected empirical possibility, which will in general be rather exceptional.

In sum, the version of the main theorem adapted to the stratified (naive instantial) case can be stated as follows. If we have a counterexample to the claim that πM^* is instantially closer to I_0 than πM, there are three mutually exclusive possibilities: it may concern a counterexample to the claim that M^* is instantially closer to I_t than M, or to the claim that M is relatively adequate, or to the reference claim associated with the theoretical frame. Hence, being instantially at least as successful can be explained by being instantially closer to the truth on the theoretical level, together with the hypothesis that the theoretical frame refers and the hypothesis that the less successful theory is at least relatively adequate.

As a consequence, we again obtain the result that IRS (and hence the HD-method) is functional for approaching the truth, not only in the plausible observational sense, but also in the realist sense. We start with the observational sense, which is directly based on the unstratified version of the main theorem. If theory $M^\star$ is (observationally) more successful than theory $M, \pi M^\star$ may still be closer to the observational truth I_0 than πM, which would explain that $M^\star$ is at least as successful, and πM cannot be closer to the observational truth I_0 than $\pi M^\star$, for otherwise M could not be less successful. But it follows from our analysis that IRS (and hence the HD-method) is also functional for approaching the truth in the following realist sense. If theory $M^\star$ is more successful than theory M we may conclude the following. First, $M^\star$ may still be closer to the theoretical truth I_t than M, which would explain that $M^\star$ is at least as successful, assuming for the instantial part in addition that the theoretical frame refers and that M is at least relatively adequate. Second, M cannot be closer to the theoretical truth I_t than $M^\star$, for otherwise M could not be less successful.

For the stratified case the situation can be summarized as follows. The theoretical or realist version of the (comparative) truth approximation hypothesis:

RTAH: $M^\star$ is closer to I than M

implies, assuming some auxiliary hypotheses and some extra observational success of $M^\star$, the observational version

OTAH: $\pi M^\star$ is closer to I_0 than πM

which in turn implies, if we suppose that there are no data mistakes, the instrumentalist comparative success hypothesis

CSH: $M^\star$ (is and) will remain more successful than M.

As a consequence, going in the other direction, IRS, prescribing when and how to choose the more successful theory, is functional for observational truth approximation, which, in turn, is functional for realist truth approximation.

It is not difficult to prove that CSH is equivalent even to OTAH, assuming that all members of I_0 can in fact be realized, and that I_0 can in fact be established as a true hypothesis, and assuming the data to be correct. This equivalence directly follows from the fact that I_0 contains all possibilities that can possibly be realized and is the strongest observational law that can possibly be established. From the equivalence it follows that the instrumentalist cognitive aim of the most successful theory in principle coincides with observational truth approximation.

Taking into account similarity between structures which gives rise to a refined notion of truthlikeness, an analogously adapted theorem for stratified theories can be obtained. This is easy to conclude on the basis of the theorem proved in Kuipers (1992). The 1992-article does not contain the later discovered adapted theorem for the naive instantial stratified case presented above. It is still to be investigated whether the analysis can also be extended to quantitative forms of naive and refined truthlikeness.

4. *Comparing Methodologies and Rules of Inference*

The foregoing analysis puts into question several points of view that are popular in philosophy of science. It almost directly follows that falsificationist methodologies, including not only realist but also empiricist versions, are not the most efficient for truth approximation. Hence, the standard explanations of the fact that scientists do not attribute dramatic impact to falsifications, namely in terms of defensible dogmatic strategies or in terms of social factors, are not necessary. The instrumentalist practice can be justified in terms of truth approximation. Moreover, the instrumentalist rule of success naturally suggests the rule to infer that the best theory is closest to the truth. This rule can be seen as a severe correction of socalled inference to the best explanation.

Falsificationist methodology and instrumentalist practice

Let us first generalize the core elements of the instrumentalist methodology to the best (available) theory, if any. The hypothesis of interest is:

> GCSH: The best theory (is and) will remain the best

and the generalized instrumentalist rule of success

> IGRS: If GCSH has passed 'sufficient tests', then choose, for the time being, the best theory.

By simple generalization of the main theorem it can be shown that IGRS is functional for approximating observational and even theoretical truth.

We arrive at the core of standard falsificationist, in particular realist and empiricist, methodology by restricting the hypothesis of interest (GCSH) and the instrumentalist rule (IGRS) to not yet falsified theories. It is important to recall that when we speak of true or false theories we always mean true or false ***as hypothesis***; when we want to refer to 'the true theory' we will speak of 'the (theoretical or observational) truth'. Let us call the best not yet falsified theory ***the best explanation***, which is suggested by the terminology of 'inference to the best explanation' (see below). Now the hypothesis of falsificationist interest is:

> FGCSH: The best explanation (is and) will remain the best of the available unfalsified theories

and the falsificationist rule of success

> FGRS: If FGCSH has passed 'sufficient tests', then choose, for the time being, the best explanation.

The falsificationist message is that convincing falsification of a theory means that the game is over for that theory and that the only option is to look for another one. There is no interest in independent HD-evaluation or comparative HD-testing of already falsified theories. To be sure, from the foregoing analysis it follows that FGRS is also functional for observational and even theoretical truth approximation: what holds for the best theory in relation to its competitors holds ipso facto when that theory has not yet been falsified, and hence has only explanatory successes and no instantial failures.

However, realist and empiricist versions of the falsificationist methodology are, though effective, not as efficient as the instrumentalist methodology in approaching their respective cognitive goals. Ironically enough, instrumentalists will reject the theoretical truth as their cognitive goal; at most they are willing to accept the observational truth as their cognitive target.

Consider first the combination of realist means and ends. Since a false theory can very well be closer to the truth than another, according to sophisticated realist epistemological convictions captured by our truthlikeness definition, the methodological restriction to unfalsified theories is an unnecessary retardation of truth approximation. While the instrumentalist methodology (unintentionally) goes, as straight to the point as possible, along a chain of true or false theories, falsificationist methodology used by the realist attempts to approach theoretical truth by expelling false theories as soon as possible. When a theory has been falsified, the only way out is to invent a better theory, at least in the sense that it avoids the same falsification. Put differently, while realist epistemology recognizes the possibility of one false theory being closer to the theoretical truth than another, it is not exploited by the falsificationist methodology. On the other hand, instrumentalist methodology does so, although its defenders are at most willing to subscribe that one aims at observational truth, that is, at the strongest observationally adequate theory with respect to the relevant domain. This goal corresponds to the ultimate aim of Van Fraassen's constructive empiricist approach.

Hence, let us now consider empiricist means and ends. On the methodological side there is, as long as the distinction between theoretical and observational is not in dispute, no serious difference with the realist methodology. By restricting itself to unfalsified theories, it amounts, like realist but unlike in strumentalist methodology, to an unnecessary retardation of approaching its own cognitive goal. For that goal, the observational truth, can be approached equally well by a chain of falsified or unfalsified theories.

To be sure, the above relativization of the methodological role of falsification, should not be understood as a plea to drop falsifiability as a criterion for the empiricalness of a theory. On the contrary, empirical theories are very much supposed to be able to score success, to be precise, explanatory success. This evidently implies falsifiability.

The above diagnosis leaves room for a straightforward realist interpretation of the overwhelming evidence provided by Kuhn, Lakatos, Laudan and many others, that scientists appear to be not very much impressed by a falsification of their theory, at least as long as they don't have an unfalsified better one. As we have seen, it is possible to justify this attitude from a realist point of view in terms of approximation towards the true theory, even if this theory is supposed to contain theoretical terms. Hence, the standard realist appeal nowadays to the Quine/Duhem-thesis in order to justify the attitude abovementioned, is fundamentally redundant and misplaced. No doubt, the thesis, that unambiguous falsification is impossible due to auxiliary hypotheses, observation theories, and the like, is right and leaves room for all kinds of dogmatic strategies to avoid the falsification of one's favorite theory. So, the point is not that such strategies do not play a role; on the contrary, they do play an important role and may even be very well justified in concrete cases. However, such considerations are not necessary to justify an independent or comparative evaluation of a falsified theory, at least as long as there is no unfalsified better theory.

It is also interesting to look at the socalled social studies of science presently flourishing. They started from and are still based on the assumption that the findings of Kuhn and others show that the behaviour of (the majority of) scientists is not that rational as was previously thought. For they apparently do not apply elementary methodological rules. However, as we have seen, this instrumentalist, if not dissolute, attitude cannot only be defended as a dogmatic strategy that may be appropriate in the case at hand, as Lakatos has pointed out. On closer inspection it may even be seen, particularly when there is no unfalsified better theory available, as a paradigmatic case of straightforward rational behaviour in order to approach the truth. To be sure, again, the diagnosis does not imply that recourse to social-psychological factors is always beside the point. However, the diagnosis raises serious doubts as to the current foundations of social studies of sciences. The overwhelming evidence of non-falsificationist behaviour has brought social studies of science so far as to neglect elementary methodology as a description of the ideal course of events, whereas social factors are supposed to explain deviations.

Our realist justification of instrumentalist behaviour may be useful as a new reference point for studying problematic social factors. Let us formulate the ideal course of nature from this perspective. From the point of truth approximation, IGRS can be justified as a prescriptive rule, and the HD-method as its drive mechanism. Intuitive versions of method and rule are seen by most scientists and instrumentalist methodologists as the hall-mark of scientific rationality. The analysis of their contribution truth approximation can be conceived of as an explication of to what many scientists are inclined to think, and others, including instrumentalists and sociologists, are inclined to doubt. Fortunately, the understanding of the whole mechanism is not relevant for the practice of applying the HD-method and IGRS.

Inference to the best explanation

So-called inference to the best explanation is frequently seen as the realist's crucial rule of inference behind the above indicated methodological rule. However, let us start its evaluation by formulating the corresponding rules suggested by our analysis.

It is important to note that IGRS neither is nor suggests a rule of inference, in the standard sense that the conclusion is (supposed to be) true when the premises are true. That is, IGRS does not suggest the conclusion that the best theory is true.

According to our analysis IGRS roughly suggests 'Choose, after sufficient comparative testing, the best theory'; this is at most a rule to provisionally conclude that the best theory is the one closest to the truth. Although the suggested rule is not a standard rule of inference, it is tempting to call it ***realist inference to the best theory***:

> RIBT: If a best theory has been chosen, then conclude, for the time being, that it is the one closest to the theoretical truth I_t (among the theories available).

The justification of RIBT lies of course in the fact that the comparative hypothesis that the best theory is closest to the truth explains why it is the best one, whereas any competing comparative hypothesis of the same kind (that is, one telling that one of the other available theories is closest to the truth) cannot explain this.

It is also plausible to formulate an empiricist version, with similar justification:

EIBT: If a best theory has been chosen, then conclude, for the time being, that its projection is closest to the observational truth I_0 (among the available theories).

It is important to stress and defend the severe corrections these two versions of IBT contain in comparison to what is standardly called 'Inference to the best explanation' (IBE), where the best explanation is conceived of as the best not yet falsified theory. See Lipton (1991) for an extensive exposition and defence of IBE. Again we will distinguish a realist and an empiricist version, both suggested by FGRS.

RIBE: If a best explanation has been chosen, then conclude, for the time being, that it is true (that I is a subset of it).

EIBE: If a best explanation has been chosen, then conclude, for the time being, that it is observationally true (that I_0 is a subset of it).

Let us first elaborate the differences between (R/E)IBE and (R/E)IBT. They are the same for both versions of the two. IBE is restricted to the case that the best theory has not yet been falsified, hence instantial failures are not taken into account. To put it differently, the best theory to which IBE can be applied is the unfalsified theory (if any) which is explanatorily the most successful (if any), which also explains the label '(best) explanation'. In contrast to IBE, IBT generally applies in the case that there is a best theory. Here the best theory is defined, as suggested before, as the theory (if any) which is instantially and explanatorily at least as successful as all other available theories, and more successful than any of the others in at least one of the two respects.

The other important difference is that IBE is a standard rule of inference in the sense that the conclusion is supposed to be true when the premises are. That is, according to IBE the best (unfalsified) theory is supposed to be true, whereas IBT only infers that the best theory is closest to the truth. As a consequence, even if the best theory is not falsified, IBT does not imply that this theory is true, rather, it still leaves much room for the possibility that it is false.

The main objections to IBE are directly related to its differences to IBT. First, in contrast to IBT, IBE is restricted to the case that the best theory has not been falsified; it cannot deal with the case in which the best theory has been falsified. Second, whereas it is selfevident that being closest-to-the-truth is relative to other (i.c. the available) theories, as IBT presupposes, it nearly is a contradiction that being-true (-as-hypothesis) is relative to other theories, as IBE presupposes.

In sum, in contrast to FGRS and IBE, IGRS and IBT do not degrade false theories from further evaluation and competition. From the perspective of truth approximation, to be precise, approximation of the strongest relevant truth, the question whether a theory is true or false is essentially inessential.

It is tempting to summarize the main reason for this claim by stating that a false theory may well be closer to the truth than a true theory. However, it should be conceded that neither naive nor refined qualitative formal explications of closer-to-the-truth (Kuipers (1992)) permit such a judgement, nor the converse judgment that a true, but not the strongest, theory is closer to the truth than a false one. This is just due to the strict character of comparative explications, leading to partial orderings of theories. However, it is evident that naive and refined quantitative explications based on numbers or distance functions will make such judgements perfectly possible.

Hence, the restriction to unfalsified theories in the (realist and empiricist version of the) FGRS/IBE-approach to theory evaluation and selection is indefensible from the truth approximation point of view. Exactly the same objection applies to all standard probabilistic approaches, notably the Bayesian. They settle all falsified theories on the uniform level of zero posterior probability, putting them aside from further evaluation, let alone from selection.

5. Novel Facts, Crucial Experiments, Idealization and Concretization

Our picture of theory evaluation has still to be confronted with several important methodological points made by Popper, Lakatos, and Nowak. We will first evaluate the emphasis Popper and Lakatos put on novel facts from the truth approximation perspective. Then it will be shown that Lakatos's relativization of crucial experiments directly follows from this perspective. It will be concluded that the present account of theory evaluation can be seen as an explication, including corrections, and justification of the methodology of Lakatos, but that it also leaves room for the idealization and concretization methodology developed by Nowak and others. Both methodologies are special cases of the instrumentalist methodology, and hence functional for truth approximation.

Novel facts and ad hoc repairments

Suppose that our favourite theory has been falsified. Now it is possible that a well-conceived change of the theory leads to a new theory which is not falsified by the counterexample for the old one. As Popper has stressed, in scientific practice it is considered to be very important that such a new theory not only avoids the problem of the old one, in which case it would just be an ad hoc repairment, but that it also leads to new test implications, that could not be derived from the old theory and which turn out to pass the corresponding tests. Popper and Lakatos even thought that this extra success, the prediction of novel facts, was the litmuss test for whether or not the new theory is possibly closer to the (relevant) truth than the old one.

From our analysis it follows immediately that it is formally possible that a new theory is closer to the truth than the old one, while it only corrects the instantial failures of the old one, without additional explanatory success. Suppose that M is a subset of I and let G be the general test implication of M (hence M is subset of G) which has been falsified. Suppose now that $M^\star$ is such that $M^\star \cap G = M$ and that $M^\star$ is, like M, a subset of I. Under these conditions $M^\star$ is closer to the truth than M only in that it avoids known instantial failures of M, but without further unintended explanatory success. For under the specified conditions M must be a subset of $M^\star$, and hence theory M implies all general hypotheses derivable from theory $M^\star$, hence all explanatory successes of $M^\star$.

A similar case is possible for theories M and $M^\star$ containing I as a subset. Let M fail to imply the established law L and let $M^\star = M \cap L$, then $M^\star$ again is closer to the truth than M, only in that it has L as additional explanatory success, but without unintended avoidance of instantial failures.

These special cases can be summarized as follows: if the theory under test happens to be stronger or weaker than the true theory, the suggested ad hoc repairments will bring one closer to the truth, but without unexpected additional success. However, if the theory is not simply stronger or weaker than the true theory, the suggested ad hoc changes of it will almost inevitably lead to new predictions of additional success, some of them coming true or false depending on whether the repaired theory is or is not closer to the truth. It is even not excluded that there are plausible general conditions under which 'almost inevitably' can be replaced by 'inevitably'. Whatever the case may be, if comparative HD-testing of a new theory is in favour of that theory, then, depending on the test route, this means either an unexpected additional instantial failure of the old theory, or an unexpected extra explanatory success of the new theory, where what is unexpected is of course determined by the old theory. In sum, ad hoc repairment of a theory will seldom be a real improvement without unexpected extra success. In other words, comparative HD-testing of an ad hoc repairment will either lead to unexpected extra successes of the new theory or extra successes of the old theory that could have been, but were not, explicitly expected before. So besides some qualifications the intuitions of Popper and Lakatos with respect to ad hoc repairments and novel facts are largely justified. Instead of a ban on ad hoc changes, they can be allowed, provided they are subjected to comparative HD-testing with the original theory.

Crucial experiments

In *Section 3* we saw that the HD-method could be applied to the comparative hypothesis 'theory $M^\star$ is closer to the truth than theory M'. The comparative hypothesis is suggested when one theory is more successful than another. Let us now, from the truth approximation perspective, look at the situation where two theories are equally successful, and hence at the idea of a so called crucial test.

Let the two theories concerned, say M and $M^\star$, be instantially as well as explanatorily equally successful before the crucial test. We may or may not in addition assume that their equal instantial success means that both theories have not yet been falsified. The methodological side of the idea of a crucial test amounts to the following. First, a crucial test typically is supposed to be a repeatable experiment, hence it refers to general test implications. More specifically, the idea is to derive from M a general test implication $G(M)$ of the form 'always when C then F' and from $M^\star$ $G(M^\star)$ of the form: 'always when C then not-F'. Let us further assume that it follows from our background knowledge that one of these general testable conditionals has to hold and that it is possible to test them with C as initial condition (see below).

Under these conditions it is excluded that the two theories remain equally successful, for the experiments will force us to accept either $G(M)$ or $G(M^\star)$. Moreover, if the experiments force us to accept, for instance, $G(M^\star)$, this not only implies that $M^\star$ is explanatorily more successful than M, but also that it is instantially more successful. The reason for this is that $G(M^\star)$, which in the beginning is a falsifying general hypothesis of M in Popper's sense, has become a *falsifying general fact* of M. Every investigated 'C-case' apparently results in not-F, in such a way that their combination not only is in agreement with $G(M^\star)$'s prediction, but also appears as a falsifying instance of $G(M)$ (and hence of M). $G(M^\star)$ summarizes and, pace Popper, inductively generalizes these falsifying instances.

The assumption that the tests can and will have C as initial test condition is quite important. If, due to physical or practical constraints, it is impossible to test $G(M)$ and $G(M^\star)$ by realizing the test condition C, the two test implications can at most be checked by tests suggested by their logically equivalents: 'always when not-F then not-C' and 'always when F then not-C'. However, if this is the case, i.e., if the initial test conditions have to be not-F and F, respectively, the tests will not lead to such convincing results. For in that case every successful test result for e.g. $G(M)$, i.e., cases of not-F and not-C, merely are neutral results with respect to $G(M^\star)$, and not falsifying ones. Similarly, positive test results with respect to $G(M^\star)$, i.e., cases of F and not-C, are neutral for $G(M)$, and not negative.

So far concerning the methodological aspects of a crucial test. What about its consequences for truth approximation? Of course, at least all conclusions we have drawn from the comparative statement that one theory is more successful than another follow: M cannot be closer to the truth than $M^\star$, and $M^\star$ can still be closer to the truth than M. Moreover, new experiments (related to an old or a new GTI of M or $M^\star$) may destroy the success dominance. This would not block the conclusion that M is not closer to the truth than $M^\star$, but it blocks the conclusion that $M^\star$ could still be closer to the truth than M. As a consequence, a crucial experiment may temporarily lead to better truth approximation perspectives for one theory compared to the other, but these perspectives may well become blocked. To be sure, the reverse perspectives cannot arise, except when the outcome of the crucial experiments is put into question or when new considerations about auxiliary hypotheses lead to the conclusion that the supposed falsifying general fact is in fact an explanatory success.

Concerning the case that both M and $M^\star$ were not yet falsified before the crucial test, the following noncomparative conclusions can be added to the above truth approximation conclusions. M is false, and $M^\star$ may still be true. Moreover, $M^\star$ may later become falsified as well, but the conclusion that M is false can only be withdrawn by reconsidering data or auxiliary hypotheses.

In several respects the present analysis is in accordance with Lakatos's analyses of crucial experiments, in which the temporary character and the revisability of the conclusions is generally accepted. Our analysis adds to this that the conclusions can unproblematically be stated in terms of (perspectives on) truth and truth approximation, and can be generalized to falsified theories. The latter point is very important as long as the theories under consideration have to be assumed to be 'born refuted', for instance due to unavoidable idealizations.

Idealization and concretization

As a consequence, we can now conclude that our instrumentalist methodological portrait is a refined and corrected version of Lakatos's methodology. The basic common feature is that theory testing is primarily a matter of comparative testing. Theories within a research program are evaluated by the HD-method, there is room for falsified theories, theory transitions can be made on the basis of comparative HD-testing, including ad hoc repairments, and such transitions are functional for truth approximation, though no guarantee for it. The most important difference is the explicit truth approximation analysis, leading to a straightforward justification of comparative testing and of leaving falsified theories in the game, i.e., the instrumentalist methodology. This allows to avoid a recourse to dogmatic strategies for saving the hard core of a program, which is unavoidable from Lakatos's sophisticated falsificationist perspective, or so it seems.

A good reason for supposing that Lakatos's theory development strategy is not the only way of justifying nonfalsificationist practice is provided by what is perhaps the most striking case of this practice: the theory development strategy elaborated by Nowak and others, called 'idealization and concretization'. According to this strategy one starts with a theory neglecting some factors of which one knows that they are relevant. Hence, one knows beforehand that the theory, suggested by strong idealization, is false. To note that, one does not need experimental testing. The same ('to be born refuted') holds for all concretizations of the theory that one obtains from successive accounting for up to that point neglected factors. Only when one thinks to have taken into account all relevant factors, it makes sense not to exclude the possibility that the theory is true. Following Lakatos, one may of course say that the hypothesis that a certain factor is negligible is an auxiliary hypothesis that can be blamed for the falsification. But the problem is that the falsification is not at all a surprising result for which one has to find reasons. In other words, there need not be any inclination to dogmatism with respect to a certain theory that one wants to save.

Consequently, the nonfalsificationist strategy of idealization and concretization does not at all presuppose a division between a hard core and auxiliary hypotheses, but neither does it exclude it. In other words, Lakatos's and Nowak's strategies of theory development can, but need not, go together. Both are specific strategies fitting very well into the general instrumentalist methodology. In both cases it follows automatically that they are functional for truth approximation as far as they succeed in generating increasingly successful theories. (The way in which this works for idealization and concretization has been explained in detail in *Section 4* of Kuipers (1992).) The strategies of Lakatos and Nowak illustrate the general message of this paper that instrumentalist methodology and realist epistemology can nicely go together.

Let us finally note that, in the line of Lakatos, the empiricist version of our methodological portrait also leaves room for a wellconceived transition of one research program (with hard or semihard core) to another. Suppose that a new research program turns out to 'have a theory' of which the observational projection is (on the basis of tests) more successful than the projection of the best known theory of an old program. Of course, this conclusion presupposes an observational level not contaminated with theoretical principles of either one of the programs. Moreover, it remains a momentary judgement, because the old program can later turn out to leave room for a theory that surpasses the success of the best known theory belonging to the new program. Whatever the case may be, the sketched program transition is functional for approaching observational truth, for the theory of the new program can be closer to observational truth than the original best theory of the old program, and the reverse is impossible.

6. *Concluding Remarks*

Although it may be conceded that *the* scientific method does not exist, this does not yet imply that any method works, as Feyerabend suggests with his slogan 'anything goes'. It is more realistic to start with a distinction (to be specified pragmatically) between two aspects of scientific research, viz. invention and testing of theories. This distinction became known as the *Context of Discovery* versus the *Context of Justification.*

Concerning the Context of Discovery it may well be that almost all conceivable methods, from inductive generalization to another night's sleep, work in certain cases: 'anything goes *sometimes*'.

Within the Context of Justification there may also not be a universal method. However, the HD-method certainly is a dominant method. Unfortunately, the term 'Context of Justification', suggests - before and after its falsificationist specification, like the terms 'confirmation' and 'corroboration' -, that the truth or falsity of a theory is the sole interest of testing. Our analysis of the HD-method and its functonality for truth approximation makes it clear that it would be much more plausible to speak of the *Context of Evaluation.* This term would in the first place refer to the separate and comparative HD-evaluation of theories in terms of explanatory successes and instantial failures, but on this basis also to the evaluation of their relative merits in approaching the (strongest relevant) truth, or at least the observational truth.

It also follows from our analysis that improving a theory is a twosided affair of designing a new theory with less instantial failures and more explanatory successes. Elsewhere, we have already studied design research programs, using their (partial) analogy with truth approximation (Kuipers & Vos & Sie (1992)). In designing one tries to make a product of which the operational profile of factual and counterfactual properties gradually approaches the intended profile of desired and undesired properties. Our present analysis suggests to view theory revision, in turn, as a special type of design with particular specifications of the desired and undesired properties, i.e., explanatory successes and instantial failures, respectively. Assuming a fixed domain, the explanatory successes to be covered and the instantial failures to be avoided can only increase in number. Of course, Lakatos's and Nowak's strategies are from this perspective specific design strategies, and there may be others as well.

Moreover, according to the standard structuralist view, it frequently occurs that the domain of intended applications is changed on grounds of theoretical considerations and empirical findings. In other words, the target may change. Hence, as already suggested in *Section 3*, it is plausible to study the extrapolation of the present analysis to the case of a changing domain by further exploring the indicated analogy with design research, where changing the target is the rule. Besides rules and heuristic strategies for improving the available theory for a fixed domain, rules and strategies for adapting the domain to the available theory will then come into the picture, too.

References

Hempel, C.G., 1966, *Philosophy of natural science*, Englewood Cliffs.

Kuipers, T., 1982, 'Approaching descriptive and theoretical truth', *Erkenntnis* 18, 343-387.

Kuipers, T., 1984, 'Approaching the truth with the rule of success', *Philosophia Naturalis* 21, 244-253.

Kuipers, T. (ed.), 1987a, 'What is closer-to-the-truth? A parade of approaches to truthlikeness', *Poznan Studies Vol. 10,* Amsterdam.

Kuipers, T., 1987b, 'A structuralist approach to truthlikeness', In: Kuipers (1987a), 79-99.

Kuipers, T., 1989, 'How to explain the success of the natural sciences', In: Weingartner, & Schurz, G. (1989), 299-341.

Kuipers, T., 1992, 'Naive and refined truth approximation', *Synthese* 93, 299-341.

Kuipers, T. & Vos, R. & Sie, H., 1992, 'Design research programs and the logic of their development', *Erkenntnis* 37, 37-63.
Kuipers, T., 1994, 'The refined structure of theories', In: Kuokkanen (1994), 3-24.
Kuokkanen, M. (ed.), 1994, *Idealization VII: Structuralism - Idealization and Approximation, Poznan Studies* 42, Amsterdam.
Lipton, P., 1991, *Inference to the best explanation*, London.
Miller, D., 1990, 'Some logical mensuration', *The British Journal for the Philosophy of Science* 41, 281-290.
Mormann, Th., 1988, 'Are all false theories equally false?', *British Journal for the Philosophy of Science* 39, 505-519.
Niiniluoto, I., 1987, *Truthlikeness*, Dordrecht.
Popper, K.R., 1959, *Logik der Forschung*, Vienna, 1934, translated as *The logic of scientific discovery*, London.
Schurz, G. (ed.), 1988, *Erklären und Verstehen in der Wissenschaft*, München.
Weingartner, P. & Schurz, G., 1989, *Philosophy of the Natural Sciences, Proceedings 13th Wittgenstein symposium, 1988*, Vienna.

Chapter 7: Probability, Confirmation and Testing from a Structuralist Perspective[1]

Bernhard Lauth

Introduction

The aim of this paper is to provide a structuralist account of confirmation and testing. The structuralist approach has been outlined in numerous publications, beginning with Sneed's (1971). § 1 summarizes some of the concepts and definitions, which are needed in this paper. The basic notions are that of a theory element and that of a theory net.[2]

In § 2, we assign a probability space $\Omega(T)$ to every theory element T, such that each 'point' of the space is a collection of potential models for T. Every measurable subset $H \subset \Omega(T)$ is called an hypothesis for T. Let $p(H/E_0, \ldots, E_{n-1})$ denote the conditional probability of hypothesis H with respect to a finite sample $E_0, \ldots, E_{n-1}$ of empirical data. The main result of this section is Theorem (2.13). The theorem states that $p(H/E_0, \ldots, E_{n-1})$ converges almost everywhere to the actual truth value (One or Zero) of H for $n \longrightarrow \infty$, provided that $\Omega(T)$ can be construed as a complete metric space, which is separable by the data $E_0, E_1, \ldots$. Similar results have been obtained by Gaifman and Snir (1982) for some types of first order languages.[3]

Theorem (2.13) motivates the following testing procedure: An hypothesis H is accepted, if $p(H/E_0, \ldots, E_{n-1})$ exceeds some arbitrarily chosen threshold $\beta, (0 < \beta < 1)$, otherwise H is rejected. Under the assumptions of the theorem, every testing procedure of this type yields correct results in the limit, when $n \rightarrow \infty$, i.e. H is confirmed in the limit, if and only if H is true.

In practice, however, the conditions of the Theorem (completeness of metric, separability by the data) cannot always be met. In § 3, I shall therefore propose a weaker type of testing procedure, which is applicable under much more general circumstances.

Assume that the data $E_0, E_1, \ldots$ is obtained by a sequence of experiments, (measurements, observations) $\mathbf{E}_0, \mathbf{E}_1, \ldots$, respectively, where $E_n =$

---

[1] This paper has been written under DFG-project La 676/1-1.

[2] Cf. Sneed (1971), Stegmüller (1979), (1986), and Balzer & Moulines & Sneed (1987).

[3] Cf. Gaifman & Snir (1982), Theorem 2.1.

the outcome of $\mathbf{E}_n$, for all $n \geq 0$. An empirical theory can entail various predictions about the outcomes of $(\mathbf{E}_n)_{n\geq 0}$. We consider each prediction as an hypothesis H about the actual data stream $(E_n)_{n\geq 0}$, which is obtained from $(\mathbf{E}_n)_{n\geq 0}$, i.e. an hypothesis is just a subset $H \subset \times_{n\geq 0}\mathbf{E}_n$.

A testing procedure for H is a mapping $\alpha_H(E_0, \ldots, E_n)$, which takes as input a finite sample $E_0, \ldots, E_n$ of data and outputs a value

$$\alpha_H(E_0, \ldots, E_n) = 1 \text{ or } \alpha_H(E_0, \ldots, E_n) = 0,$$

where '1' means that the hypothesis is to be accepted and '0' means that H is rejected. A testing procedure α_H is *reliable*, if the values $\alpha_H(E_0, \ldots, E_n)$ converge to the actual truth value of H, i.e. if $lim_{n\to\infty}\alpha_H(E_0, \ldots, E_n) = 1$ iff H is true, and $lim_{n\to\infty}\alpha_H(E_0, \ldots, E_n) = 0$ otherwise. The main result is formulated in Theorem (3.6) and Corollary (3.7). The theorem shows that an almost everywhere reliable testing procedure can be found for each measurable hypothesis $H \subset \times_{n\geq 0}\mathbf{E}_n$. The proof is based on Doob's martingale convergence theorem for conditional expectations.

Inductive procedures, which operate on data sequences $E_0, E_1, \ldots$ and infer correct results in the limit (when $n \longrightarrow \infty$) have been extensively studied in formal learning theory. Their theoretical foundations were developed by E. M. Gold in his (1967) paper on language identification in the limit. Some of the basic ideas are foreshadowed in Putnam (1963) and (1965). Gold analyzed methods for inductive identification of formal languages in automata theory. Similar techniques can be used for the inductive extrapolation of recursive functions (cf. L. Blum and M. Blum (1975), D. Angluin and C. Smith (1983), D. Osherson, M. Stob, S. Weinstein (1986) for a survey). Inductive procedures for structure identification in first order languages are provided by E. Shapiro (1981), D. Osherson, S. Weinstein (1986), (1989), K. Kelly, C. Glymour (1989), (1990) and B. Lauth (1993), (1994). To some extent, the above methods and accounts give some substance to Popper's idea of scientific progress by approaching the truth (cf. Popper (1963), and (1984)).

1. Theory Elements and Theory Cores.

According to the structuralist view, complex empirical theories like Newtonian mechanics and Maxwell's electrodynamics are to be considered as connected networks of different, interrelated 'theory elements'.[4]

Each theory element T is characterized by its core$K(T)$ and its range of intended applications $I(T)$, such that $T =< K(T), I(T) >$. Basically, the theory core consists of the following components:

(1) $M(T)$: the models of T

[4] Cf. Balzer & Moulines & Sneed (1987), chap. IV.

(2) $M_p(T)$: the potential models of T

(3) $M_{pp}(T)$: the partial-potential models of T

(4) $C(T)$: the constraints of T.

Thus, the core has the form $K(T) =< M(T), M_p(T), M_{pp}(T), C(T), \ldots >$, where the dots indicate additional components, usually included in the definition of a core (like topological structures and external links).

A potential model for theory element T is a set theoretical structure (k+n-tuple) of the form $x =< D_1, \ldots, D_k, R_1, \ldots, R_n >$. $D_1, \ldots, D_k$ are non-empty sets (the 'base sets' of the structure) and $R_1, \ldots, R_n$ are relations (functions, operations) on these sets. Each relation R_i has some relation type σ_i associated with it[5]. The number of base sets and the types of the relations determine the structure type of x: x is called a structure of type $\tau =< k, \sigma_1, \ldots, \sigma_n >$, if x consists of k base sets $D_1, \ldots, D_k$ and n relations $R_1, \ldots, R_n$ such that $R_1, \ldots, R_n$ are of type $\sigma_1, \ldots, \sigma_n$, respectively.

The collection of all structures of type τ is denoted by $Str(\tau)$. Note that, if there exist $i_1, \ldots, i_k \in \{1, \ldots, n\}$ for all $i = 1, \ldots, n$ such that $R_i \subset D_{i1} \times \ldots \times D_{ik}$, then x is an L-structure for some many sorted first order language L with relation symbols $\mathbf{R}_1, \ldots, \mathbf{R}_n$ (where k is the arity of R_i). Similarly, we have that if there exist $j, i_1, \ldots, i_k \in \{1, \ldots, n\}$ for all $i = 1, \ldots, n$ such that $R_i : D_{i1} \times \ldots \times D_{ik} \longrightarrow D_j$, then x is an L-structure for some many sorted language L with function symbols $\mathbf{f}_1, \ldots, \mathbf{f}_n$. Finally, x is an L-structure for a one sorted first order language, if the above assumptions hold true with $k = 1$.

We associate a structure type τ with every theory element T such that $M(T) \subset M_p(T) \subset Str(\tau)$ and $C(T) \subset Po(M_p(T)) \subset Po(Str(\tau))$, hence each model of T is a structure of type τ. Under the assumptions of the preceeding paragraph, we can assign a first order language $L(T)$ to structure type τ such that $Str(\tau)$ consists of all L-structures for $L(T)$. $M(T)$ is an elementary class of L-structures, if there is some collection S of first order axioms, such that $M(T) = \{x \in Str(\tau)/x \models S\}$, i.e. the models of T are those structures of type τ, which satisfy the axioms of T. Constraints can be considered as 'second order' axioms, i.e. they determine

---

[5]k-relation types are defined inductively: (i) For $i = 1, \ldots, k$, $[i]$ is a k-relation type; (ii) If $\sigma, \sigma_1, \sigma_2$ are k-relation types, then $\sigma_1 \times \sigma_2$ and $Po(\sigma)$ are k-relation types, too. For every k-relation type σ and every k-tuple $D_1, \ldots, D_k$ of 'base sets' we define the echelon set $\sigma(D_1, \ldots, D_k)$: (i) If $\sigma = [i]$, then $\sigma(D_1, \ldots, D_k) = D_i$; (ii) If $\sigma = \sigma_1 \times \sigma_2$, then $\sigma(D_1, \ldots, D_k) = \sigma_1(D_1, \ldots, D_k) \times \sigma_2(D_1, \ldots, D_k)$; (iii) If $\sigma = Po(\sigma')$, then $\sigma(D_1, \ldots, D_k) = Po(\sigma'(D_1, \ldots, D_k))$. We say that R_i is a relation of type σ_i, if $R_i \in \sigma_i(D_1, \ldots, D_k)$.

the relationships between different models of T. (Example: Newtons's second law and Maxwell's laws are the basic axioms of classical mechanics and electrodynamics, respectively, i.e. each model of these theories must satisfy the corresponding axioms. The constraints require that these axioms be invariant with respect to Galilei and Lorentz transformations, respectively.[6])

Now, let $X \subset M_p(T)$ be a collection of potential models. We say that $X \in Cn_{th}(T)$, if X satisfies all axioms and constraints of T, i.e. if $X \subset M(T)$ and $X \in C(T)$. Different parts of an empirical theory (different theory-elements) may have a different range of intended applications, (e.g. the intended applications of classical mechanics include, among other things, the motion of the planets in the solar system and the motion of charged particles in magnetic and electric fields when relativistic or quantum effects can be neglected). Actually, the range of intended applications of a theory T may change over time, hence $I(T)$ should be considered as a function of time.[7] Within the structuralist framework, the intended applications of a theory element T are represented by partial potential models of T, hence $I(T) \subset M_{pp}(T)$. According to Sneed (1971), partial-potential models are obtained by cutting off all 'T-theoretical' components of some potential model of T.[8] Let $y = r(x)$ be the partial potential model, which is obtained by omitting the T-theoretical components of $x \in M_p(T)$. A collection $X \subset M_p(T)$ is called a theoretical extension for $I(T)$, if for each $y \in I(T)$ there is some $x \in X$ such that $y = r(x)$. We let $X \in e(I(T))$, if X is a theoretical extension for $I(T)$. The Ramsey sentence for T claims that all intended applications of T can be extended to full potential models in such a way that all axioms and constraints of T are satisfied, i.e.: $\exists X \in Cn_{th}(T) : X \in e(I(T))$.

In the present context, however, I prefer a generalized notion of partial models, which does not depend on Sneed's distinction between T-theoretical and T-non-theoretical terms of a theory T. This generalized notion of partial models was introduced by W. Balzer (1985) and has been adopted in Stegmüller (1986).[9] I use it in a slightly modified version: According to this version, y is called a substructure of $x \in Str(\tau)$, if y is obtained by replacing all relations of x by subsets of these relations in such a way that the resulting structure is again of type τ. We write $y \subset x$, if y is a substructure of x. Thus, let $x =< D_1, \ldots, D_k, R_1, \ldots, R_n >$ be a structure of type τ. Then $y \subset x$ iff there exist $R'_1, \ldots, R'_n$, such that

[6] The representation of physical invariances by suitable constraints has been analyzed in detail in T. Bartelborth (1988).

[7] See B. Lauth (1991) for such an account.

[8] The standard example are masses and forces in Newtonian mechanics.

[9] Stegmüller (1986), p. 159 ff. and Balzer (1985), p. 27.

$y =< D_1, \ldots, D_k, R'_1, \ldots, R'_n >, R'_1 \subset R_1, \ldots, R'_n \subset R_n$ and $y \in Str(\tau)$. $\subset$ is a partial ordering of $Str(\tau)$. In the following, $M_{pp}(T)$ denotes the collection of all substructures of potential models of T, hence

(1.1) Definition: $M_{pp}(T) := \{y \subset x / x \in M_p(T)\}$.

We assume that $I(T) \subset M_{pp}(T)$ is closed under substructures, i.e. if $y \in I(T)$ and $y' \subset y$, then $y' \in I(T)$, too. A structure $y =< D_1, \ldots, D_k, R_1, \ldots, R_n >\in M_{pp}(T)$ is called a *finite* substructure (of some $x \in M_p(T)$), if $R_1, \ldots, R_n$ are finite relations or functions (no matter, whether the base sets $D_1, \ldots, D_k$ are finite or not).

Finite substructures can represent 'empirical data' in a very compact form, i.e. the finite relations $R_1, \ldots, R_n$ of some substructure can represent the outcomes of a finite number of measurements or observations. Note that partial models in Sneed's sense can be subsumed - in a slightly modified version - under the generalized definition (just replace all T-theoretical relations by the empty set). In general, however, substructures may contain 'data' about T-theoretical quantities as well as non-theoretical quantities. From the preceeding definitions we can immediately infer:

(1.2) $M(T) \subset M_p(T) \subset M_{pp}(T) \subset Str(\tau)$.

A collection $X \subset M_p(T)$ of potential models is called a theoretical extension for $E \subset M_{pp}(T)$ iff $\forall y \in E \; \exists x \in X : y \subset x$ and vice versa: $\forall x \in X \exists y \in E : y \subset x$. We write $X \in e(E)$, if X is a theoretical extension of E. In general, we let $E \in e(E_0)$, for $E, E_0 \subset M_{pp}(T)$, iff $\forall y_0 \in E_0 \; \exists y \in E : y_0 \subset y$ and $\forall y \in E \; \exists y_0 \in E_0 : y_0 \subset y$. Accordingly, the Ramsey sentence for theory element T states that $\exists X \in Cn_{th}(T) : X \in e(I(T))$, i.e.: all substructures in $I(T)$ can appropriately be extended to potential models of T such that the extended structures satisfy the axioms and constraints of T.

Complex empirical theories like Newtonian mechanics and Maxwell's electrodynamics contain a large number of different axioms and constraints, each of which may have a different range of intended applications[10]. Within a structuralist frame, Newtonian mechanics or Maxwell's theory are considered as theory nets, where each part of the theory is represented by some particular theory element T. Theory nets are usually obtained by successive 'core specialization' from a single 'basic theory ele-

[10] Example: Newton's first and second law, the third law (actio = reactio), the law of gravitation, Coulomb's law, the Lorentz force law, Hooke's law, Stoke's law and other 'special' laws of classical mechanics.

ment' T_0 : *A* theory element T is called a specialization of $T_0, T\sigma T_0$, iff $Cn_{th}(T) \subset Cn_{th}(T_0)$ and $I(T) \subset I(T_0)$, i.e.: T is obtained by adding new axioms and constraints to the theory core of T_0, thereby restricting T_0's range of intended applications. A theory net is a finite collection N of theory elements which is partially ordered by the σ–relation.[11] Note that, by the definition of σ, the models of all theory elements in a theory net are structures of the same type, since $M_p(T) = M_p(T')$ holds for all $T, T' \in N$.

Theory nets may occur as parts of still larger units, which are called theory holons. A theory holon is a collection of theory elements (where the models are not necessarily of the same type), which are interrelated by 'external links' (cf. Stegmüller, loc. cit. p. 271 ff and Balzer, Moulines, Sneed (1987), chap. VIII.1). In the case of classical mechanics, a theory holon may include theory elements for Newtonian, Hamiltonian or Lagrange mechanics on different levels of specialization along with an appropriate collection of external links.

2. Theory Elements and Probability Measures.

Let $x = < D_1, \ldots, D_k, R_1, \ldots, R_n >$ and $x' = < D'_1, \ldots, D'_k, R'_1, \ldots, R'_n >$ be structures of type τ. We define $Basis(x) := < D_1, \ldots, D_k >$, hence $Basis(x) = Basis(x')$ iff $< D_1, \ldots, D_k > = < D'_1, \ldots, D'_k >$. For collections $X, X' \subset M_{pp}(T)$ we let $Basis(X) = \{Basis(x)/x \in X\}$, hence $Basis(X) = Basis(X')$ iff there exists $x' \in X'$ for every $x \in X$ such that $Basis(x) = Basis(x')$ and vice versa. For example, we have that $Basis(y) = Basis(x)$, if y is a substructure of x (cf. definitions in § 1). Accordingly we get $Basis(X) = Basis(E)$, if $X \in e(E)$ is a theoretical extension of E, where $X \subset M_p(T)$ and $E \subset M_{pp}(T)$.

(2.1) Definition: Let T be a theory element. We let $\Omega(T) =$
$\{X \subset M_p(T)/Basis(X) = Basis(I(T))\}$.

By its definition, $\Omega(T)$ consists of all possible theoretical extensions for $I(T)$. Intuitively, we consider $\Omega(T)$ as a collection of possible 'worlds' or possible states of empirical systems which can be represented by structures of type τ. Moreover, we define expressions like $p(T)$ or $p(T/E)$ (= the absolute probability of T and the conditional probability of T given E), such that

[11] Cf. Stegmüller (1979), p. 91. Some examples, which have been analyzed within the structuralist framework, are the theory net of Newtonian particle mechanics, classical (equilibrium) thermodynamics and classical electrodynamics (cf. Balzer & Moulines (1981), Balzer & Moulines & Sneed (1987), chap. IV and Bartelborth (1988).

$$p(T) := p(\{X \in \Omega / X \in Cn_{th}(T)\}),$$

i.e: $p(T)$ is the probability that any collection $X \in \Omega$ satisfies the axioms and constraints of T. Similarly, we consider $p(E)$ as an abbreviation for $p(\{X \in \Omega/E(X)\})$, where $\{X \in \Omega/E(X)\}$ is some measurable subset of Ω (see below for details). Accordingly, the conditional probability $p(T/E)$ is the ratio $p(T/E) = p(\{X \in \Omega/X \in Cn_{th}(T) \wedge E(X)\})/p(\{X \in \Omega/E(X)\})$.

Of course, probabilities like $p(T)$ and $p(E)$ must be defined with respect to some suitable measure space (σ-algebra) over $\Omega(T)$. We take the σ-algebra to be generated by the collection of all those subsets of Ω, which are elementarily definable in ZF (ZF = the Zermelo-Fraenkel set theory).

To be more precise, let $L = L(ZF)$ be the language of ZF, Var_L the variables of L, Fml_L the collection of all L formulas and let h denote any function with $Dom(h) = Var_L$. $L(ZF)$ consists of a single binary relation symbol $\dot{\in}$. For any L-formula A, we define $\models A[h]$ by induction over the length of A: For atomic formulas $v \dot{\in} w$, where $v, w \in Var_L$, we let $\models v \dot{\in} w[h]$ iff $h(v) \in h(w)$. For equations $v \doteq w$ we let $\models v = w[h]$ iff $h(v) = h(w)$. We adopt the usual conventions for Boolean combinations. Finally, we let $\models \exists v A[h]$ iff there exists some function h' with $Dom(h') = Dom(h) = Var_L$ and $h'(v') = h(v')$ for all $v' \neq v$ such that $\models A[h']$.

A collection $\Omega_0 \subset \Omega$ is elementarily definable by some ZF formula A iff there exists exactly one variable v occuring free in A and an assignment function h with $Dom(h) = Var_{L(ZF)}$ such that $X \in \Omega_0$ iff $X = h(v)$ and $\models A[h]$ holds for all $X \in \Omega$. We denote Ω_0 by X_A, if Ω_0 is definable by formula A and write $A(X)$, iff $X \in X_A$. Accordingly, we let the expression $p(A)$ denote $p(\{X \in \Omega/A(X)\})$. Thus we get

(i) $X_A := \{X \in \Omega/h : h$ is a function, $Dom(h) = Var_{L(ZF)}, X = h(v), \models A[h]\}$

(ii) $A(X) \leftrightarrow X \in X_A$

(iii) $p(A) = p(\{X \in \Omega/A(X)\})$.

In what follows we suppose that T is such that $\{X \in \Omega/X \in Cn_{th}(T)\}$ is definable by some suitable ZF formula A. Now, let $V = V(\Omega)$ denote the collection of all subsets X_A, which are elementarily definable by some ZF formula A. $V(\Omega)$ is countable, since $Fml_{L(ZF)}$ is countable.

(2.2) Corollary: $V = V(\Omega)$ is a Boolean algebra.

Proof: Let X_A and X_B be elements of V. Then $X_A \cup X_B$ and $X_A \cap X_B$ are elements of V, too, since $Fml_{L(ZF)}$ is closed under $\wedge$ and $\vee$. Moreover, the complement $X_A^C := \Omega - X_A$ is in V, for each $X_A \in V$, since $X_A^C = \{X \in \Omega / \neg A(X)\}$. Finally, there are elements $\mathbf{0}, \mathbf{1} \in V$, namely $\mathbf{0} =$ the empty set and $\mathbf{1} = \Omega$. Die distributive laws for $\cup$ and $\cap$ are trivial. □

(2.3) Corollary: V is a Boolean ring.

Proof: $\emptyset \in V$. If $X_A, X_B \in V$, then $X_A - X_B = X_A \cap X_B^C \in V$ and $X_A \cup X_B \in V$. □

In general, however, V is not a σ–algebra, since $\bigcup_{i \geq 0} X_{A_i}$ is not necessarily an element of V for any sequence $X_{A_0}, X_{A_1}, \ldots, X_{A_i} \in V$. Hence, we must extend V in order to get a probability space for $\Omega(T)$:

(2.4) Definition: Let T be a theory element. $(\Omega, \mathbf{A}, p)$ is a probability space for T iff the following holds:

(1) $\Omega = \Omega(T) = \{X \subset M_p(T) / Basis(X) = Basis(I(T))\}$

(2) $\mathbf{A} = \mathbf{A}(V)$ is the smallest σ-field, which contains all elements of $V = V(\Omega)$ (hence $\mathbf{A}$ is the σ-field generated by V)

(3) p is a probability measure on $(\Omega, \mathbf{A})$.

Note that $\mathbf{A}(V)$ is equal to the Dynkin system, which is generated by V, since V is closed under finite intersections.[12]

A measure on a Boolean ring V is a non-negative and σ- additive function p, such that $Dom(p) = V$ and $p(\emptyset) = 0$. By a well known theorem of Caratheodory,[13] each measure on a Boolean ring can be extended to a measure on the σ-field which is generated by V. The extension is uniquely determined, if there exists an increasing sequence $X_1 \subset X_2 \subset X_3 \subset \ldots$ in V such that $\Omega = \bigcup_{n \geq 1} X_n$ and $p(X_n) < +\infty$ for all $n \geq 1$.[14] This condition is trivially satisfied, if measure p is normalized ($p(\Omega) = 1$).

(2.5) Theorem: Let $p : V \longrightarrow [0,1]$ a non-negative, $\sigma - additive$ function on $V(\Omega)$, such that $p(\emptyset) = 0$ and $p(\Omega) = 1$. p determines a unique extension, which is a probability measure on $\mathbf{A}(V)$.

The proof is obvious from the above remarks combined with Corollary 2.3. □

[12] Cf. Bauer (1978), p. 20, Proposition 2.4.
[13] Cf. H. Bauer, loc cit., p. 31, Proposition 5.2.
[14] Cf. H. Bauer, loc. cit. Proposition 5.7.

Next, we define a topology on $\Omega(T)$ by taking the Boolean algebra $V(\Omega)$ as a basis of open sets:

(2.7) Theorem: Let T be a theory element, $\Omega = \Omega(T)$ and $V = V(\Omega)$ as before. V is the basis for a topology on $\Omega(T)$.

Proof: Recall that $Q \subset Po(X)$ is a basis for a topology on X, if (1)$X = \bigcup Q$, and (2) if, for all $A, B \in Q$ and every $x \in A \cap B$, there is some $C \in Q$, such that $x \in C \subset A \cap B$. For the present purpose, it suffices to choose $C = A \cap B$, since V is a Boolean algebra (hence V is closed under finite intersections). □

(2.8) Definition: Let T be a theory element and $\Omega = \Omega(T), V = V(\Omega)$ as before. We let $O(V)$ denote the topological space generated by V, i.e: $O(V)$ is the intersection of all topologies S on $\Omega(T)$ such that $V \subset S$.

Note that every open (closed) set is a countable union (intersection) of some elements of V, since V itself is countable. Thus the open sets have the form $\bigcup\{X_i/i \in \mathbf{N}\}$ with $X_i \in V$; ($\mathbf{N}$:= the natural numbers). The closed sets are the complements $(\bigcup X_i)^c = \bigcap X_i^c$, i.e. they have the form $\bigcup\{X_i/i \in \mathbf{N}\}$ with $X_i \in V$. The elements of V are clopen, since $X_i^c \in V$, for every $X_i \in V$.

(2.9) Definition (Borel sets): Let $O = O(V)$ denote the topology which is generated by V. We let $\mathbf{B} = \mathbf{A}(T)$ denote the σ-algebra which is generated by O (= the collection of all Borel sets of $O(V)$).

We show that $\mathbf{B} = \mathbf{A}(T)$ coincides with the σ-algebra generated by V, i.e. $\mathbf{B} = \mathbf{A}(T) = \mathbf{A}(V)$:

(2.10) Theorem: Let $O = O(V), V = V(\Omega)$ and $\mathbf{B} = \mathbf{A}(T)$ as before. Then: $\mathbf{B} = \mathbf{A}(T) = \mathbf{A}(V)$.

Proof: First note that $\mathbf{B} = \mathbf{A}(T) = \mathbf{A}(O(V))$, by the definition of $\mathbf{B}$. $\mathbf{A}(V) \subset \mathbf{A}(O(V))$ is trivial, since $V \subset O(V)$. Hence, it suffices to show that $\mathbf{A}(O(V)) \subset \mathbf{A}(V)$. Indeed: $\mathbf{A}(V)$ contains all open sets of $O(V)$, since $O(V)$, by its definition, consists of all unions of elements of V. All of these unions are countable, however, since V itself is countable. But all countable unions of elements of V are contained in $\mathbf{A}$(V). Thus we get that $V \subset O(V) \subset \mathbf{A}(V)$, i.e. $\mathbf{A}(V)$ is the smallest σ-algebra, which contains all elements of $O(V)$. For otherwise, there would be a smaller σ-field

$\mathbf{A}' \neq \mathbf{A}(V)$ with $V \subset O(V) \subset \mathbf{A}' \subset \mathbf{A}(V)$– contradicting the assumption that $\mathbf{A}(V)$ is the smallest σ- algebra, which contains all elements of V. □

We note that Ω, with the above topology, is separable, since O has a countable basis.[15] We say that $\Omega(T)$ is separable by a sequence $A_0, A_1, \ldots$ of ZF formulas, if, for all $X, X' \in \Omega$ with $X \neq X'$ there is some $n \geq 0$ such that $A_n(X)$ but not $A_n(X')$ or vice versa. If $\Omega(T)$ is separable by a sequence of ZF formulas, then $O(V)$ is a Hausdorff space, i.e. there exist disjoint open environments $U, U' \in O(V)$ for all $X, X' \in \Omega, X \neq X'$, such that $X \in U, X' \in U', U \cap U' = \emptyset$. (Choose $n \geq 0$ so that $U = \{X \in \Omega / A_n(X)\}$ and $U' = \{X \in \Omega / \neg A_n(X)\}$). Under these assumptions we can define a metric on Ω as follows:

(2.11) Theorem: Let Ω be separable by a sequence A_0, $A_1, \ldots$ of ZF formulas. We define a mapping $d : \Omega \times \Omega \longrightarrow \mathbf{R}$ such that $d(X, X') = (i+1)^{-1}$, if A_i is the first formula in $A_0, A_1, \ldots$ which separates X, X',[16] and $d(X, X') = 0$, if no such formula exists. d is a metric on $\Omega(T)$.

Proof: (1) The symmetry $d(X, X') = d(X', X)$ is trivial. (2) By assumption (separability of Ω), $d(X, X') = 0$ can obtain only if $X = X'$. (3) The only non trivial property is the triangle inequality: $d(X, Z) \leq d(X, Y) + d(X, Z)$. Let A_i be the first formula in $A_0, A_1, \ldots$ which separates X and Z. Similarly let A_j be the first formula which separates X and Y. By the definition of d we get $d(X, Z) = (i+1)^{-1}$ and $d(X, Y) = (j+1)^{-1}$. Two cases can obtain: 1st case: $j \leq i$: Then $d(X, Y) = (j+1)^{-1} \geq (i+1)^{-1} = d(X, Z)$, and a fortiori $d(X, Z) \leq d(X, Y) + d(Y, Z)$. 2nd case: $i < j$: Then A_j must occur after A_i in $A_0, A_1, \ldots$. Hence $d(X, Y) < d(X, Z)$, since A_j is the first formula, which separates X and Y. In that case, Y and Z are also seprarated by A_i, since Y and X agree with respect to all formulas $A_k, k < j$, but X and Z disagree with respect to A_i. On the other hand, Y and Z must agree with respect to all formulas $A_k, k < i$, since X and Z agree on A_k and X and Y also agree on A_k (for $k < i < j$). Hence, $d(Y, Z) = d(X, Z)$ and we have again $d(X, Z) \leq d(X, Y) + d(Y, Z)$.□

Next we show that the metric topology which is induced by d coincides with the topology $O(V)$, which is generated by V:

(2.12): Theorem: Let $O = O(V)$ and $V = V(\Omega)$ as before and let $d : \Omega \times \Omega \longrightarrow \mathbf{R}$ denote the above metric on Ω. Then $O(V)$ is equal to

[15] Cf. Lipschutz (1977), Proposition 9.5.

[16] This means that A_i is the first formula in $A_0, A_1, \ldots$ such that $A_i(X)$, but $\neg A_i(X')$ or vice versa.

the metric topology O' which is induced by d.

Proof: d induces $O(V)$, if and only if the collection of all open balls $S(X,r) := \{X' \in \Omega / d(X',X) < r\}$ is a basis for $O(V)$, i.e: if (1) every union of open spheres is contained in $O(V)$, such that $O' \subset O(V)$, and if (2) every open set of $O(V)$ is a union of open balls, hence $O(V) \subset O'$. (1) $O' \subset O(V)$: Let $r = (i+1)^{-1}$. Then $S(X,r) = \{ X' \in \Omega/d(X',X) < r\}$ consists of all collections $X' \in \Omega$, which agree with X wrt all formulas $A_0, \ldots, A_i$, hence $S(X,r) = \{X' \in \Omega/\forall j \leq i : A_j(X') \leftrightarrow A_j(X)\} = \cup\{X_j/j \leq i\}$ where $X_j = \{X' \in \Omega/A_j(X')\}$, if $A_j(X)$ and $X_j = \{X' \in \Omega/\neg A_j(X')\}$, otherwise. Since $\neg A_j$ is a ZF formula we have $X_j \in V$ in either case. Thus $S(X,r)$ is a finite intersection of elements of V, and $S(X,r) \in O(V)$, since $O(V)$ is closed under finite intersections. By its definition, O' consists of all unions of open balls, hence $O' \subset O(V)$. (2) $O(V) \subset O'$: It suffices to show that each element of V is a union of open balls, since $O(V)$ is defined to conists of all unions of elements of V. We show that $X_{A_i} = \bigcup\{S(X,r)/A_i(X), r = (1+i)^{-1}\}$ holds for every formula A_i in $A_0, A_1, \ldots$ Indeed: $S(X,r)$ consists of all collections $X' \in \Omega$ which agree with X wrt all formulas $A_0, \ldots, A_i$, since otherwise $d(X',X) \geq r = (i+1)^{-1}$. Thus we get the following result: If $X' \in \bigcup\{S(X,r)/A_i(X), r = (i+1)^{-1}\}$, then $A_i(X')$ and hence $X' \in X_{A_i}$. And vice versa: If $A_i(X')$ holds, then $X' \in \bigcup\{S(X,r)/A_i(X), r = (i+1)^{-1}\}$, since$X' \in S(X',r)$. □

A sequence $(X_n)_{n\geq 0}$ of collections $X_n \in \Omega$ is a Cauchy sequence, if, for each $\varepsilon > 0$ there exists $n_0 \geq 0$ such that $d(X_n, X_m) < \varepsilon$ holds for $n, m > n_0$. A metric space is complete, if every Cauchy sequence in Ω converges to some element X of Ω, i.e. any open environment of X contains all but finitely many members of the sequence. A complete metric space with countable basis is called a polish space.[17]

Now let Ω be a polish space and let $E_0, E_1, \ldots$ be a sequence of ZF formulas which separates Ω. For each collection $X \in \Omega$ we define E_i^X such that $E_i^X = \{X' \in \Omega/E_i(X')\}$, if $E_i(X)$ and $E_i^X = \{X' \in \Omega/\neg E_i(X')\}$, otherwise. Accordingly we get $\bigcap_{i<n} E_i^X = \{X' \in \Omega/\forall i < n : E_i(X') \leftrightarrow E_i(X)\}$.

Now we are in a position to investigate the conditional probabilities $p(T/E_0, \ldots, E_{n-1})$ in the limit, when $n \longrightarrow \infty$. The following theorem shows that the conditional probabilities $p(T/\bigcap_{i<n} E_i^X)$ converge almost everwhere to 1, if $X \in Cn_{th}(T)$, and to 0, otherwise. The proof is adapted from Gaifman and Snir (1982) which have derived similar results for some types of first order languages. Henceforth, we let $1_T : \Omega \longrightarrow \{0,1\}$ denote the characteristic function for T, i.e.: $1_T(X) = 1$, if $X \in Cn_{th}(T)$ and

[17] Cf. H. Bauer, loc. cit. Definition 41.2.

$1_T(X) = 0$, otherwise, for all $X \in \Omega$.

(2.13) Theorem: Let $(\Omega, \mathbf{A}, p)$ be a probability space for theory element T. Let $d : \Omega \times \Omega \longrightarrow \mathbf{R}$ be a complete metric for Ω, such that $\mathbf{A}$ is the family of all Borel sets of the metric topology. Finally let $(E_i)_{i \geq 0}$ be a countable sequence of ZF formulas which separates Ω. Under these assumptions,

$$p(T/\bigcap_{i<n} E_i^X) \longrightarrow 1_T(X)$$

holds p-almost everywhere in Ω.

Proof: Recall from measure theory that for each non-negative (or integrable) real valued random variable X defined on Ω and every σ-field $\mathbf{D} \subset \mathbf{A}$ there exists a non-negative (or integrable) random variable $X_0 =: E(X/\mathbf{D})$, which is $\mathbf{D}$-measurable and which satisfies the equation

(1) $\int_C X_0 dp = \int_C X dp$

for each $C \in D$. The existence of X_0 follows from the Radon-Nikodym theorem (cf. H. Bauer, (1978), 17.10). X_0 is p-almost everywhere uniquely determined, i.e. every function X_0, which satisfies equation (1) is p-almost everywhere equal to X_0. $X_0 = E(X/\mathbf{D})$ is called the conditional expectation of X given $\mathbf{D}$ (cf. Bauer, loc.cit., 54.1 and 54.2).

Let $\{B_i/i \in I\}$ be a countable collection of pairwise disjoint subsets of Ω, such that $\Omega = \bigcup B_i$ and let $\mathbf{D}$ be the σ-field generated by the B_i. The conditional expectation of X given $\mathbf{D}$ is $X_0 := \sum_{i \in I} E_{B_i}(X) 1_{B_i}$, where $E_{B_i}(X) := \int X dp_{B_i}$ and $p_{B_i}(A) = p(A/B_i)$ is the conditional probability of A given B_i. Indeed: By its definition, X_0 has the constant value $E_{B_i}(X)$ on every B_i. Hence $\int_{B_i} X_0 dp = E_{B_i}(X)p(B_i) = \int_{B_i} X dp$ for all $i \in I$, and therefore $\int_C X_0 dp = \int_C X dp$ holds for all $C \in \mathbf{D}$, since every $C \in \mathbf{D}$ is a union of some of the B_i. If $X = 1_A =$ the characteristic function of A (where $1_A(w) = 1$, if $w \in A$, and $1_A(w) = 0$, otherwise), then $p(A \cap B_i) = \int_{B_i} 1_A dp = E(1_A/\mathbf{D})(w)p(B_i)$ holds for all $w \in B_i$. Accordingly we get

(2) $E(1_A/\mathbf{D})(w) = p(A \cap B_i)/p(B_i) = p(A/B_i)$,

for $w \in B_i$, provided that $p(B_i) > 0$. For the proof of the theorem we need the following facts: (1) If $\mathbf{D}_0 \subset \mathbf{D}_1 \subset \mathbf{D}_2 \subset \ldots \mathbf{A}$ is an increasing sequence of σ-fields over Ω such that $\mathbf{A} = \bigcup_n \mathbf{D}_n$, then $E(X/\mathbf{D}_n) \longrightarrow X$ holds almost everywhere in Ω, for every integrable random variable X.

(2) In a polish space we have: A collection $X \subset \mathbf{A}$ of Borel sets, which separates Ω, generates the σ-field of all Borel sets, hence $\mathbf{A} = \mathbf{A}(X)$ (cf. the Martingale convergence theorems in Bauer, loc. cit. §60).

Now, let $\mathbf{D}_n$ be the σ-field which is genarated by $E_0, \ldots, E_{n-1}$, i.e. let $\mathbf{D}_n$ be the smallest σ-field, which contains all collections $\{X \in \Omega / E_i(X)\}$ for $i = 0, \ldots, n-1$. According to our previous definitions we have $+E_i := E_i^X = \{X' \in \Omega / E_i(X')\}$, if $E_i(X)$ and
$-E_i := E_i^X = \{X' \in \Omega / \neg E_i(X')\}$, otherwise. Thus $\mathbf{D}_n$ is equal to the σ-field which is generated by all intersections of the form $\bigcap_{i<n} \pm E_i$. The family of all collections $\bigcap_{i<n} \pm E_i$ is a partition of Ω. Hence

(3) $p(T/\bigcap_{i<n} E_i^X) = E(1_T/\mathbf{D}_n)(X)$,

provided that $p(\bigcap_{i<n} E_i^X) > 0$, (cf. equation (2), above). The union of all intersections $\bigcap_{i<n} \pm E_i$ of p-measure Zero is a measure Zero set, too, hence (3) holds p-almost everywhere. By assumption, the sequence $(E_i)_{i \geq 0}$ separates Ω, hence $\mathbf{A} = \mathbf{A}(\bigcup_n \mathbf{D}_n)$, i.e. $\bigcup_n \mathbf{D}_n$ generates the σ-field of all Borel sets of Ω, hence

(4) $E(1_T/\mathbf{D}_n)(X) \longrightarrow 1_T(X)$

holds p-almost everywhere. The theorem is now immediate from (3) and (4). □

It is easy to verify that the proof of Theorem (2.13) remains valid, if $p(T/\bigcap_{i<n} E_i^X)$ is replaced by $p(H/\bigcap_{i<n} E_i^X)$, where H is any hypothesis $H \in \mathbf{A}(V)$. Under the assumptions of the theorem we therefore get that $p(H/\bigcap_{i<n} E_i^X) \longrightarrow 1_H(X)$ holds p-almost everywhere in Ω, where $1_H(X) = 1$, if $H(X)$ and $1_H(X) = 0$, otherwise. Similarly, we get that the difference $p(H/\bigcap_{i<n} E_i^X) - p'(H/\bigcap_{i<n} E_i^X)$ of the conditional probabilities for different measures p, p' must converge to Zero, provided that both measures take on the value 0 on the same hypotheses:

(2.14) Corollary: Under the assumptions of the preceeding theorem, $| p(H/\bigcap_{i<n} E_i^X) - p'(H/\bigcap_{i<n} E_i^X) | \longrightarrow 0$, holds p-almost everywhere, for all probability measures p, p' on $\mathbf{A}$ such that $p'(H) = 0 \leftrightarrow p(H) = 0$ holds for all hypotheses $H \in \mathbf{A}$.

The proof is immediate from the preceeding theorem. Actually, we could prove a stronger claim, namely that the supremum of the deviations $| p(H/\bigcap_{i<n} E_i^X) - p'(H/\bigcap_{i<n} E_i^X) |$ taken over all $H \in \mathbf{A}$ must approach Zero for $n \longrightarrow \infty$. But the proof is rather lengthy, and we don't need it

in the sequel.

3. Probabilities and Testing Procedures.

The above results motivate a straightforward testing procedure for scienhtific theories: A theory T is accepted, if $p(T/E_0, \ldots, E_{n-1})$ exceeds some arbitrarily chosen threshold $\beta, (0 < \beta < 1)$, otherwise T is rejected. Under the assumptions of Theorem (2.13), all testing procedures of this type yield correct results in the limit, when $n \longrightarrow \infty$, i.e. T is confirmed in the limit, if and only if T is true.

In practice, however, the assumptions of the theorem (completeness of metric, separability by the data) cannot always be met.(A simple cardinality argument shows, for example, that Ω cannot be separated by a countable sequence of data, if $| \Omega |> 2^{|\aleph_0|}$). In this section, I will therefore propose a different type of testing procedures, which is weaker in some respects, but which is applicable under much more general circumstances. I shall first sketch the basic ideas in an informal way, before I proceed with technical details.

Thus let $\mathbf{E}_0, \mathbf{E}_1, \ldots, \mathbf{E}_n, \ldots$ denote a sequence of experiments (measurments or observations). Each experiment is supposed to have a specific range of possible outcomes. To simplify notation, I shall denote the collection of all possible outcomes of an experiment n also by $\mathbf{E}_n$. Moreover, we will assume that each $\mathbf{E}_n$ is equipped with a metric $d_n : \mathbf{E}_n \times \mathbf{E}_n \longrightarrow \mathbf{R}$. which provides a measure for the 'distance' between different possible outcomes of experiment n. (If the experiment consists in measuring the value of some physical quantity, just take the natural metric of the reals, for that purpose).

An empirical theory may entail a variety of predictions on the outcomes of $\mathbf{E}_0, \mathbf{E}_1, \ldots$. We consider each such prediction as an *hypothesis* H about the actual data stream $E_0, E_1, \ldots$, which is obtained from $\mathbf{E}_0, \mathbf{E}_1, \ldots$, i.e. an hypothesis is just a subset $H \subset \times_{n \geq 0} \mathbf{E}_n$. A testing procedure for H is a mapping $\alpha_H(E_0, \ldots, E_{n-1})$, which takes as input a finite sample $E_0, \ldots, E_{n-1}$ of empirical data ($E_i \in \mathbf{E}_i$) and outputs a value $\alpha_H(E_0, \ldots, E_{n-1}) = 1$ or $\alpha_H(E_0, \ldots, E_{n-1}) = 0$, where '1' means that H is accepted, and '0' means that H is rejected. The procedure is *reliable*, iff the values $\alpha_H(E_0, \ldots, E_{n-1})$ converge to the actual truth value of H, i.e. if $lim_{n\rightarrow\infty} \alpha_H(E_0, \ldots, E_{n-1}) = 1$ holds for every data sequence $e = (E_i)_{i \geq 0}$ with $e \in H$ and $lim_{n\rightarrow\infty} \alpha_H(E_0, \ldots, E_{n-1}) = 0$, for every sequence $e \notin H$.

The main result of this section is Theorem (3.7). The theorem states that one can define an almost everywhere reliable testing procedure for each measurable hypothesis $H \subset \times_{n \geq 0} \mathbf{E}_n$, if some very weak assumptions are satisfied. The proof is based on Doob's martingale convergence theorem for conditional expectations.

In order to make these ideas precise, we must first explain the notion of an experiment in some more detail. In our present, structuralist, framework, experiments are considered (roughly speaking) as intended applications of some theory T. Intended applications are represented by partial models (i.e. by substructures of potential models) of T. The reader should recall the generalized definition of partial models from §1. To simplify matters, we make the assumption that the partial models of T have the form $x =< D_1, \ldots, D_k, f_1, \ldots, f_n >$, where $f_1, \ldots, f_n$ are functions with $rng(f_i) \subset \mathbf{R}$.

3.1. Definition: Let $x, x' \in M_{pp}(T), x =< D_1, \ldots, D_k, f_1, \ldots, f_n >$, $x' =< D'_1, \ldots, D'_k, f'_1, \ldots, f'_n >$. We let $x \approx x'$ iff $Basis(x) = Basis(x')$, i.e. $D_1 = D'_1, \ldots, D_k = D'_k$, and if $Dom(f_1) = Dom(f'_1), \ldots, Dom(f_n) = Dom(f'_n)$. For collections $X, X' \subset M_{pp}(T)$ we let $X \approx X'$ iff for every $x \in X$ there exists $x' \in X'$ such that $x \approx x'$ and vice versa: for each $x' \in X'$ there exists $x \in X$ such that $x \approx x'$.

3.2. Corollary: $\approx$ is an equivalence relation. □

Let $x =< D_1, \ldots, D_k, f_1, \ldots, f_n >$ be a finite structure of type τ, such that $Dom(f_1), \ldots, Dom(f_n)$ are finite sets. Intuitively speaking, the elements $x' \in [x]_\approx$, represent all possible values of the functions $f_1, \ldots, f_n$ for their respective domains. $\mathbf{E} \subset Po(M_{pp}(T))$ is called an experiment for T, iff there is some finite number of finite substructures $E \subset I(T)$, such that $\mathbf{E} = [E]_\approx$.[18] Each $E \in \mathbf{E}$ is called a possible outcome of $\mathbf{E}$. $\mathbf{E}_0, \mathbf{E}_1, \ldots$ is a monotonic sequence of experiments for T, if there is some $E_{i+1} \in \mathbf{E}_{i+1}$ for each outcome $E_i \in \mathbf{E}_i$, such that $E_{i+1} \in e(E_i)$ and vice versa: for each $E_{i+1} \in \mathbf{E}_{i+1}$ there is some $E_i \in \mathbf{E}_i$ such that $E_{i+1} \in e(E_i)$. Intuitively speaking, empirical data are 'accumulated' in a monotonic sequence of experiments, no data is 'lost'.

Now, let $\mathbf{E}_i, i = 0, 1, \ldots$, be a monotonic sequence of experiments for T and $\Gamma = \times_{i \geq 0} \mathbf{E}_i$. For every sequence $e = (E_0, E_1, \ldots)$ with $E_i \in \mathbf{E}_i$, the finite initial segment of length n is denoted $e[n]$, hence $e[n] = (E_0, \ldots, E_{n-1})$. We agree that $e[0] = \emptyset$. Henceforth, we let $SEQ_\Gamma = \{e[n]/n \geq 0, e \in \Gamma\}$ denote the collection of all finite initial segments of

[18] Thus we get $Basis(E) = Basis(E')$ for all $E' \in [E]_\approx$. Intuitively, we think of $E' \in [E]_\approx$ as a large, but finite collection of numerical values, which are obtained from a finite number of empirical systems, each of which is represented by some substructure $x \in E$.

some data sequence $e \in \Gamma$.[19]

An empirical theory can entail various predictions on the outcomes of $\mathbf{E}_0, \mathbf{E}_1, \ldots$. We consider each such prediction as an *hypothesis* about the actual data stream, which will be obtained from $\mathbf{E}_0, \mathbf{E}_1, \ldots$. Technically speaking, we can identify every hypothesis with the collection of all data streams $E_0, E_1, \ldots$, for which the hypothesis holds true, i.e. we let $H \subset \times_{n \geq 0} \mathbf{E}_n$. H is *true* for some sequence $e \in \times_{n \geq 0} \mathbf{E}_n$ iff $e \in H$, otherwise H is *false*.

3.3. Definition: (Testing procedures): Let $\mathbf{E}_0, \mathbf{E}_1, \ldots$ be a monotonic sequence of experiments for theory element T.
(1) A testing procedure for $H \subset \times_{n \geq 0} \mathbf{E}_n$ is a mapping $\alpha_H(E_0, \ldots, E_{n-1})$, which takes as input a finite sequence $E_0, \ldots, E_{n-1}$ of outcomes $E_i \in \mathbf{E}_i$ and outputs a value $\alpha_H(E_0, \ldots, E_{n-1}) = 1$ or $\alpha_H(E_0, \ldots, E_{n-1}) = 0$, where '1' means that H is accepted and '0' means that H is rejected.
(2) Hypothesis H is *confirmed in the limit* by the outcomes $E_0, E_1, \ldots$, iff $lim_{n \to \infty} \alpha_H(E_0, \ldots, E_{n-1}) = 1$, i.e. if there exists $N \geq 0$ such that $\alpha_H(E_0, \ldots, E_{n-1}) = 1$ holds for all $n \geq N$.
(3) Testing procedure α_H is *reliable*, iff H is confirmed in the limit, iff H is true, such that
$lim_{n \to \infty} \alpha_H(E_0, \ldots, E_{n-1}) = 1$ iff $e \in H$ and $lim_{n \to \infty} \alpha_H(E_0, \ldots, E_{n-1}) = 0$ iff $e \notin H$ holds true for every data sequence $e = (E_n)_{n \geq 0} \in \times_{n \geq 0} \mathbf{E}_n$.

An n-tuple $(E_0, \ldots, E_{n-1})$ is called a 'positive' outcome for H, if $\alpha_H(E_0, \ldots, E_{n-1}) = 1$; otherwise, the outcome is 'negative'. Now let T be an empirical theory and let $(\mathbf{E}_n)_{n \geq 0}$ be a monotonic sequence of experiments for T. One particular hypothesis about $(\mathbf{E}_n)_{n \geq 0}$ would be
H: *All outcomes of* $(\mathbf{E}_n)_{n \geq 0}$ *agree with the predictions of* T.

In our present, structuralist, context, such a prediction would read $H = \{(E_n)_{n \geq 0} \in (\mathbf{E}_n)_{n \geq 0} / \exists X \in Cn_{th}(T) \forall n \geq 0 : X \in e(E_n)\}$.

A theory T is called *empirically adequate* wrt $e = (\mathbf{E}_n)_{n \geq 0}$ iff the above hypothesis holds true for e, i.e. iff $e \in H$. In practice, however, we would not expect theories to be perfectly adequate in this sense for several reasons. One reason is imprecision of measurement. (Data, which are obtained by meauring phsysical quantities will differ from theoretically predicted values, even if T is true). Another reason is the existence of idealizing and/or simplifying assumptions within a theory. (For example, a dynamical model for a physical system could never account for all interactions between the system and the rest of the universe). A somewhat

[19] Note, however, that SEQ_Γ also contains non monotonic sequences of outcomes $E_0, E_1, \ldots$ where $E_{i+1} \notin e(E_i)$for some $i \geq 0$.

more realistic hypothesis, which accounts for these problems, would be H': *'All outcomes of* $(\mathbf{E}_n)_{n\geq 0}$ *approximately agree with the predictions of* T'.

A theory T is called ϵ – *adequate*, iff the 'distance' between the actual outcomes of $(\mathbf{E}_n)_{n\geq 0}$ and the predictions of T is less than ϵ. In order to make this idea precise, we need a measure for the distance between different sequences $e, e' \in \times_{n\geq 0}\mathbf{E}_n$. Such a measure can be obtained under some very weak assumptions. Moreover, the measure will be such that an almost everywhere reliable testing procedure can be found for every measurable hypothesis $H \in \times_{n\geq 0}\mathbf{E}_n$.

(3.4) Definition: Let $(\mathbf{E}_i)_{i\geq 0}$ be a monotonic sequence of experiments for T and let $d_i : \mathbf{E}_i \times \mathbf{E}_i \longrightarrow \mathbf{R}$ be a metric, for $i = 0, 1, \ldots$, such that every $\mathbf{E}_i$, considered as a metric space, has a countable basis. We define a measurable space $(\Gamma, \mathbf{A})$, such that $\Gamma = \times_{n\geq 0}\mathbf{E}_n$ and $\mathbf{A} = \mathbf{B}(\Gamma)$ = the σ-algebra of all Borel sets of Γ with respect to the product topology.

By its definition, a projection $\pi_i : \Gamma \longrightarrow \mathbf{E}_i$ maps every sequence $E_0, E_1, \ldots$ in Γ onto its i-th component E_i. Let $\mathbf{A}_n = \mathbf{B}(\mathbf{E}_n)$ denote the family of all Borel sets of $\mathbf{E}_n$. π_n is $\mathbf{A} - \mathbf{A}_n -$ measurable, if $\pi_n^{-1}[B] \in \mathbf{A}$ holds for every Borel set $B \in \mathbf{A}_n$. By definition, the family of all open sets of $\mathbf{E}_n$ generates $\mathbf{A}_n$, hence it suffices to show that $\pi_n^{-1}[B] \in \mathbf{A}$ holds for every open set $B \in \mathbf{A}_n$.[20] By assumption, π_n is continuous, i.e. $\pi_n^{-1}[B]$ is an open set of the product topology on Γ, and therefore $\pi_n^{-1}[B]$ is a Borel set, too. As a consequence, all projections are actually $\mathbf{A}$- $\mathbf{A}_n$-measurable. Thus (i) $\mathbf{A}(\pi_n : n \geq 0) \subset \mathbf{A}$, where $\mathbf{A}(\pi_n : n \geq 0)$ is the smallest σ -algebra $\mathbf{D}$ on Γ, such that all projections are $\mathbf{D} - \mathbf{A}_n$-measurable. On the other hand, each open set of the product topology is a member of $\mathbf{A}(\pi_n : n \geq 0)$ by the following reason: Γ has a countable basis,[21] where each member of the basis is an element of $\mathbf{A}(\pi_n : n \geq 0)$. Hence every open set B is a countable union of basic neighborhoods, hence $B \in \mathbf{A}(\pi_n : n \geq 0)$, since $\mathbf{A}(\pi_n : n \geq 0)$ is closed under countable unions. By definition, $\mathbf{A} = \mathbf{B}(\Gamma)$ is the smallest σ -algebra on Γ, which contains every open set, hence (ii) $\mathbf{A} \subset \mathbf{A}(\pi_n : n \geq 0)$. From (i) and (ii) we infer that $\mathbf{A} = \mathbf{B}(\Gamma) = \mathbf{A}(\pi_n : n \geq 0)$.

We note that the product topology on Γ can be induced by a metric d, since Γ is a countable product of metric spaces $\mathbf{E}_0, \mathbf{E}_1, \ldots$. A suitable metric is, for instance,

---

[20] Cf. H. Bauer, loc cit., Prop. 7.2.

[21] Γ has a countable basis, since Γ is a countable product of spaces $\mathbf{E}_i$, each of which has a countable basis (cf. cf. Lipschutz, loc cit., p. 178, Exc. 22(i))

$d((E'_i)_{i\geq 0},(E_i)_{i\geq 0}) = \sum_{n\geq 0} 2^{-n}(d_n(E_n, E'_n)/(1+d_n(E_n, E'_n)))$.[22]

The open sphere $u_\varepsilon((E_i)_{i\geq 0}) = \{(E'_i)_{i\geq 0}/d((E'_i)_{i\geq 0},(E_i)_{i\geq 0}) < \varepsilon\}$ is an open Borel set, for every sequence $(E_i)_{i\geq 0}$. Accordingly, the hypothesis

$$H := T^\epsilon = \bigcup\{u_\varepsilon((E_i)_{i\geq 0})/\exists X \in Cn_{th}(T)\forall i \geq 0 : X \in e(E_i)\}$$

is an open set, hence $T^\epsilon \in \mathbf{B}(\Gamma)$. Intuitively speaking, T^ϵ contains all possible outcomes of the experiments $\mathbf{E}_0, \mathbf{E}_1, \ldots$, which do not significantly deviate from T's predictions, where 'significant' means a distance $\geq \epsilon$. Similarly,the hypotheses
$T^{\epsilon,n} := \bigcup\{u_\varepsilon((E_i)_{i\geq 0})/\forall i \leq n \exists X \in Cn_{th}(T) : X \in e(E_i)\}$, and $T^{\epsilon,\infty} := \bigcup\{u_\varepsilon((E_i)_{i\geq 0})/\forall i \geq 0 \exists X \in Cn_{th}(T) : X \in e(E_i)\}$
are measurable sets (for all $\varepsilon > 0$ and all $n \geq 0$), $T^{\epsilon,n}, T^{\epsilon,\infty} \in \mathbf{B}(\Gamma)$. We obviously have: $T^\epsilon \subset T^{\epsilon,\infty} \subset T^{\epsilon,n}$, for $\forall n \geq 0$. We note some obvious properties of $T^{\epsilon,n}, T^{\epsilon,\infty}$:

3.5. Corollary: Let p be a probability measure on $\mathbf{B}(\Gamma)$: The following holds for all $\varepsilon > 0, n \geq 0$:

(i) $T^{\epsilon,\infty} = \bigcap_{n\geq 0} T^{\epsilon,n}$, and $T^{\epsilon,n} = \bigcap_{i\leq n} T^{\epsilon,i}$;

(ii) $T^{\epsilon,n}$ is a monotonically decreasing sequence, i.e. $T^{\epsilon,n+1} \subset T^{\epsilon,n}$;

(iii) $p(T^{\epsilon,n+1}) \leq p(T^{\epsilon,n})$ and $p(T^{\epsilon,n+1}/T^{\epsilon,n}) \geq p(T^{\epsilon,n+1})$ and $p(T^{\epsilon,n}/T^{\epsilon,n+1}) = 1$;

(iv) $p(T^{\epsilon,n})$ is monotonically decreasing for $n = 0, 1, \ldots$,

(v) $p(T^{\epsilon,\infty} \mid T^{\epsilon,n}) \geq p(T^{\epsilon,\infty})$;

(vi) $p(T^{\epsilon,\infty}/T^{\epsilon,n})$ is monotonically increasing for $n = 0, 1, \ldots$

The proof is obvious from the above definitions. □

For every sequence $e = (E_n)_{n\geq 0} \in \Gamma$ and every $n \geq 0$ let $\pi_{\leq n}(e) = (E_0, \ldots, E_{n-1})$ denote the projection of $(E_n)_{n\geq 0}$ onto its initial segment of length n. Moreover, we let $p(E_0, \ldots, E_{n-1}) := p(\{e \in \Gamma/\pi_{\leq n}(e) = (E_0, \ldots, E_{n-1})\}$.

Theorem (3.7) states, roughly speaking, that the conditional probability $p(T^\epsilon/E_0, \ldots, E_{n-1})$ almost everywhere converges to 1, if the hypothesis $H = T^\epsilon$ is true, and to 0, otherwise. The main problem for a proof of

[22] Cf. Lipschutz, loc. cit. p. 178, Prop. 12.12.

this result is the fact that, in most cases, the probability $p(E_0, \ldots, E_{n-1}) = 0$, so that the conditional probability $p(T^\epsilon/E_0, \ldots, E_{n-1})$ is not defined. (The reader should recall that $p(E_0, \ldots, E_{n-1}) = 0$ is the probability that the first n experiments yield outcomes $E_0, \ldots, E_{n-1}$, respectively. This probability must vanish for all but countably many outcomes $(E_0, \ldots, E_{n-1}) \in \mathbf{E}_0 \times \ldots \times \mathbf{E}_{n-1}$). Fortunately, the tool-box of measure theory provides some means to circumvent this difficulty. The solution that I want to propose, is to replace the conditional probabilities by conditional expectations. So let 1_{T^ϵ} denote the characteristic function of $H = T^\epsilon$, i.e. $1_{T^\epsilon}(e) = 1$, if $e \in T^\epsilon$, and $1_{T^\epsilon}(e) = 0$, otherwise. Moreover, let $\mathbf{A}(\pi_0, \ldots, \pi_{n-1})$ be the smallest σ-algebra $\mathbf{D}$ on Γ, such that all projections $\pi_i, i \leq n$ are measurable wrt $\mathbf{D}$. We obviously have that $\mathbf{A}(\pi_0, \ldots, \pi_{n-1}) \subset \mathbf{A} = \mathbf{B}(\Gamma)$, since $\mathbf{A}$ is defined to be the smallest σ-algebra on Γ, such that all $\pi_i, i = 0, 1, \ldots$ are measurable wrt $\mathbf{A}$.

Now let $E^{\pi_{\leq n}}(1_{T^\epsilon}) := E(1_{T^\epsilon})/\mathbf{A}(\pi_0, \ldots, \pi_{n-1}))$ denote the conditional expectation on 1_{T^ϵ} given $\mathbf{A}(\pi_0, \ldots, \pi_{n-1}))$. $E^{\pi_{\leq n}}(1_{T^\epsilon})$ is a random variable on Γ, which is almost everywhere uniquely determined by the following equation $\int_B E^{\pi_{\leq n}}(1_{T^\epsilon})dp = \int_B 1_{T^\epsilon}dp$ (cf. Bauer, loc.cit., chap. X, for details).

The fundamental relationship between conditional probabilities and conditional expectations is exhibited by the following

(3.6) Lemma: For all sequences $e = (E_n)_{n \geq 0} \in \Gamma$ we have:
$E^{\pi_{\leq n}}(1_{T^\epsilon})(e) = p(T^\epsilon/E_0, \ldots, E_{n-1})$ provided that $p(E_0, \ldots, E_{n-1}) > 0$

Proof: The proof is based on the following Lemma (*) of measure theory:
(*) Let $Y : \Gamma \longrightarrow \Gamma'$ be a mapping from Γ into some measure space Γ' with σ-algebra $\mathbf{A}'$ and let $X : \Gamma \longrightarrow \mathbf{R}$. Let $\mathbf{A}(Y)$ denote the smallest σ-algebra $\mathbf{A}$ on Γ, such that Y is $\mathbf{A}$-$\mathbf{A}'$-measurable. X is measurable wrt $\mathbf{A}(Y)$ iff some $\mathbf{A}'$-$\mathbf{B}(\mathbf{R})$-measurable function $g : \Gamma' \longrightarrow \mathbf{R}$ exists such that $X = g \circ Y$ (cf. Bauer, loc. cit., p. 298).

We use the lemma with $Y = \pi_{\leq n}$ and $X = E^{\pi_{\leq n}}(1_{T^\epsilon})$, hence $\Gamma' = \mathbf{E}_0 \times \ldots \times \mathbf{E}_{n-1}$. The Lemma then states that some measurable, real valued function $g : \mathbf{E}_0 \times \ldots \times \mathbf{E}_{n-1} \longrightarrow \mathbf{R}$ exists, such that

(1) $E^{\pi_{\leq n}}(1_{T^\epsilon})(e) = g \circ \pi_{\leq n}(e)$

holds for all data sequences $e \in \Gamma$. g is uniquely determined, since $rng(\pi_{\leq n}) = \mathbf{E}_0 \times \ldots \times \mathbf{E}_{n-1}$ (cf. Bauer, loc.cit., p. 300). Let $p_{\pi_{\leq n}}$ be the induced measure of p wrt the mapping $\pi_{\leq n}$, i.e. $p_{\pi_{\leq n}}$ is a probability measure on $\mathbf{E}_0 \times \ldots \times \mathbf{E}_{n-1}$ such that $p_{\pi_{\leq n}}(B) = p(\pi_{\leq n}^{-1}(B))$ for every $B \in \mathbf{A}_0 \times \ldots \times \mathbf{A}_{n-1}$. We get:

(2) $\int_B g dp_{\pi_{\leq n}} = \int_{\pi_{\leq n}^{-1}(B)} g \circ \pi_{\leq n} dp = \int_{\pi_{\leq n}^{-1}(B)} E^{\pi_{\leq n}}(1_{T^\epsilon}) dp = \int_{\pi_{\leq n}^{-1}(B)} 1_{T^\epsilon} dp$ for all $B \in \mathbf{A}_0 \times \ldots \times \mathbf{A}_{n-1}$.

By our assumptions, each $\mathbf{E}_i$ is a metric space, hence each singleton set $\{E\}$ with $E \in \mathbf{E}_i$ is closed, whence $\{E\} \in \mathbf{A}_i$. In a similar way, one can define a metric for the finite product $\mathbf{E}_0 \times \ldots \times \mathbf{E}_{n-1}$, which generates the product topology, such that $\{(E_0, \ldots, E_{n-1})\} \in \mathbf{A}_0 \times \ldots \times \mathbf{A}_{n-1}$ holds for each n-tuple $(E_0, \ldots, E_{n-1}) \in \mathbf{E}_0 \times \ldots \times \mathbf{E}_{n-1}$. If we let $B = \{(E_0, \ldots, E_{n-1})\}$ in (2) we get:

(3) $\int_{\{(E_0, \ldots, E_{n-1})\}} g dp_{\pi_{\leq n}} = g(E_0, \ldots, E_{n-1}) p(E_0, \ldots, E_{n-1}) = \int_{B'} 1_{T^\epsilon} dp,$

where $B' := \{e \in \Gamma / \pi_{\leq n}(e) = (E_0, \ldots, E_{n-1})\}$. Note that the value $g(E_0, \ldots, E_{n-1})$ is well defined, even if $p(E_0, \ldots, E_{n-1}) = 0$.
If $p(E_0, \ldots, E_{n-1}) > 0$, we get :

(4) $g(E_0, \ldots, E_{n-1}) = \int_{B'} 1_{T^\epsilon} dp / p(E_0, \ldots, E_{n-1}) = p(T^\epsilon / E_0, \ldots, E_{n-1})$

i.e: $g(E_0, \ldots, E_{n-1})$ is equal to the conditional probability of T^ϵ given that $\pi_{\leq n}(e) = (E_0, \ldots, E_{n-1})$. The Lemma is now immediate from (1) and (4). □

Note that the conditional expectation has a well-defined value, even when $p(E_0, \ldots, E_{n-1}) = 0$. We now consider the limit of the conditional expectations, when n tends to infinity. So let $\mathbf{C}_n, n = 0, 1, \ldots$ be a monotonic sequence of subalgebras $\mathbf{C}_n \subset \mathbf{A}$ such that $\mathbf{C}_n \subset \mathbf{C}_{n+1}$ for all $n \geq 0$. A sequence $X_n, n = 0, 1, \ldots$ of $\mathbf{C}_n$-measurable random variables is called a *martingale* wrt $(\mathbf{C}_n)_{n \geq 0}$ iff $\int_B X_n dp = \int_B X_{n+1} dp$ holds for all $B \in \mathbf{C}_n$ and all $n \geq 0$. It is easy to see that the sequence $E^{\pi_{\leq n}}(1_{T^\epsilon})$ actually is a martingale wrt the particular sequence $\mathbf{C}_n := \mathbf{A}(\pi_0, \ldots, \pi_{n-1})$, since $\int_B E^{\pi_{\leq n}}(1_{T^\epsilon}) dp = \int_B E^{\pi_{\leq n+1}}(1_{T^\epsilon}) dp = \int_B 1_{T^\epsilon} dp$ for all $B \in \mathbf{C}_n$ by the fundamental property of conditional expectations. We therefore can apply Doob's martingale convergence theorem, in order to obtain the following

(3.7) Theorem: Let $(\mathbf{E}_n)_{n \geq 0}$ be a monotonic sequence of experiments for theory element T, such that each $\mathbf{E}_n$ is a metric space with countable basis and let p be a probability measure on $\mathbf{A} = \mathbf{B}(\Gamma), \Gamma = \times_{n\geq 0}\mathbf{E}_n$. Then

$$lim_{n\to\infty} E^{\pi\leq n}(1_{T^e}) = 1_{T^e}$$

holds p-almost everywhere in Γ, i.e: the conditional expectation of 1_{T^e} converges almost everywhere to the actual truth value of $H = T^e$.

Proof: By Bauer, Prop. 60.01, every martingale $(X_n)_{n\geq 0}$ with $sup(E(\mid X_n \mid)) < \infty$ converges almost everywhere to an integrable random variable $X_\infty : \Gamma \longrightarrow \mathbf{R}$, such that

(1) $lim_{n\to\infty} E^{\pi\leq n}(1_{T^e}) = X_\infty$ p-almost everywhere.

By Bauer 60.4 and 60.3, the sequence $(X_n)_{n\in\mathbf{N}\cup\{\infty\}}$, which is obtained by adding X_∞ to the sequence $X_n := E^{\pi\leq n}(1_{T^e}), n = 0, 1, \ldots$, is a martingale wrt the sequence $(\mathbf{A}_n)_{n\in\mathbf{N}\cup\{\infty\}}$ of σ-algebras, where $\mathbf{A}_n := \mathbf{A}(\pi_0, \ldots, \pi_{n-1})$ and $\mathbf{A}_\infty := \mathbf{B}(\Gamma)$. Thus

(2) $X_n := E^{\pi\leq n}(1_{T^e}) = E^{\pi\leq n}(X_\infty)$ p-almost everywhere for $n = 0, 1, \ldots$

By Bauer 60.6, the variable X_∞ is almost everywhere uniquely determined by (2), whence

(3) $1_{T^e} = X_\infty$ p-almost everywhere.

From (1) and (3) we infer that the conditional expectations $E^{\pi\leq n}(1_{T^e})$ converge almost everywhere to X_∞, which is almost everywhere equal to 1_{T^e}, hence $E^{\pi\leq n}(1_{T^e}) \longrightarrow 1_{T^e}$ almost everywhere. □

Now let H be any measurable hypothesis $H \subset \times_{n\geq 0}\mathbf{E}_n, H \in \mathbf{B}(\Gamma)$. A testing procedure α_H for H is called *almost everywhere reliable*, if the values $\alpha_H(E_0, \ldots, E_{n-1})$ converge almost everywhere to the actual truth value of H, i.e. if $lim_{n\to\infty}\alpha_H(E_0, \ldots, E_{n-1}) = 1_H(e)$ for almost all data sequences $e \in \times_{n\geq 0}\mathbf{E}_n$. The proof of the preceeding theorem shows that an almost everywhere reliable testing procedure exists for each measurable hypothesis H, (and, in particular, for $H = T^e$):

(3.8) Corollary: Let p denote any probability measure on $\mathbf{B}(\Gamma)$,with $H \in \mathbf{B}(\Gamma)$ and $0 < \beta < 1$. We define a testing procedure α_H for H, such that $\alpha_H(E_0, \ldots, E_{n-1}) = 1$ iff $E^{\pi\leq n}(1_H)(e) > \beta$ and $\alpha_H(E_0, \ldots, E_{n-1}) = 0$, otherwise, for all $e = (E_i)_{i\geq 0} \in \Gamma$. Then α_H is almost everywhere reliable.

Proof: We have shown in Theorem (3.7) that $lim_{n\to\infty} E^{\pi\leq n}(1_{T^e})(e) = 1_{T^e}(e)$ holds p-almost everywhere. It is easy to see that the theorem and the proof remain valid, if $H = T^e$ is replaced by an arbitrary hypothesis $H \in \mathbf{B}(\Gamma)$, hence $lim_{n\to\infty} E^{\pi\leq n}(1_H) = 1_H$ almost everywhere. Accordingly, there exists some subset $\Gamma_0 \subset \Gamma$ with $p(\Gamma_0) = 1$, such that $lim_{n\to\infty} E^{\pi\leq n}(1_H)(e) = 1$ for all sequences $e = (E_n)_{n\geq 0}$ in Γ_0 with $e \in H$. Similarly, $lim_{n\to\infty} E^{\pi\leq n}(1_H)(e) = 0$ for all $e \in \Gamma_0$ with $e \notin H$. Assume now that $e \in H$. Then there exists $N \geq 0$ such that $E^{\pi\leq n}(1_H)(e) > \beta$ for all $n \geq N$, whence $\alpha_H(E_0, \ldots, E_{n-1}) = 1$ for $n \geq N$. By a similar reasoning there exists $N \geq 0$ such that $\alpha_H(E_0, \ldots, E_{n-1}) = 0$ for $n \geq N$, if $e \notin H$. In both cases we get that $lim_{n\to\infty} \alpha_H(E_0, \ldots, E_{n-1}) = 1_H(e)$ for all $e \in \Gamma_0$, i.e. α_H is almost everywhere reliable. □

References

Angluin, D. & Smith, C. H., 1983, 'Inductive Inference, Theory and Methods' , *Computing Surveys* 15, 237-269.

Balzer, W., 1985, *Theorie und Messung*, Berlin/New York.

Balzer, W. & Moulines, C. U., 1981, 'Die Grundstruktur der klassischen Partikelmechanik und ihre Spezialisierungen', *Zeitschrift für Naturforschung* 36a, 600-608.

Balzer, W. & Moulines, C. U. & Sneed, J. D., 1987, *An Architectonic for Science*, Dordrecht.

Balzer, W. & Lauth, B. & Zoubek, G., 1993, 'A Model for Science Kinematics', *Studia Logica* 52, 519-548.

Bartelborth, T., 1988, *Eine logische Rekonstruktion der klassischen Elektrodynamik*, Frankfurt/Bern.

Bauer, H., 1978, *Wahrscheinlichkeitstheorie und Grundzüge der Maßtheorie*, Berlin/New York, 3rd edition.

Blum, M. & Blum, L., 1975, 'Toward a Mathematical Theory of Inductive Inference', *Information and Control* 28, 125-155.

Carnap, R., 1950, *Logical Foundations of Probability*, Chicago.

Carnap, R. & Jeffrey, R. C. eds., 1971, *Studies in Inductive Logic and Probability*, Vol. I, Berkeley, Los Angeles, London.

Earman, J. ed., 1983, *Testing Scientific Theories, Minnesota Studies in the Philosophy of Science*, Vol. X, Minneapolis.

Gaifman, H., 1964, 'Concerning Measures on First Order Calculi', *Israel Journal of Mathematics*, 2, 1-18.

Gaifman, H., 1979, 'Subjective Probability, Natural Predicates and Hempel's Ravens', *Erkenntnis* 14, 105-147.

Gaifman, H. & Snir, M., 1982, 'Probabilities over Rich Languages, Testing and Randomness' , *Journal of Symbolic Logic*, 47, 495-548.

Glymour, C., 1980, *Theory and Evidence*, Princeton.
Gold, E. M., 1967, 'Language Identification in the Limit', *Information and Control* 10, 447-474.
Goodman, N., 1965, *Fact, Fiction and Forecast*, Indianapolis, 2nd edition.
Hempel, C. G., 1945, 'Studies in the Logic of Confirmation', *Mind* 54, 1-26 and 97-121.
Jeffrey, R. C. ed., 1980, *Studies in Inductive Logic and Probability*, Vol. II, Berkeley.
Kelly, K. & Glymour, C., 1989 , 'Convergence to the Truth and Nothing but the Truth', *Philosophy of Science* 56.
Lauth, B., 1991, 'Theory Evolution and Reference Kinematics', *Synthese* 88, 279-307.
Lauth, B., 1993, 'Inductive Inference in the Limit for First Order Sentences', *Studia Logica* 52, 491-517.
Lauth, B., 1994, 'An Abstract Model for Inductive Inference', *Erkenntnis* 40, 87-120.
Lauth, B., 1995, 'Inductive Inference in the Limit of Empirically Adequate Theories', *The Journal of Philosophical Logic* 24, 525-48.
Lipschutz, S., 1977, *Allgemeine Topologie*, London/New York.
Osherson, D. N. & Stob, M. & Weinstein, S., 1986, *Systems That Learn*, Cambridge.
Osherson, D. N. & Weinstein, S., 1986, 'Identification in the Limit of First Order Structures', *Journal of Philosophical Logic* 15, 55-81.
Popper, K., 1984, *Objektive Erkenntnis*, Hamburg, 4th edition.
Putnam, H., 1963, 'Degree of Confirmation and Inductive Logic', In, Schilpp, A. 1963, 761-783.
Putnam, H., 1965, 'Trial and Error Predicates and the Solution to a Problem of Mostowski', *Journal of Symbolic Logic*, 30. No.1, 49-57.
Schilpp, A. ed., 1963, *The Philosophy of Rudolf Carnap*, La Sall, Ill.
Shapiro, E., 1981, *Inductive Inference of Theories from Facts*, Report YLU 192, Department of Computer Science, Yale University.
Sneed, J. D., 1971, *The Logical Structure of Mathematical Physics*, Dordrecht, 2nd edition 1979.
Stegmüller, W., 1979, *The Structuralist View of Theories*, Berlin/Heidelberg/New York.
Stegmüller, W., 1986, *Probleme und Resultate der Wissenschaftstheorie und Analytischen Philosophie*, Vol. II, Theorie und Erfahrung, Berlin/Heidelberg/New York.

Chapter 8: Theoretical Terms: Recent Developments

Wolfgang Balzer [1]

Introduction

The theme of theoretical terms exhibits a progressive development within general philosophy of science, and in particular within the structuralist approach. Without attempting at completeness let me briefly recall the history. Logical empiricists started from a 'non-theoretical' or 'empirical' observation language the terms and sentences of which should be accessible to 'direct' experience. All other terms of scientific language were called 'theoretical'. A distinction between theoretical and non-theoretical (=observational) terms was thus drawn relative to some all-embracing language comprising all of ordinary and scientific language. This view was motivated by the goal of making theoretical, scientific statements more 'secure' by some kind of definition or reduction to observational ones. It was felt that observational statements were objective enough to provide a basis for agreement in science since they could be checked by every person using her own senses. This objectivity would then be transferred to theoretical statements by defining them in observational terms.

It soon became clear that this distinction was rather arbitrary and vague, for most 'observation' statements in science cannot be checked directly by means of the human senses: they involve instruments and/or interpretations. The fixation to first order logic prevented every serious attempt to investigate relations of definability in existing empirical theories; and other more trivial examples of the 'all copper conducts electricity' type were seen to escape standard, first-order definability. Logical investigations of definability in a broader setting did not improve the situation because they still assumed the distinction itself as given, and thus operated with sweeping notions of observability, or standard interpretation of O-terms.[2] These problems led to the replacement of the bipartition by a stratified construction in which, starting again from a neutral observation

---

[1] I am indebted to Ulrich Gähde and Theo Kuipers for helpful comments on an earlier draft, to Dorothea Lotter for correcting my English and to Jörg Sander for assistence in processing the manuscript.

[2] Like (Lewis,1970), who takes a theoretical term to be a 'term introduced by a given theory T at a given stage in the history of science' (p.428). His account moreover rests on the descriptively very problematic assumption that all 'good' theories have exactly one realization.

language, 'theoretical terms' of successively higher order were introduced.[3] But this did not eliminate the problems. The remaining 'observational' sentences fell outside of scientific language, and there was no idea of how to define the comprehensive class of non-observational terms by the few and elusive observational ones. Also, it was argued that the notion of a neutral, non-theoretical observation language was a phantom.[4] It was argued that each scientific term is 'theory-laden' insofar as theoretical assumptions play a central role in the determination of its meaning.

The result of this development may be stated as follows. It is futile to characterize 'sound' empirical theories (as opposed to empty or 'metaphyiscal' ones) by using observability or empiricity as meta-theoretical primitives. Of course, this does not mean to give up the attempt to characterize empirical science as distinct from other social phenomena like metaphysics or religion. It means that observability and empiricity will have to be explicated by means of a comprehensive, and probably rather complex model of science. Whether there is a distinction between theoretical and observational terms in science is not a matter of apriori assumption but of empirical, meta-scientific study.

Consequently, attention was shifted to the study of real science, and the question was now whether scientific structures and practice indicate at all the presence of the distinction. One tried to understand theoretical terms by means of the role(s) they play in scientific theories and in the application of such theories.[5] J.D.Sneed was the first to suggest a definite distinction between the 'theoretical' and the 'non-theoretical' terms of a given theory by means of the roles they play in that theory.[6] According to Sneed, a term $\bar{t}$ is *theoretical in* theory T iff every determination of (a realization of) that term presupposes that T has been successfully applied beforehand.[7] Whereas previous distinctions rested on some positive or implicitly presupposed characterization of observability, Sneed's criterion for theoretical terms is free from such a strong philosophical assumption. On the basis of his criterion it became possible for the first time to investigate questions of 'empiricity' or non-theoreticity and 'reduction of security' of scientific statements on an entirely empirical basis given by data about theories and their development. No wonder that Sneed's distinction served as a paradigm for subsequent research attempting at the explication and elimination of pragmatic elements in Sneed's original formulation.

[3] (Hempel,1974).
[4] E.g. (Hanson,1958).
[5] (Putnam,1962), see also (Putnam,1965).
[6] (Sneed,1971).
[7] An attempt to capture this notion in a modal theoretic setting is found in (Miroiu,1984).

Stegmüller suggested to explicate Sneed's 'presupposition' as 'logical implication' and arrived at the following.[8] $\bar{t}$ is *T-theoretical* iff for every determination of (a realization of) $\bar{t}$, the sentences describing $\bar{t}$'s determination according to each existing exposition of T logically imply the existence of some successful application of T, i.e. some system satisfying the laws of T. The next step was to investigate the range of 'every determination' to be considered in this formulation. 'f-determining models' and 'methods of determination' were introduced as special kinds of models and model classes in order to explicate the notion of determination.[9] This left just one pragmatic component in Sneed's criterion, namely the reference to determinations of $\bar{t}$ as they are described in existing expositions of the theory. This reference could not be replaced by formal requirements because using a purely formal characterization of a class of determinations would turn practically every term into a non-T-theoretical one according to Sneed's criterion.[10]

In addition to this criterion of theoreticity, Moulines proposed a pragmatic condition for T-non-theoreticity.[11] Term $\bar{t}$ is T-non-theoretical if there exists an acknowledged method of determination for $\bar{t}$ in some theory T' different from T plus some link from T' to T which permits the transfer of realizations of $\bar{t}$ from T' into T.

In discussing Sneed's criterion, Kamlah and Tuomela proposed alternative criteria which -like Sneed's- start from a given theory but draw a different distinction.[12] According to Kamlah and Tuomela term $\bar{t}$ is T-theoretical, roughly, if *there exists* some $\bar{t}$-determining model (or measuring model for $\bar{t}$) in T which is also a proper model of T. Very roughly, the formulations of Sneed, Stegmüller, Balzer & Moulines have the form: 'For all x: if x is a measuring model for $\bar{t}$, then x is a model of T', whereas Kamlah's and Tuomela's versions have the form: 'There exists some x such that x is a measuring model for $\bar{t}$ and also a model of T'. Obviously, these two versions are not equivalent. They contain some common core, though. If in the 'for all' version $\bar{t}$ is T-theoretical in a non-trivial way (i.e. if the 'if' part is true for some x), then $\bar{t}$ also is T-theoretical according to the second, 'there exists', version.

A different route was taken by U.Gähde who proposed two purely formal conditions for the distinction of theoretical terms which avoided any pragmatic notion. These conditions put special emphasis on the cooperation between the fundamental laws of empirical theories and more specia-

[8] (Stegmüller,1973), Sec.VIII.

[9] (Balzer & Moulines,1980). See also (Balzer,1979), p.29ff.

[10] A detailed account of this development is given in (Zoglauer,1993.).

[11] (Moulines,1985), (Moulines,1992).

[12] (Kamlah,1976), (Tuomela,1973).

lized laws, auxiliary hypotheses etc. in the determination of the values of theoretical terms (functions). Although for most empirical theories these two requirements significantly narrow down the set of possible candidates for the theoretical/non-theoretical dichotomy, in general they do not determine such a distinction uniquely: they formally admit more than one distinction among a theory's terms.

Starting from Gähde's work, I proposed a weaker criterion which also is fully formal and yields a unique distinction (see Sec.3 below) within the theory's vocabulary.[13] This criterion was criticized by Hoyningen-Hüne, Kuokkanen, and -most severely- by G.Schurz for yielding unintended and thus inadequate results.[14] In the following, after a brief presentation of the main intuitions found in the literature, and of Gähde's two innertheoretical conditions and my purely formal account in Secs.2 and 3, I will take up with this criticism and defend my formal approach to theoreticity.

2. Presystematic Meanings of Theoreticity

1) The logical empiricist meaning. This presupposes the notion of observational terms, i.e., terms which can be determined by 'direct' observation. Theoreticity then is the property of being non-observational. Theoretical terms are those terms whose referents cannot be 'directly' observed.

2) The operationalist meaning. This presupposes the notion of operational definition. Theoretical terms are those which cannot be operationally defined. Normatively, this view implies that they should be abandoned from science.

3) The organization meaning. Theoretical terms are terms which simplify the task of theoretical systematization. They are used to organize scientific knowledge in an efficient way. They are only partially defined by means of the theory in which they were first introduced.

4) The Sneedian meaning. Theoretical terms are those which within a given theory can only be determined by presupposing that theory. That is, they cannot be determined externally.

5) The contextual meaning. This view focuses on the existence of terms which are not T-theoretical. Such terms indicate the way in which T is not self-contained. They refer to T's external theoretical context.

[13] (Balzer,1985), see also (Balzer,1986) for a discussion of Sneed's account.
[14] (Hoyningen-Hüne,1988), (Kuokkanen,1988), especially pp.54, (Schurz,1990).

6) The holistic meaning. Theoretical terms are terms which function in binding together the web of theories such as to form one (or few) unseparable unit(s).

7) The instrumental meaning. Theoretical terms, like theories, are mainly instruments for achieving predictions and applications by means of scientific knowledge. They are mere tools for managing large sets of data.

8) The formalist meaning. This view centers on purely formal properties of the terms of axiomatized theories.[15]

9) The ontological meaning. Theoretical terms are constructs that allow reference to very small entities, or entities which are not directly accessible.

Of course, these meanings are not independent from each other. The formalist meaning, for instance, goes together with the organization meaning and the instrumental meaning, Sneedian meaning goes together with holistic and instrumental meaning. Contextual meaning goes together with formalist and instrumental meaning, etc. I do not want to elaborate on each meaning. Instead, I want to use these views as a basis for clustering. Indeed, on the basis of the above nine items, two different broad views of theoreticity can be seperated which I want to call the philosophical and the meta-scientific view.

The *philosophical* view of theoreticity makes the distinction of theoretical and non-theoretical terms dependent on some previous philosophical distinction or assumption. The characterization and the existence of theoretical terms is derived from some 'more fundamental' philosophical presupposition. Logical empiricists presuppose observational terms as philosophically primary, theoretical terms are only 'admitted' as terms which are found in science and are not observation terms. Operationalists start from the normative idea of founding science on operational definitions. Theoretical terms are then only 'admitted' as those terms which occur in science but are not properly introduced by operational definitions. The ontological meaning presupposes the existence of some kinds of entities, and takes theoretical terms as a means for referring to such entities. The holistic meaning may be seen as a borderline case of the philosophical view. If it is taken in its radical form 'every term depends inseparably on every other term' then theoretical terms are needed to explain why this should be so.

[15] An interesting view of this kind is found in (Gaifman,Osherson & Weinstein,1990), who argue that theoretical terms may facilitate theory revision.

The *meta-scientific* view, on the other hand, does not start from broad philosophical assumptions. Rather, it takes scientific phenomena and developments as data on the basis of which a distinction of scientific terms in two categories can be drawn and evaluated. Whether there is such a distinction is contingent on the actual forms of scientific theories and practice. The organization meaning, the Sneedian meaning, the contextual, intrumentalist, and formalist meaning are of this nature. In all these approaches, the main interest in theoretical terms is in their role in science, in the way how theoretical terms function in the organization, application and development of scientific knowledge. In its moderate form, the holistic meaning finds its place here, too.

With respect to this distinction of philosophical versus meta-scientific accounts of theoreticity two things should be noted. First, there is a kind of natural, scientific presupposition relation among the two views. Philosophy should not proceed entirely from apriori assumptions.[16] Therefore, it is natural that philosophical distinctions should take into account the views resulting from more empirical approaches, if there are such. In the case at hand, philosophy should take serious the findings of philosophy of science, and study whether the meta-scientific accounts of theoreticity are compatible with the philosophical views, but should not expect that meta-scientific accounts are scheduled along prior philosophical assumptions.

The second point to be noted is that structuralism as an empirically oriented program of meta-science of course attaches more weight to the meta-scientific view of theoreticity than to the philosophical one. This observation also is in line with the historical development of these views as indicated in Sec.1. The philosophical view was exposed to severe criticism and has been largely abandoned. One implication of this is that one should not expect to get an explanation of why empirical theories are 'empirical' or even 'useful' simply in terms of their employing observational terms.

3. Gähde's Innertheoretical Conditions of Theoreticity

In his pioneering work[17] U.Gähde formulated two purely formal conditions for theoretical terms. In the example of classical particle mechanics these conditions led to a distinction between theoretical and non-theoretical functions which coincided with pre-systematic views on that dichotomy. In the case of other empirical theories, however, they did not provide a unique distinction. Thus, in a revised version presented in (Gähde,1990)

[16] I take this as the result of empiricists' struggle against metaphyics. Philosophy without empirical basis has no place within science.

[17] (Gähde,1983).

they were regarded as necessary conditions for theoreticity only. In the following, by Gähde's account I will always mean this revised account.

His account rests on the assumption that a theory at a certain stage of it's development is given in the form of a specialization net, consisting of a basic theory-element plus an array of specializations of the basic element.[18]

His first innertheoretical condition states that a function which is theoretical with respect to T cannot be uniquely determined by T-non-theoretical functions with the help of T's axioms alone. This condition contains a negative statement about the relation of a theoretical term and the theory it comes from: the theory's axioms alone are *not* sufficient for it's determination. Thus T-theoretical functions are left fundamentally underdetermined by T's basic laws. This property guarantees for the flexibility of an empirical theory to adjust to new sets of data and circumstances. In particular, it accounts for the fact that fitting new data may require some modification of special laws but usually can be achieved without affecting the theory's basic law(s). If theoretical terms were uniquely determined by T's fundamental principles alone this variability would not exist. Every change of the theory would have to be a change of the basic law(s).

In contrast to the first, negative statement, his second condition establishes a positive bond between a theoretical term and the theory it *comes from*. According to this condition, a theoretical term -though underdetermined within the theory-net's basic element- can be uniquely determined in some models of a suitable specialization of this basic element, i.e. in some models of a specialized theory-element. In (Gähde,1990) different versions of variable strength of these two conditions are discussed. In any of these variants the positive role of theoretical terms is revealed. They can be determined by a specific cooperation between the fundamental axioms and special laws of 'their' theory, and so their definite values can be propagated and used in an entire net of theories through the well known means of constraints, links and intertheoretical relations.

The special requirements added in order to obtain unique determinability have to be appropriately restricted. It is intuitively clear, that by adding suitable requirements to the axioms of the basic theory-element, *every* term can be made uniquely determinable in some specialization, and thus *every* term could be made to fulfill the second innertheoretical condition if arbitrary requirements were allowed for specialization. The point in (Gähde,1983) was to admit only those requirements (special laws etc.) which are invariant under the transformations characteristic for the basic theory-element. In other words: in order to determine the values of a theoretical term only those additional formulas were admitted which

[18] Compare (Balzer,Moulines,Sneed,1987), Chap.4, for these notions.

have the right properties of invariance as do the theory's basic law(s). In the reformulation in (Gähde,1990) the choice of special requirements in one version even is restricted to those occurring in special theory-elements which *de facto* belong to the theory-net in question at a certain stage of it's development. Note that the latter condition usually implies the former jamell one which refers to invariances: usually, only those specialized theory-elements will be included in a theory-net which fulfill the same invariance principles as does the net's basic element. Relativizing the special requirements to a given theory-net at a given time also allows to account for the following. The notion that theoretical terms *come from some empirical theory* T (and thus are theoretical with respect to T) includes the aspect that their values can be determined with the help of this theory. Now, what one can do with some theory (i.e. theory-net) depends crucially on the set of laws included in this net at a certain stage of it's development. If a theory-net develops, new specialized theory-elements will be included and values of some terms may become determinable which had not been determinable before. This may have impacts on the question of theoreticity, as mirrored in Gähde's revised second condition.

It is crucial for an adequate understanding of Gähde's innertheoretical conditions that they are imposed on all theoretical terms simultaneously. This has an important methodological consequence with respect to their application. Suppose that $r_1, ..., r_n$ are all the non-logical terms that occur in the basic element of some theory-net corresponding to theory T at a certain stage of it's development. Let us call each structure from which the interpretations of the allegedly theoretical terms are removed a *restriction*, and each full structure obtained from such a restriction an *extension*. In order to find out which of the terms of $r_1, ..., r_n$ are T-theoretical we have to take up one bipartition of this set of terms at a time, and see whether it satisfies the two conditions. For a comparatively strong version of the two requirements we thus have to carry out the following checks. In a first step we have to find out for all restrictions whether or not they have (up to transformations of scale) exactly one extension in the set of models of the net's basic element. In a second step we have to check if there exist a restriction and a specialized theory-element such that this restriction has exactly one extension (again up to scale transformations) in the class of models of this specialized theory-element.[19] If the bipartition does not satisfy the two requirements, we check the next one, and so on.

In general there is no guarantee that the two conditions will single out exactly one bipartition for each empirical theory. Therefore, in

[19]For alternative formulations of the two conditions with different strength cf. (Gähde,1990).

(Gähde,1990), they are treated as necessary conditions only. In that paper, a rather sceptical view is expressed with respect to the question whether or not innertheoretical conditions (that refer only to the logical structure of the theory in question) can provide such a unique and adequate distinction. He thinks it more likely that *inter*theoretical conditions are needed in addition. As it is an open question whether or not such intertheoretical conditions can be formulated in a purely formal way, one cannot speak of his conditions, so Gähde holds, as constituting a *purely formal (sufficient) criterion* for theoreticity.

His two innertheoretical conditions, however, are purely formal requirements, and, as he has shown, they suffice to account both for the flexibility of empirical theories to adapt new applications and for their defending themselves against configating observational data. Furthermore, they are responsible for the occurrence of certain holistic phenomena as mirrored in the Ramsey-sentence of empirical theory-nets.[20]

4. Balzer's Criterion

My own proposal is obtained by three alterations of Gähde's account. On the one hand, I agree with Gähde's methodological reason for his first condition. However, I see three reasons for dropping the condition. First, the flexibility of adjustment of theories to new data which is achieved by this condition might also be obtained in a different way, which is best illustrated by the example of classical particle mechanics, CPM. In CPM, total force F might be used as a primitive of a peculiar kind, namely a primitive which is explicitly defined in the theory by Newton's second law. In order to treat special laws, the special force-functions $f_1, ..., f_n$ might be introduced by existential quantification, and the adjustment to changing data could be achieved by the additional stipulation $F = \Sigma f_i$. All kinds of adjustment in this setting could be exercized by changing, or introducing new, special forces. Another case to the point is Moulines' treatment[21] of Gibbsean thermodynamics where 'equilibrium' is explicitly defined. Second, there are some theories which happen to contain definable primitives which nevertheless are *the* central, 'theoretical' primitives of those theories. A second example besides that of thermodynamics just mentioned is the theory of institutions in which 'power' has such a status.[22] In such cases, by sticking to the condition, we had to conclude that either the respective term is not theoretical or that the theory's presentation is

[20] The latter issue is dealt with in detail in Gähde's contribution to this volume.
[21] (Moulines,1975).
[22] Compare (Balzer,1993), in particular Chap.10.

flawed from a meta-theoretical point of view. Both possibilities are not very attractive, so put it mildly. Third, in empirical theories, it usually is rather easy to see whether a term is definable or not, and this also holds for the weaker condition of being uniquely determined in a model. So the condition is not of primary importance for the *practical* determination of theoreticity. These considerations, together with the aim of simplicity led me to omit Gähde's first condition.

Second, Gähde's insistance on determining *all* theoretical terms by means of non-theoretical terms only, is quite restrictive. Though the methodological reason for this insistance is compelling, it rules out interesting, 'weaker' cases in which one theoretical term can be determined only relative to another, or in which there are no 'non-theoretical' terms left.[23] Of course, in the first case one would expect that a chain of 'relative' determinations can ultimately be combined, and covered by one model of an appropriate specialization in which all the previous determinations are achieved in one step. The issue is formally involved, though, and my more liberal version certainly is easier in application.

Third, in the old version of Gähde's innertheoretical conditions as presented in (Gähde,1983) it was assumed that for the basic theory-element a class of invariance transformations is given. In the CPM example this class is undisputed: the class of Galilei transformations. In other examples, however, this class may be controversial, and it even may be unknown or not investigated, as for instance in the case of thermodynamics. For a general application of the conditions it therefore is necessary to clarify the nature of these transformations. My proposal is to determine for each term its range of invariance in the models of the basic theory-element, and take the resulting invariances as a standard for restricting the choice of admissible special laws.

As I want to prove some formal theorems below let me introduce at least those technical definitions pertaining to my definition of theoreticity needed there.[24]

A *species of structures* is a class Σ of $\langle k, m, n, \tau_1, ..., \tau_n \rangle$-structures such that $\Sigma \neq \emptyset$ and Σ is invariant under canonical transformations, i.e. for all $x = \langle D_1, ..., D_k, A_1, ..., A_m, R_1, ..., R_n \rangle$ and all $\psi = \langle \psi_1, ..., \psi_k \rangle$, $\psi_i : D_i \rightarrow D'_i$ bijective: $x \in \Sigma$ iff $x^\psi \in \Sigma$. The definitions of potential models and models are as usual.[25] A class M_p of potential models is given

[23] The latter case is perhaps exemplified by general relativity.

[24] The definitions of $(k+m)$-types τ, echelon sets $\tau(D_1, ..., D_k, A_1, ..., A_m)$ of type τ over sets $D_1, ..., D_k, A_1, ..., A_m$, typified classes S of $\langle k, m, n, \tau_1, ..., \tau_n \rangle$-structures, and of the ψ-transport R^ψ and x^ψ of a relation R and a structure x can be looked up in (Balzer,1985).

[25] See (Balzer,Moulines,Sneed,1987), Chap.2.

as the intersection of characterizations of the different relations, and a class M of models as a species of structures which is not a class of potential models.

If $T = \langle M_p, M\rangle$ is a theory with structures of type $\langle k, m, n, \tau_1, ..., \tau_n\rangle$ then for $j \leq n$, $\bar{R}_j = \{R_j / \exists x(x \epsilon M_p \wedge R_j = pr_{k+m+j}(x))\}$ is called the *j-th term* of T.[26] $\bar{R}_j$ is called *real-valued* iff for all $R \epsilon \bar{R}_j$ there is D such that $R : D \rightarrow \mathbb{R}$ or $R : D \rightarrow \mathbb{R}^3$. I assume that $\mathbb{R}$ and $\mathbb{R}^3$ are auxiliary base sets, i.e. they occur among the sets $A_1, ..., A_m$ in structures x of the above form. For $R, R' \epsilon \bar{R}_j$ I use '$R \sim R'$' as an abbreviation for '$Dom(R) = Dom(R') \wedge Range(R) = Range(R') \wedge \exists \alpha \epsilon \mathbb{R}^+ \exists \beta \epsilon \mathbb{R} \forall c \epsilon Dom(R)(R(c) = \alpha \cdot R'(c))$' if $\bar{R}_j$ is real-valued, and for '$R = R'$' else. By x_{-j} I denote the result of deleting R_j in structure x, and by $x_{-j}[R']$ I denote the result of replacing R_j by R', provided $R' \epsilon \tau_j(D_1^x, ..., D_k^x, A_1^x, ..., A_m^x)$.

D1 Let $T = \langle M_p, M\rangle$ be a theory with structures of type $\langle k, m, n, \tau_1, ..., \tau_n\rangle$ and let $j \leq n$.

a) If $x, y \epsilon M_p$ then $x \simeq_j y$ iff $x \epsilon M \wedge y \epsilon M \wedge x_{-j} = y_{-j}$.

b) If $B \subseteq M$ then B is called *M-$\bar{R}_j$-invariant* iff $\forall x, y(x \epsilon B \wedge x \simeq_j y \Rightarrow y \epsilon B)$.

c) $\bar{R}_j$ is *T-theoretical* iff there is B so that

1) $B \subseteq M$ is a species of structures
2) B is M-$\bar{R}_j$-invariant
3) $\forall x \forall R, R'(x_{-j}[R] \epsilon B \wedge x_{-j}[R'] \epsilon B \Rightarrow R \sim R')$.

In other words: term $\bar{R}_j$ is theoretical in theory T iff there is an admissible specialization B in which $\bar{R}_j$ is uniquely determined up to the invariance of scale $\sim$, or definable in B up to $\sim$. Admissibility of B is expressed in requirement D1-c-2. B has to have the same degree of invariance for term $\bar{R}_j$ which this term has in the original model class M.

In contrast to Gähde's conditions, this definition allows for the case, in which the unique determination of $\bar{R}_j$ in B is possible only by using other terms in x_j which also may be theoretical in T. In principle, this allows for theoretial terms which have no relation to non-theoretical terms at all. In particular, in some cases *all* terms of T may turn out as T-theoretical.

In first-order logic, the above definition is naturally related to the notion of definability. D1-c-3 says that the functions R are uniquely determined up to equivalence $\sim$, that is, the $R_{/\sim}$ are uniquely determined. By

[26] An alternative proposed by B.Lauth, to define a term as a function $\tau : M_p \rightarrow \mathcal{D}$ where $\mathcal{D}$ is a suitable echelon set, is more flexible, for instance in treating defined terms, but is neglected here for reasons of simplicity.

Beth's well known definability theorem this means that the term denoting $R_{/\sim}$ is definable in B. D1 then may be read as stating that the term denoting the $R_{/\sim}$ is definable in some *invariant* (in the sense of D1-c-2) sub-theory B of the theory of M. If, in the first-order case each model is covered by just one invariant sub-theory then the term under consideration is piecewise definable.[27] In general, however, first-order logics is not an adequate frame for the representation of empirical theories, and what is left from the first-order theory of definability in our set-theoretic frame essentially is the condition of unique determination.

5. Some Formal Results

For the following discussion I state three theorems. The first expresses an equivalent characterization of theoretical terms in the sense of Sec.3.[28] If M is a class of models and $j \leq n$ then $x \in M$ is called a *measuring model* for $\bar{R}_j$ (in M) iff $\forall R'(x_{-j}[R'] \epsilon M \Rightarrow R_j^x \sim R')$.

Theorem 1 A term $\bar{R}_j$ is T-theoretical according to D1 with $T = \langle M_p, M\rangle$ iff there exists a measuring model x for $\bar{R}_j$ in M.

Proof: '$\Rightarrow$': If $\bar{R}_j$ is T-theoretical then by D1 there exists some non-empty $B \subseteq M$. Let $x \in B$, so $x \in M$. We show that x is a measuring model for $\bar{R}_j$. Let R' be given such that $x_{-j}[R'] \epsilon M$. By assumption, $x \epsilon B$, and by the definition of x_{-j}, $x_{-j}[R'] \simeq_j x$. This, by D1-c-2 and D1-b, implies $x_{-j}[R'] \epsilon B$. So $x \epsilon B$ and $x_{-j}[R'] \epsilon B$. From D1-c-3 we therefore obtain $R' \sim R_j^x$.
'$\Leftarrow$': It is easy to check that for any canonical transformation ψ:
(1) $(x_{-j}[R])^\psi = (x^\psi)_{-j}[R^\psi]$,
and to prove by induction with respect to τ that
(2) the function mapping each

$$R \epsilon \tau(D_1, ..., A_m) \textit{ into } R^\psi \in \tau(D_1^\psi, ..., D_k^\psi, A_1, ..., A_m)$$

is bijective (see, for instance, (Balzer,1985), T2-b.).

Now let $B = \{x \epsilon M / \forall R'(x_{-j}[R'] \epsilon M \Rightarrow R_j^x \sim R')\}$. By assumption there is a measuring model for $\bar{R}_j$, so $B \neq \emptyset$. We show that B is a species of stuctures. Let $x \epsilon B$ and let ψ be a canonical transformation of x. We have to show that $x^\psi \epsilon B$ which, by the fact that M itself is a species of

[27] See (Tuomela,1973), Chap.IV, for a description of some standard, first-order notions of definability.

[28] This result was communicated to me independently by H.Rott and H.-J.Schmidt. I state Schmidt's version which is the stronger one.

structures, reduces to $\forall R'((x^\psi)_{-j}[R']\epsilon M \Rightarrow R' \sim R_j^{(x^\psi)})$. Let R' and ψ be given such that $(x^\psi)\epsilon M$. By the definition of canonical transformations, $(R_j)^{(x^\psi)}\epsilon\tau_j(D_1^\psi, ..., D_k^\psi, A_1, ..., A_m)$, so $R'\epsilon\tau_j(D_1^\psi, ..., D_k^\psi, A_1, ..., A_m)$. From (2) we obtain that R' has the form R_o^ψ with $R_o\epsilon\tau_j(D_1, ..., A_m)$. By (1), $(x^\psi)_{-j}[R'] = (x^\psi)_{-j}[R_o^\psi] = (x_{-j}[R_o])^\psi\epsilon M$. As M is a species of structures, this yields: $x_{-j}[R_o]\epsilon M$, so $x\epsilon M$ and $x_{-j}[R_o]\epsilon M$. From this and $x\epsilon B$ we obtain with the definition of B: $R_o \sim R_j^x$. As $\mathbb{R}$ and $\mathbb{R}^3$ are auxiliary base sets, $\sim$ is preserved under canonical transformations, so: $R' = R_o^\psi \sim (R_j^x)^\psi = R_j^{(x^\psi)}$, so $R' \sim R_j^{(x^\psi)}$. As, trivially, $B \subseteq M$, we have proved D1-c-1).

B also is M-$\bar{R}_j$-invariant. Let $x\epsilon B$, $y\epsilon M$, and $x \simeq_j y$. By the definition of B, (3) $\forall R'(x_{-j}[R']\epsilon M \Rightarrow R_j^x \sim R')$. Now let $y_{-j}[R']\epsilon M$. From the assumption that $x \simeq_j y$ we obtain $y_{-j}[R'] = x_{-j}[R']$, and from this by (3): $R' \sim R_j^x$. Taking R_j^y for R' in (3) we obtain $y\epsilon M \Rightarrow R_j^x \sim R_j^y$. This yields $R' \sim R_j^x \sim R_j^y$. So $\forall R'(y_j[R']\epsilon M \Rightarrow R_j^y \sim R')$, i.e. $y \epsilon B$.

Finally, R_j in B is uniquely determined up to $\sim$ by the definition of B. □

The next theorem roughly states that each T-non-theoretical term can be made T-theoretical by adding suitable axioms to those of T. There are, however, formal exceptions to this which -though empirically irrelevant-have to be specified in order to obtain a strict proof. I say that term $\bar{R}_s$ *admits fixpoints* in M iff there is $x \epsilon M$, and a canonical transformation ψ of x such that 1) ψ is not identity, 2) for all $r \leq k$: $\psi(D_r) = D_r$, and 3) $R_s^x = R_s^\psi$.

Theorem 2 If $\bar{R}_j$ is not T-theoretical and M is such that not all $\bar{R}_s$, $s \leq n, s \neq j$ admit fixpoints in M then there exists a class M' of structures such that $\bar{R}_j$ is $\langle M_p, M \cap M' \rangle$-theoretical.

Proof: Let $x\epsilon M$, and define M' as the class of all x^ψ which can be obtained from x by some canonical transformation ψ. Obviously, M' is a species of structures, and so is $B := M \cap M'$ which proves D1-c-1. Also, B is trivially B-$\bar{R}_j$-invariant (D1-c-2). We show D1-c-3. Let $x\epsilon B$ and $x_{-j}[R']\epsilon B$. If $x_{-j}[R']$ has the form x^ψ it follows that, for all $i \leq k$: $\psi(D_i^x) = D_i^x$, and, for all $s \leq n, s \neq j$: $R_s^x = (R_s)^{(x^\psi)}$. As, by assumption, not all $\bar{R}_s$ admit fixpoints in M it follows that ψ is the identical function, i.e., for all $i \leq k$: $\psi_i : D_i \rightarrow \psi_i(D_i)$ is identity. This yields: $R_j^x = R'$, and so $R_j^x \sim R'$. □

Theorem 3 Let $R \approx R'$ be defined by $\exists x(x_{-j}[R]\epsilon M \wedge x_{-j}[R']\epsilon M)$. If, in D1-c, $\sim$ is replaced by $\approx$ then every term is T-theoretical in every theory T.

Proof: Take $B = M$ in D1-c. c-1 and c-2 are trivial, and c-3 follows from the definition of $\approx$. □

Note that, in general, $\sim$ need not be compatible with $\approx$.

6. Theoreticity's Meta-Theoretical Locus

The results of Sec.2 may be sharpened by regarding meta-science decidedly as an empirical enterprise. If meta-science is empirical, it should result in empirical theories of the kind which meta-science, and in particular the structuralist approach, has studied as objects and characterized theoretically.

Now one crucial point of the structuralist notion of an empirical theory is its containing -among other things- a set of intended applications. Intended applications are those real systems to which practitioners intend to apply the theory. Commitment to such systems anchors the theory in the real world. Without such commitment, no standard of confirmation would be available, and the acceptance of a 'theory' would be *entirely arbitrary*, a matter of pure decision for or against it without any empirical reason.[29]

Besides providing a basis against which the theory can be checked the intended applications also focus the kind of arguments to be used for or against modifications of the theory. In the absence of intended applications, scientists have to retreat to intuitions, goals, plausibilities and the like in order to convince one another, usually without much success. The history of several discussions within logical empiricism was beset by the problem of ignoring the need of specifying intended applications. For instance, there were endless discussions about all kinds of examples and counterexamples of explanations. As the array of 'explanations' occurring in the real world of science is very and perhaps even inconsistently comprehensive, 'all possible' explanations do not form a sound basis or set of intended applications for a meta-scientific theory of explanation. On the other hand, if intended applications are present, they will of course have strong priority over mere fuzzy attitudes or intuitions. In discussing theoreticity I would postulate a definite priority of intended applications over philosophical intuitions. An intuition cannot override a distinction drawn in a well specified meta-theoretical model when this model is based on a well specified set of intended applications. To admit for the contrary would ultimately mean to admit the dominance of philosophy over empirical science.

[29] The discussion of immunity of an empirical theory even in the light of a given set of intended applications, for instance in (Stephan,1990), Sec.3.3, is suited to put this kind of empirical basis in the right perspective. A theory may get into serious trouble at least when its application to the *paradigmatic* intended applications becomes problematic.

In structuralism, theoreticity is approached from the meta-scientific point of view -as opposed to the philosophical view. Structuralism provides a theory elaborated in great detail, and intended to apply to various arrays of scientific theories and disciplines. In this setting, theoreticity is checked by the role which scientific terms play in the structuralist model of science, and by the historical occurrences of scientific theories which form the intended applications of the structuralist model of science. I do not want to rule out reference to intuitions and the like, but I want to insist on the correct weighing of such elements against arguments arising from the proper empirical setting just specified. In particular, I do not say that general criteria in the style of logical empiricism should be abandoned. But I want to emphasize the proper weight of such endeavors in the discussion of theoretical terms.

The domain of intended applications for the structuralist model comprises theories of varying degree of comprehensiveness, ranging from isolated over *theoretical hypotheses*[30] to comprehensive theories. Typical examples for the first category are the ideal gas law or Ohm's law, for the second Newton's second law or the second fundamental law of thermodynamics. I have no good example for the third type (perhaps a Grand Unified Field Theory), but I do not want to exclude the possibility of such theories. Theoretical hypotheses generate theories which usually differentiate into whole theory-nets.

With respect to the presence of theoretical terms the question is whether these are assumed to occur in *all* intended applications of the metatheoretical model. Sneed in his original account does not assume that every theory contains empirical terms, his model provides room for the extreme cases where all or no terms are T-theoretical. But, he insists that the *distinction* between theoretical and non-theoretical terms is decisive and is part of every empirical theory. The same holds with respect to Gähde's two conditions. For him, the theoretical/non-theoretical distinction is fundamental both for the logical reconstruction of empirical theories as well as for the formulation of the Ramsey-sentence. Stegmüller even made this distinction *the* crucial point and achievement of the structuralist approach, which it certainly is not.[31] According to Sneed the reconstruction of each

[30] These terms are defined more precisely in (Balzer,1987), pp.219.

[31] I avoid discussion of the 'problem of theoretical terms' which, in my view, also was overemphasized by Stegmüller. This problem arises if we want to test a theory containing theoretical terms in Sneed's sense without involvement of that theory in the test procedure. Clearly, this notion of a test is very strong. In scientific practice, other more liberal ways of testing -like bootstrapping- are used. Stegmüller's overemphasis of 'theory-independent' testing has led to critical overreactions in the opposite direction. (Hoyningen-Hüne,1988), for instance, declares the Sneed-Stegmüller problem of theoretical terms a *Scheinproblem* without recognizing that the bootstrap idea he of-

theory affords clarification which of its terms are theoretical and which are not; the non-theoretical terms then make up the 'partial potential models' among which we find the theory's 'intended applications'. This account has a severe *practical* drawback for there are theories for which it is very difficult to find out which of its terms are theoretical. This may be so because the criterion chosen to rule the distinction is pragmatic,[32] or it may be so for reasons of theoretical complexity, as in the examples of wave mechanics and of Einstein's theory of general relativity. It is a severe obstacle for reconstruction to insist on the clarification of the distinction in such cases, an obstacle which in practice prevents application of the structuralist model to such theories -theories which are rather clear cut in all other respects. In addition, the extreme cases of all or no terms being T-theoretical lead to strange situations wrt T's Sneedian empirical claim: 'All intended applications can be extended to proper models'. If all terms are T-theoretical, this empirical claim reduces to a statement about the cardinality of the base sets, if no term is T-theoretical, then the full range of terms has to be determined or measured for the full range of arguments, before the empirical claim is checked.

In (Balzer,1982), Sec.V, I proposed to generalize Sneed's theory-concept such that the distinction between theoretical and non-theoretical terms is no longer presupposed and needed for a full reconstruction of the theory. As my proposal does not seem to have been recognized, I repeat it here. The simple idea is to generalize Sneed's notion of partial potential models to arbitrary sub-structures.[33] In that paper, I pointed out that this generalization does not depreciate the distinction, but only *avoids* it in cases where it would be cumbersome. If it is feasible to draw the distinction in a convincing way, then it should be included in the reconstruction of the theory under investigation in the original Sneedian way. To put it differently: my proposal was to treat the distinction between theoretical and non-theoretical terms as giving rise to a meta-theoretical specialization. In general, i.e. in the empirical claim of the meta-theory which is made about *all* intended applications ('i.e. all empirical theories fall under the structuralist notion of a theory'), the distinction is eliminated from the notion of a theory. There is, however, an important and interesting subdomain of intended applications (= empirical theories) for which such a distinction is made.

Two of the three types of intended applications mentioned above clearly cannot be expected to contain an interesting distinction between theo-

fers as more adequate to scientific practice also provides for the possibility of Sneedian circularities.

[32] Witness the endless discussions about the CPM-theoreticity of mass and force.

[33] See (Balzer,1982), Appendix, for details.

retical and non-theoretical terms. Isolated hypotheses do not contain a distinction. According to Sneed's criterion all their terms are non-theoretical, according to my own criterion all their terms are theoretical. In any case: there is no distinction among their terms, neither formally, nor from the perspective of measurement. It is futile to look for theoretical as opposed to non-theoretical terms in isolated hypotheses.[34] Also, there is some reason why very comprehensive theories -if they exist- might contain no proper distinction. Such theories might 'contain' *all* methods of measurement for *all* their terms, and, formally, all their terms might be in similar positions relative to each other.

The distinction between theoretical and non-theoretical terms of a theory is interesting only for the array of 'theoretical hypotheses', that is, roughly, of theories which give rise to theory-nets but are not all-comprehensive. It is only in the domain of such theoretical hypotheses that we meet theories which on the one hand introduce new terms for the first time, terms which turn out as theoretical, and on the other hand also contain non-theoretical terms which rely on previously existing theories or methods of measurement.

7. *Criticism 1: Theoreticity and Specialization*

G.Schurz has attacked the formal approach to theoreticity. In (Schurz,1990) he presents not less than ten formalized 'theories' in which application of the formal criterion leads to a distinction between theoretical and non-theoretical terms which according to Schurz is at odds with presystematic expectation. Also, he puts forward three general criteria of adequacy for theoreticity which are not satisfied by my formal account. I will consider his criticism under three different headings and argue that the problems he points out do not arise because of a flaw in my formal criterion of theoreticity. Rather they arise because of misconceptions of the meta-theoretical locus of theoreticity and they can be accommodated by locating them at their proper positions in the structuralist model.

A first class of 'counterexamples' consists of 'theories' which are specializations in the stucturalist sense, that is, they consist of some basic axioms valid in many theory-elements of a theory-net, plus some more special axiom(s) which hold true in a special domain of application only. Schurz's examples are: the harmonic oscillator, a special system of CPM in which inertial frames are made explicit, two stationary cases of quantum mechanical atoms, and plastic collisions (in which the particles stick

[34] I mentioned the case of the ideal gas law as a problematic one already in (Balzer,1985a), pp.149, but there I did not draw a definite conclusion from it, as I do now.

with each other after the collision). All these examples under structuralist reconstruction in fact are specializations. The basic theory-element for the harmonic oscillator and for CPM with inertial frames is that for CPM, given essentially by Newton's second law, the basic theory-element for various treatments of atoms is Schrödinger's wave mechanics[35] (as long as the Pauli principle and/or spin do not play a role), the basic theory-element for plastic collisions is that of 'general' collision mechanics (CCM) given essentially by the law of conservation of total momentum. Application of my formal criterion to these cases yields, in respective order, position, position, the spectrum of eigenvalues, and velocity, as theoretical terms which presystematically are expected to be non-theoretical in their respective theories.

With respect to the *basic* theory-elements mentioned, each term indeed *is* non-theoretical in the sense of Sec.3; this has been formally proved for position in CPM and velocity in CCM.[36] As shown by Schurz, these terms are, respectively, theoretical *in specializations* of the basic theory-elements. Is that paradoxical, as he claims? I do not think so. In fact, a look at Theorem 2 above shows that nothing else should be expected, and that the production of lots of examples is quite redundant in this case. Each T-non-theoretical term can be made T'-theoretical when T' is obtained from T by adducing suitable axioms.

This was clear from the beginning, and nobody has claimed that theoreticity is invariant under specialization or remains fixed during the historical development. If a problem is seen this must come from what I called presystematic expectations. Why do we expect, for instance, that position should be T-non-theoretical in all specializations of CPM? There is a very clear answer: because they are specializations of the basic theory-element, and because in this basic theory-element they are non-theoretical. If this answer is correct then theoreticity in special laws or specializations is not checked by applying the formal criterion to the special law. It is checked by application of the formal criterion to the basic theory-element *of which* the special law is a specialization. There are three general arguments in favour of this view.

First, the notion of a theory or empirical theory has turned out to be difficult to characterize. There are several different explications of that notion, ranging from the logicians' deductively closed set of statements over theory-elements to theory-nets and theory-evolutions.[37] In applying

[35] See (Zoubek,1987) for a very brief structuralist account.

[36] (Balzer,1985). For the quantum mechanical case the distinction has not been investigated. I conjecture that the state function (Ψ), and hence the spectrum, turns out as non-theoretical.

[37] Compare (Balzer,Lauth,Zoubek,1993).

the notion of a theory, i.e. in identifying some scientific output as a theory, priority is given to theory-nets. Scientific theories primarily are theory-nets (or more complex entitites like theory-evolutions). Restriction to 'smaller' formal entities is admissible only in the special cases where there is just one isolated law for which there is no acknowledged specialization. In cases where special laws exist and are acknowledged, it is *not* the case that each special law 'is', or gives rise to, a theory. Consequently, in these cases we cannot ask for theoreticity with respect to one special law. Theoreticity is relative to the *theory*, i.e. to the whole theory-net.

Second, my formal criterion of theoreticity is closely linked to invariances. The method of measurement (B in D1) to exist for the theoretical term has to have the same degree of invariance with respect to that term as that term has in the basic models. In a theory-net, the proper invariance is that determined by the basic theory-element. All specializations are assumed to have -and usually do have- the same invariances associated with them, namely those given by the basic theory-element.[38] These common invariances are just one way of expressing what is common to all the specializations of one net, one way to express that all special laws have some 'same form' in common. Binding theoreticity to invariances thus naturally leads to relativize it to theory-nets rather than to single specializations.

Third, this relativization of invariance to basic theory-elements mirrors scientific practice. I think we can agree that physicists hold, for instance, Hooke's law to be Galilei-invariant, which means that the law holds irrespective of which inertial frame is chosen for the description of the system. *Formally*, Galilei-invariance of CPM means that the Galilei-transform of each model of CPM again is a model. However, the specialization of CPM by Hooke's law formally is *not* Galilei-invariant. If s satisfies Hooke's law $f = -k \cdot s$, and $v \neq 0$ then $s + vt$ does *not* satisfy this law. Similar examples are found in other theory-nets. Of course, the specializations could be formulated in a way to have the same invariance as the basic theory-element; then each term non-theoretical in the basic element would provably remain non-theoretical in the specialization.[39] This view at scientific practice shows that the invariance of the basic theory-element has clear priority over that of some special law.

---

[38] This justifies, by the way, the structuralist treatment of special laws as *including* the basic law, a treatment which *prima facie* seems at odds with the way special laws are formulated in the textbook literature.

[39] Which means that 'Schurz-type' examples of the kind under discussion are no longer possible.

8 Criticism 2 : Invariances of scale

In my formal definition of theoreticity described in Sec.4 reference is made to two kinds of invariances. On the one hand, in order to be theoretical a term is required to be uniquely determined up to some given invariance of scale $\sim$ (compare D1-c-3 above). Second, the method of determination (measurement) B used in that determination has to be invariant under the same class of transformations as are the theory's models. This is expressed by the method of measurement's being M-R_i-invariant (compare D1-c-2), or invariant under relation $\simeq_i$. Whereas $\simeq_i$ is explicitly defined in terms of the model class M, $\sim$ is defined 'externally' to be given by linear transformations or identity, respectively. In contrast to this invariance of scale, $\simeq_i$ might be called a *theoretical* invariance, for it is an invariance of the theory (the theory's axioms) under certain transformations.

In (Balzer,1985a), Sec.11, I proceeded in a more general way, and first introduced the notion of a method of measurement with 'arbitrary' invariance instead of invariance of scale. Only afterwards I restricted my considerations to invariances of scale, and I did not offer arguments for the distinction between invariance of scale and other, theoretical, invariance. Taking up this route, Schurz[40] puts forward two theorems showing that in CPM, position s can be determined uniquely up to Galilei invariance (which he expresses by saying that s is 'CPM-theoretical with Galilean invariance'). He then tries to argue that Galilei invariance is '*the proper* scale invariance for s and v in classical physics'[41], and that therefore the formal criterion is inadequate (because according to it all terms in CPM would be CPM-theoretical wrt Galilei invariance).

This point has not been discussed in detail before, and, since it is important for my distinction, I will show that Schurz's argument is basically misled, being based on a superficial understanding of invariances of scale.

First, there is the simple formal observation expressed in Theorem 3 above that by appropriately weakening the invariance of scale $\sim$ in the definition of uniqueness *every* term can be made T-theoretical.[42] So in order to obtain a *distinction* it is central to somehow constrain $\sim$ to be 'well behaved'. Even if no other argument were available, I think it is an interesting observation that by cutting at linear transformations in CPM a formal distinction along the lines of D1-c above arises. Even this purely formal distinction would show that *there is* some distinction between linear transformations and the 'rest', including Galilei transformations in CPM.

[40] (Schurz,1990), Secs.9 and 10.
[41] Op. cit. p.188.
[42] This I mentioned in (Balzer,1985a) in passing, p.51.

Second, this distinction, and in general: the distinction between invariances of scale and theoretical invariances, is not only a purely formal one, it also is meaningful. To show this let me begin by noting that statements about the invariance of a function f are used to express that 'something' remains identical when f is replaced by an admitted variant. To say simply that f is invariant under transformations of a given mathematical form (like linear or Galilei transformations) bears little content. The proper syntax of statements of invariance is this: '*f in its property σ* is invariant under transformations of kind ϕ', where ϕ is defined beforehand. In general, the property σ of f consists of f's occurring in a system x which satisfies an axiomatically fixed set of assumptions. So the proper syntax is: 'f in its property of occurring in a model x of axioms A is invariant under transformations of the kind ϕ'.

I submit that this property is different for invariance of scale and theoretical invariance. In invariance of scale, f is a 'scale', i.e. a numerical representation of an empirical relation where representation is governed by a system of fundamental measurement, like for instance an extensive system.[43] The system x therefore is a model of some axioms for fundamental measurement, and the axioms A are those required to govern the empirical relations in that system as well as the connection of the latter with f. In theoretical invariance, f is a function occurring in the models of an empirical theory T, x is a model of T and A are the axioms of T. Though I was eager to bring out the similarity, I think the difference also is clearly visible. The difference between invariance of scale and theoretical invariance amounts to that between systems of fundamental measurement and empirical theories. It may be held that systems of fundamental measurement are special kinds of empirical theories, and I don't object to this.[44] But I insist that there is a clear difference between the two. In the class of empirical theories, there is a clear distinction between those 'theories' which capture methods of fundamental measurement and the 'rest', consisting of empirical theories proper. In general, therefore, we are able to draw a distinction between invariances of scale and theoretical invariances, and we can exclude theoretical invariances for $\sim$.

Applied to the case of CPM, the position function s, and Galilei invariance the question was whether Galilei invariance of s is an invariance of scale (as claimed by Schurz) or a (CPM-)theoretical invariance. This question turns out to be touching the borderline between empirical philo-

[43] There is a large literatur on such systems, I mention only the paradigmatic, authoritative work of (Krantz et al.,1971). A simple kind of extensive systems there is defined on pp.71.

[44] Indeed, this claim follows from my thesis that the structuralist view of measurement is a generalization of representationalist approaches to measurement, see (Balzer,1992).

sophy of science which leaves science unchanged and foundational work in a particular field. Considered from the purely descriptive, meta-scientific view the situation is this. Suppose that Galilei invariance of s were an invariance of scale, that is, s were invariant under Galilei transformations in its property of being a numerical representation in some system of fundamental measurement. Then the task is to find some such system in real-life science. But such a system does not occur in physics, i.e. in physical literature. There is no acknowledged system of fundamental measurement for s in which s is Galilei invariant, so Galilei invariance is not an invariance of scale. Though this reply seems quite satisfactory as far as empirical science of science is concerned, there is an alternative which should not be overlooked.

The alternative consists in actively producing a new system of fundamental measurement for s in which s, in fact, *is* Galilei invariant. One may wish to point to Suppes' system[45] in which a Galilei invariant position function is produced as the numerical representation of a system of axioms for empirical relations. This is a nice formal result for the representationalist approach. There are, however, three reasons why it does not affect my claim that Galilei invariance of s in CPM is not an invariance of scale.

The first reason is that the construction of such a fundamental system is a piece of work in the foundations of physics.[46] As the definition of theoretical terms is intended to be a descriptive enterprise reflecting what is going on in science when philosophers of science don't interfere, the production of new foundational results either has to be left out of consideration or will completely change the perspective and meaning of the distinction.

Second, this attitude is backed by the conjecture that an analogous construction is possible for *each* scientific term. More specifically, let $\bar{t}$ be some term which is not T-theoretical, and let ϕ be the transformations under which $\bar{t}$ is theoretically invariant in T. Then a system of fundamental measurement for $\bar{t}$ can be constructed in which $\bar{t}$ comes out exactly as invariant under transformations of the type of ϕ. The representationalist program, in fact, aims at producing such systems for all scientific terms. But this means: all non-theoretical terms can be made theoretical by the strategy we studied in the example of CPM: the theoretical invariance is turned into a scale invariance which then can be used in the criterion of

[45] (Krantz et al.,1989), Sec.13.8. The proofs of the representation theorem we are discussing here are not given in that work, but are announced to appear in another book.

[46] It is of course highly satisfactory for a philosopher of science to achieve such foundational results, but the boundary between descriptive meta-science and active, new foundational work remains unaffected by such activities.

theoreticity, and makes the term 'invariantly definable' so that D1-c above is satisfied and the term becomes theoretical in a variant of the theory in which originally it was considered as non-theoretical.

Third, and most importantly, we may look at the actual historical development of mechanics. If Galilei invariance were the invariance of scale of position then the following historial development should have taken place. First, the position function should have emerged as a numerical representation of 'empirical' space-time relations as captured by some axioms for fundamental measurement. After that, it then should have been discovered that these axioms were Galilei invariant. But this is not what happened. The view of position as a numerical representation of space-time relations was certainly implicit in the development but it started from empirical relations in which conceptually no room was provided for frames of reference moving relative to each other.[47] In the beginning the representationalist view of position was basically a-dynamic, and the transformations thus characteristic for position were dilatations, translations, and rotations. What really happened was that Galilei invariance was not discovered as an invariance of the position function. It was discovered as an invariance of physical laws. It was empirically discovered that certain physical processes develop practically identically when described from Galilei-transformed frames of reference. Generalizing this to all physical processes the foundation was led for the transition from Aristotelean to Newtonian physics. If physical processes obey the same law when described from different frames being in constant motion relative to each other then the law should be invariant under Galilei transformations, and force should change velocity, not position. This shows that Galilei invariance developed as a theoretical invariance, and not as an invariance of a representing function.

Still, it might be said that the development of physics went on, and after full appreciation of Galilei invariance the notion of position changed its meaning. 'Position relative to a frame of reference in a (essentially) geometric system' now became 'position relative to an inertial system', and so dynamics became part of the notion of position. I don't object, but simply note that this move brings us from classical to relativistic mechanics. Theoreticity of position in relativistic mechanics, however, is not at stake.

To put it differently, in order to show that Galilei invariance is an invariance of scale for position in CPM one had to show that this kind of invariance is essential or constitutive for the meaning of the notion of position. While I do not want to deny that the meaning of position in

[47] Compare (Balzer,1983) for a detailed systematic discussion.

relativity, in fact, becomes dependent on certain theoretical transformations, I deny this for the classical theory. The simple argument is that a full appreciation of inertial systems began only in the last century.[48]

There is a third feature by which invariances of scale may be distinguished from theoretical invariances which becomes visible if we look more closely into the functioning of fundamental measurement. The idea of fundamental measurement is to generate quantities out of qualitative notions, to assign numbers to certain qualitative features of a distinguished class of objects. In order to do so *in concreto* it is *always* necessary to use some units of measurement. Length, for instance, is fixed in this way as the number of length units needed to produce a rod equal to (in length) the measured object. But the concrete numerical value arising in this way *changes* when other units are chosen, and there is no reason to prefer one system of units over another one -at least within certain limits. It is precisely this undesired dependence of the numbers on the particular units which is eliminated by means of invariances of scale. The transformation of a numerical representative corresponds precisely to a change of the system of units. In this way, linear transformations are distinguished as corresponding to changes of units in many different kinds of measurement captured by foundational systems.

The important point to note is that in all these cases each change of scale corresponds to a change of just one or two concrete units (plus copies of them). In the foundational literature, more complex cases have been studied but the pattern described always remains the same. In applying this to the case of position in CPM we have to look for the units needed in order to generate or construct a concrete position function out of qualitative data. Of course, there are the familiar units for measuring spatial distances and durations but their change will lead to dilatations of positions only. In addition to these familiar units a new kind of entity is needed to generate positions, namely coordinate systems or frames of reference. Although coordinate systems may be regarded as functioning analogously in the generation of positions as do ordinary units in the generation of quantities there is a clear difference between the two. Whereas an ordinary unit simply is an *object*, that is, an element of a base set of the corresponding model, a coordinate system is a *structure*, consisting of many such objects *plus* theoretical assumptions (the objects must not be co-planar, and must not move relative to each other).

In general, invariances of scale can be justified as just described by reference to 'objects' only. If the notion of an object is felt to be unclear, it may be relativized to the theory from which the candidate term stems.

[48] Compare (Lange,1885).

Relative to such a theory there is a very clear and precise notion of its objects. The objects of T are just the elements of the principle base sets occurring in the intended systems of T, i.e. the sets $D_1, ..., D_k$.[49] A frame of reference is clearly not an object of CPM in this sense, and therefore invariance of the position function is not an invariance of scale.

9. Criticism 3: Other Topics

There are two more criticisms to be mentioned. The first consists in pointing out examples of isolated hypotheses in which my formal criterion does not lead to a distinction. As clarified in Sec.6 this is a problem concerning the domain of application of the distinction rather than its content. A borderline case between isolated and theoretical hypotheses is Jeffrey's decision theory in which preferences turn out as formally theoretical while theoreticity of utility and probability have not yet been investigated in that theory. Probability clearly is theoretical, but for utility this can only be conjectured. If my conjecture is correct then in this theory there is no formal distinction either.[50] This corresponds to the somewhat strange role of preferences being *explicitly* definable in that theory.

A second criticism is couched in the form of general criteria of adequacy. Schurz puts forward three such criteria.[51] First, an obviously empirical term should not come out as T-theoretical. The 'counterexamples' to this criterion have already been discussed in the two preceeding sections. They come up by specializations and incorrect use of invariances, and thus are not convincing. Second, a term theoretical in some pre-theory of T should not come out as T-theoretical. Again, possible counterexamples come up in the two areas of specialization and incorrect invariances. There is some agreement that position is theoretical in a pre-theory of CPM, namely kinematics. From this a conflict with the condition stated arises only if position is turned into a mechanics-theoretical term by way of admitting Galilei invariances as invariances of scale, which was shown to be inadmissible. In cases of specialization, I would not want to call them pre-theories -though the issue is involved and cannot be dealt with in detail here.

A third 'condition of adequacy' is that theoreticity should be able to distinguish between contentful and contentless terms, which distinction is

---

[49] Compare (Balzer,Moulines,Sneed, 1987), p.10.

[50] One also might use the well known representation theorem for utility and probability as an argument for the theoreticity of these functions. I avoid this because of the four constants involved in the corresponding Gödel-Bolker transformations. While two of them have a rather clear meaning as units, the other two have not. See (Sneed,1982) for details.

[51] (Schurz,1990), Sec.3 and Sec.14.

explained by an example of a term producing a trivial Ramsey sentence. Now the existence of trivial Ramsey sentences is certainly not a feature one would want to rule out, because this feature is shared by some of the most admired physical core theories, like that of classical mechanics or thermodynamics.[52] Moreover, I do not see any reason why such a distinction should be ruled by theoreticity. In general, whether a theory is empirically trivial or not cannot be checked by asking this question for each of its terms. A term may be 'contentful' simply by contributing to a non-trivial Ramsey sentence. Schurz's proposal that every contentful term of T should be measurable in some specialization of T by 'empirical' terms alone lacks clarification of what is meant by a contentful term.

These discussions of general criteria based on intuitions, though, should not be overemphasized in the light of an existing theory (namely structuralist meta-theory) based on data, as I argued in Sec.6. To do so would mean to fall back into the mistakes of logical empiricism. In the presence of a (meta-) theory plus data, criticism may proceed without reference to general criteria of adequacy -which are nothing but a more fuzzy kind of theory without data. The present discussion to its greatest extent proceeded in the more concrete way, which confirms my initial claim about a progressive development.

Summarizing, the criticisms discussed have proved valuable for the clarification of several features of theoreticity. While they were not able to shake the basic account, they were instrumental in delineating the proper range of application of the distinction, and clarifying its role within the structuralist model of science. One aim of the structuralist approach is to analyze and define the notion of an empirical theory. In this attempt, we cannot renounce the formal distinction between theoretical and non-theoretical terms.

References

Balzer, W., 1979, 'Logische versus physikalische Definitionen in der physikalischen Begriffsbildung', in W.Balzer & A.Kamlah (eds.), *Aspekte der physikalischen Begriffsbildung*, Braunschweig/Wiesbaden, Vieweg, 13-36.

Balzer, W., 1982, 'Empirical Claims in Exchange Economics', in Stegmüller et al. (1982), 16-40.

Balzer,W., 1983, 'The Origin and Role of Invariance in Classical Kinematics', in D.Mayr & G.Süssmann (eds.), *Space, Time, and Mechanics*, Dordrecht, Reidel, 149-69.

[52] See (Balzer,Moulines,Sneed,1987), Chaps.3 and 4.

Balzer,W., 1985, 'On a New Definition of Theoreticity', *Dialectica* 39, 127-45.

Balzer,W., 1985a, *Theorie und Messung*, Berlin etc., Springer.

Balzer,W., 1986, 'Theoretical Terms, A New Perspective', *The Journal of Philosophy* 83, 71-90.

Balzer,W. 1987, 'Reference and the Development of Theories', in P.Weingartner & G.Schurz (eds.), *Logic, Philosophy of Science and Epistemology*, Proceedings of the 11th International Wittgenstein Symposium, Wien, Hölder-Pichler-Tempski, 213-224.

Balzer,W. 1992, 'The Structuralist View of Measurement, An Extension of Received Measurement Theories', in C.W.Savage & P.Ehrlich (eds.), *Philosophical and Foundational Issues in Measurement Theory*, Hillsdale NJ, Erlbaum, 93-117.

Balzer,W., 1993, *Soziale Institutionen*, Berlin, de Gruyter.

Balzer,W. and Moulines,C.U., 1980, 'On Theoreticity', *Synthese* 44, 467-94.

Balzer,W., Moulines,C.U. and Sneed, J.D., 1987, *An Architectonic for Science*, Dordrecht, Reidel.

Balzer,W., Lauth,B. and Zoubek, G., 1993, 'A Model for Science Kinematics', *Studia Logica* 52, 519-48.

Gähde,U., 1983, *T-Theoretizität und Holismus*, Frankfurt/Main-Bern, Peter Lang.

Gähde,U., 1990, 'On Innertheoretical Conditions for Theoretical Terms', *Erkenntnis* 32, 215-33.

Gaifman,H, Osherson,D.H. and Weinstein,S., 1990, 'A Reason for Theoretical Terms', *Erkenntnis* 32, 149-59.

Hanson,N.R., 1958, *Patterns of Discovery*, Cambridge.

Hempel,C.G., 1974, 'The Meaning of Theoretical Terms, A Critique of the Standard Empiricist Construal', in P.Suppes, L.Henkin, A.Joja, and G.C.Moisil, (eds.), *Logic, Methodology and Philosophy of Science IV*, Proceedings of the 1971 International Congress of LMPS, Amsterdam, North Holland

Hoyningen-Hüne,P., 1988, 'Diskussionsbemerkung zum Beitrag von Wolfgang Balzer', in P.Hoyningen-Hüne & G.Hirsch (eds.), *Wozu Wissenschaftsphilosophie?*, Berlin, de Gruyter, 76-83.

Kamlah,A., 1976, 'An Improved Definition of 'Theoretical in a Given Theory'', *Erkenntnis* 10, 349-359.

Krantz,H.D., Luce,R.D., Suppes,P. and Tverski,A., 1971, *Foundations of Measurement*, Vol. I, New York-London, Academic Press.

Krantz,H.D., Luce,R.D., Suppes,P. and Tverski,A., 1989, *Foundations of Measurement*, Vol. II, Now York-London, Academic Press.

Kuokkanen,M. 1988, *Critical Studies in the Structuralist Theory of Science*, Dissertation, University of Helsinki, Philosophy Department.
Lange,L., 1885, 'Über die wissenschaftliche Fassung des Galileischen Beharrungsgesetzes', Ber.kgl.Ges.Wiss., Math.-Phys.Kl.
Lewis,D., 1970, 'How to Define Theoretical Terms', *The Journal of Philosophy* 67, 427-446.
Miroiu,A., 1984, 'A Modal Approach to Sneed's 'Theoretical Functions'', *Philosophia Naturalis* 21, 271-286.
Moulines,C.U., 1975, 'A Logical Reconstruction of Simple Equilibrium Thermodynamics', *Erkenntnis* 9, 101-30.
Moulines,C.U., 1985, 'Theoretical Terms and Bridge Principles, A Critique of Hempel's (Self-)Criticisms', *Erkenntnis* 22, 97-117.
Moulines,C.U.,1992, 'Structures, Links, and Holons', in H.Westmeyer (ed.), *The Structuralist Program in Psychology*, Seattle etc., Hogrefe & Huber, 13-26.
Putnam,H. 1962, 'What Theories are Not', in E.Nagel, P.Suppes and A.Tarksi (eds.), *Logic, Methodology and Philosophy of Science*, Proceedings of the 1959 International Congress of LMPS, Stanford, University Press, 240-51.
Putnam,H. 1965, 'Craig's Theorem', *The Journal of Philosophy* 62, 251-60.
Schurz,G., 1990, 'Paradoxical Consequences of Balzer's and Gähde's Criteria of Theoreticity. Results of an Application to Ten Scientific Theories', *Erkenntnis* 32, 161-214.
Sneed,J.D., 1971, *The Logical Structure of Mathematical Physics*, Dordrecht, Reidel, second edition 1979.
Sneed,J.D., 1982, 'The Logical Structure of Bayesian Decision Theory', in Stegmüller et al. (1982), 201-22.
Stegmüller,W. 1973, *Theorie und Erfahrung*, Zweiter Halbband, Berlin etc., Springer.
Stegmüller,W., Balzer,W. and Spohn,W. (eds.), 1982, *Philosophy of Economics*, Berlin etc., Springer.
Stephan,E., 1990, *Zur logischen Struktur empirischer Theorien*, Berlin etc., Springer.
Tuomela,R., 1973, *Theoretical Concepts*, Wien-New York, Springer.
Zoglauer,T., 1993, *Das Problem der theoretischen Terme*, Braunschweig, Vieweg
Zoubek,G., 1987, 'Korrespondenzprinzipien der Quantentheorie. Ihre Rekonstruktion und und Bewertung im Rahmen des strukturalistischen Theorienkonzepts', in P.Weingartner and G.Schurz (eds.), *Logic, Philosophy of Science and Epistemology*, Proceedings of the 11th International Wittgenstein Symposium, Wien, Hölder-Pichler-Tempski, 234-7.

Chapter 9: Holism and the Empirical Claim of Theory-Nets

Ulrich Gähde [1]

1. Introduction

Since the pioneering work of Pierre Duhem[2] *holism* has been one of the key terms in 20th century philosophy of science. Moreover, due to the work of W.V.O. Quine,[3] it became one of the most debated issues in contemporary epistemology in general. Quine's position, however, relates to the totality of our knowledge, and its special implications concerning the holistic aspects of scientific theories remain rather vague and opaque. It is one of the attractive features of the structuralist metatheoretical concept that it enables a significantly more detailed and precise approach to certain aspects of the holistic nature of science. The following considerations focus on two holistic aspects of empirical theories.

The first aspect of holism - which was at the center of Duhem's considerations - originates from the following fact: When an empirical theory is applied to a (e.g., physical) system, numerous fundamental and special laws, auxiliary hypotheses, boundary conditions etc. are generally involved. These statements form a complex which, as a whole, is confronted with the data base. If a conflict between this theoretical complex and observational data occurs, then - at least at first sight - this complex has failed as a whole, and it is a nontrivial question how a certain hypothesis can be identified as *the* cause of the conflict.

The second aspect of holism arises from the fact that, in general, applications of empirical theories are not described in isolation. Instead, their theoretical descriptions might be interrelated in numerous ways. As a consequence, conflicts between *theory and experience* that occur in the treatment of a certain system can often be eliminated by means of modifications which are carried out 'somewhere else', i.e. in the theoretical description of other systems: Conflicts between an empirical hypothesis and observational data do not have to be solved where they first occurred. Thus, in case of such a conflict one may not only ask which laws, auxiliary hypotheses etc. used in the description of a certain application

---

[1] I thank T. Bartelborth, C.U. Moulines and F. Mühlhölzer for helpful comments on earlier versions of this paper. I am also grateful to Ms. A. Baker/ZiF (Bielefeld) for emending my English formulations.

[2] Duhem (1906).

[3] See e.g. Quine & Ullian (1978).

of the theory have to be modified. In addition, the demarcation of the set of applications affected by a certain conflict with measuring data might constitute a major problem as well.

The first aspect of holism might be enforced by the second: Let us assume that, in the theoretical description of a certain system, a conflict arises between *theory and experience*, and that attempts are made to eliminate this conflict by modifying the description of some other application. In case of the latter, however, once again different fundamental and special laws, auxiliary hyotheses etc. may exist where a modification might start. In other words: When the set of applications affected by some conflict grows, so the set of affected hypotheses will grow larger as well.

In the following considerations we will focus on how these two aspects of holism are expressed by the empirical claim of theory-nets. In the discussion of that claim we will proceed according to a modular construction system: In section 2 we start with an analysis of the simple case of the empirical claim associated with a single theory-element. In section 3 we present the established view of how the empirical claims of different theory-elements can be combined so as to provide the Ramsey-sentence of a specialization-net with more than one element. In section 4, we argue that this version of the empirical claim of theory-nets is inadequate: it has to be supplemented by a consistency requirement. We present a formal explication of this requirement and show that the revised version of the empirical claim enables a better and more detailed understanding of certain phenomena in the dynamics of empirical theories. Section 5 contains an analysis of the relation between this empirical claim, formal intratheoretical conditions imposed on the set of theoretical terms, and the occurrence of the two aspects of holism. In section 6 we show that the revision of the empirical claim associated with theory-nets has an attractive formal consequence: it can be reformulated in such a way that it takes exactly the form of the empirical claim of a single theory- element. The concluding section 7 gives an outlook on topics for future research.

2. Empirical Claim of a Theory-Element

Let $T_0 = \langle K_0, I_0 \rangle$ be a theory-element with core

$$K_0 = \langle M_p^0, M^0, M_{pp}^0, GC^0, GL^0 \rangle.$$

Here, M_p^0 denotes the set of all potential models, M^0 the set of models, and M_{pp}^0 the set of partial potential models. GC^0 has to be interpreted as the cross section of all constraints belonging to T_0, GL^0 as the corresponding cross section of all links. I_0 is the set of intended applications.[4] We may think of T_0 as the basic element of a (trivial) theory-net that consists of one theory-element only.

For the formulation of the empirical claim associated with the theory-element T_0, suitable restriction functions r, $\bar{r}$, and $\bar{\bar{r}}$ have to be defined. The task of these functions is to cut off the theoretical functions from potential models.

(α) Let $r : M_p^0 \rightarrow M_{pp}^0$ be a function such that for all potential models $x \in M_p^0$, $x = \langle D_1, \ldots, D_k, A_1, \ldots, A_l, n_1, \ldots, n_m, t_1, \ldots, t_n \rangle$, $r(x) = \langle D_1, \ldots, D_k, A_1, \ldots, A_l, n_1, \ldots, n_m \rangle$.

(β) Let $\bar{r} : Po(M_p^0) \rightarrow Po(M_{pp}^0)$ be a function such that for all sets of potential models $X \in Po(M_p^0)$ $\bar{r}(X) = \{r(x) \mid x \in X\}$.

(γ) Let $\bar{\bar{r}} : Po(Po(M_p^0)) \rightarrow Po(Po(M_{pp}^0))$ be a function such that for all $\bar{X} \in P(Po(M_p^0))$ $\bar{\bar{r}}(\bar{X}) = \{\bar{r}(X) \mid X \in \bar{X}\}$.

With the help of these restriction functions, we may now formulate the empirical claim associated with T_0. Let us start with the simple case in which no bridge structures (constraints or links) occur, i.e., in which $GC^0 = Po(M_p^0)$ and $GL^0 = M_p^0$. The empirical claim associated with this theory-element may be then formulated as follows:

$$\forall z[z \in I_0 \rightarrow z \in \bar{r}(M^0)]$$

or

$$I_0 \in \bar{\bar{r}}[Po(M^0)].$$

It can be interpreted as follows: By adding suitable theoretical functions, each intended application $z \in I_0$ can be extended into the set M_0 of models of T_0. The formulation of the empirical claim mirrors the first aspect of holism mentioned above: The definition of the set-theoretical predicate that defines the theory-element's set M_0 of models will, in general, include numerous statements. In most cases, it will contain several *characterizations* (that describe the formal features of the nonlogical terms used within that theory-element) and one or more genuine axioms. For the majority of empirical theories, this holds already for the basic element of the corresonding theory-net. If more specialized theory-elements are considered, special laws, and eventually further auxiliary hypotheses, will get involved. Together these statements form holistic complexes that are

[4]I presuppose the basic concepts of the structuralist metatheoretical approach as explained in the first chapter of this book.

confronted with the (nontheoretical) data base as a whole - in accordance with Duhem's basic idea.

Whereas the first aspect of holism refers to applications of empirical theories regarded in isolation, the second aspect relates to correlations that might exist between the theoretical descriptions of different systems. Within the structuralist approach, this aspect is mirrored by so-called bridge structures.[5] Internal bridge structures (constraints) describe the connections that exist between different applications of one and the same theory-element, whereas external bridge structures (links) give an account of the correlation between applications of different theory-elements (or even between different theory-nets). Together, constraints and links enable a detailed and precise picture of the network formed by different applications of empirical theories, as well as of the transfer of information between them. These bridge structures now have to be introduced into the formulation of the empirical claim. Let us start with introducing constraints. Let the global constraints GC^0 of T_0 be defined as the cross section

$$GC^0 = \bigcap C_i^0$$

of all constraints operating on M_p^0. The corresponding empirical claim can then be formulated as follows:

$$I_0 \in \bar{\bar{r}}[Po(M^0) \cap GC^0].$$

It states that the intended applications of I_0 can be extended into the set M^0 of models such that the internal bridge structures combined in GC^0 are fulfilled. These constraints tie together the theoretical extensions of the partial models contained in I_0 and may thus generate a correlation between the theoretical functions occurring in the corresponding models.

According to our initial assumption, the theory-net in question consists of one theory-element T_0 only. Thus there can be no links which connect T_0 with other theory-elements of the same net. But there may be (several) links that connect T_0 with theory-elements of other nets. Let

$$\{L_1^0, \ldots, L_m^0\} \quad (m \in \mathbb{N})$$

be the set of all links that connect M_p^0 with the sets of potential models of other theory-elements and are relevant for T_0. Furthermore, for all $j \leq m$ let

$$\lambda_j^0 = \{x \in M_p^0 \mid \exists \bar{a} \exists x' \exists \bar{a}' [\langle x, \bar{a}, x', \bar{a}' \rangle \in L_j^0]\}.^6$$

[5] See the chapter by Moulines & Polanski in this book.

[6] For the definition of a concrete link, cf. *Architectonic*, p. 61,78 f.

Then, the global link

$$GL^0 = \bigcap\{\lambda_1^0, \ldots, \lambda_m^0\}$$

is a subset of M_p^0. The empirical claim of a theory-element T_0 in which both constraints and links occur may be written as follows:

$$I_0 \in \bar{\bar{r}}[Po(M^0) \cap GC^0 \cap Po(GL^0)].$$

With the content of this theory-element defined by

$$Cn(T_0) = \bar{\bar{r}}[Po(M^0) \cap GC^0 \cap Po(GL^0)]$$

we thus obtain

$$I_0 \in Cn(T_0).$$

Even for the simple case of the empirical claims of theory-elements some cautioning remarks have to be added with respect to the general validity of the standard formulation given above:

1. We have assumed that all values of the nontheoretical functions are known. This, however, will be a fictitious assumption in most cases. There are several reasons for this claim: In the first place, nontheoretical functions are normally defined for an infinite number of arguments. (One example would be the position function used to describe the motion of objects in classical physics.) By contrast, only a finite number of measurement data are available. Second, there may be practical circumstances that further restrict the set of nontheoretical data. The observation of the orbits of comets illustrates this. Usually only small fractions of these orbits can be observed: due to the small size of the objects, it is difficult to determine their position when they are outside the orbit of Jupiter - and, of course, they cannot be observed at all when they are behind the sun. A similar situation arises in many experimental situations: instead of 'complete' partial models (to express it paradoxically), only a limited data base will be available. In order to cope with this (rather common) case of incomplete data sets, Balzer/Lauth/Zoubek (1993) have introduced the concept of substructures of potential models.[7]

2. We have assumed that the intended applications of empirical theories are exclusively described by functions which are nontheoretical with respect to these theories. This is trivially true at the beginning of a theory's

[7]For each potential model x a substructure is defined as a tupel x' such that each component of x' is a(n) (eventually empty) subset of the corresponding component of x.

development: When a new theory is applied for the first time, its intended applications have to be described in terms of some pre-theory (or with the help of everyday language). For example, when classical mechanics was first applied to astronomical systems, its intended applications were described by specifying the corresponding sets of objects and the position function (or, to be more precise, values of the position function that had been determined by observation before). The same may hold when a theory is applied to a new range of phenomena which have previously not been dealt with (by the same theory). Again, all values of theoretical functions are unknown and thus cannot be utilized for the description of these applications. Let us call applications of this type, which are exclusively described by nontheoretical functions, *primary applications* of that theory.

The situation may be different, however: the theory may be applied to a phenomenon, aspects of which have already been described previously with the help of the very same theory. An example: as is well known, Newtonian mechanics has been used to describe Jupiter's orbit around the sun. In the course of this theoretical description, the mass of this planet was determined. When several centuries later Amalthea - a small moon of Jupiter - was discovered and its motion around the planet analyzed, Jupiter's mass (and thus a value of a theoretical function) was assumed as already known. The same often occurs in the application of empirical theories: certain values of theoretical functions are known beforehand. A proposal how the empirical claim of theory-elements may be generalized to take account of these secondary applications is described in Balzer/Lauth/Zoubek (1993).[8]

3. So far, we have focused on the strict version of the empirical claim of theory-elements and have left phenomena of approximation unconsidered. In nearly all applications of empirical theories, however, processes of approximation play a decisive role. Following Architectonic, an approximative version of the empirical claim of theory-elements may be stated as follows:

$$I_0 \in \widetilde{Cn(T_0)}.$$

[8]There, however, the authors use a concept of partial models that is highly misleading: They identify the set M_{pp} of partial models with the set of all substructures of M_p (p. 522). Thus, the distinction between theoretical and nontheoretical terms (which is essential both for the structuralist concept in general, and for the considerations at issue, in particular) is underhand given up in the definition of M_{pp}. By contrast, the differentiation between primary applications (which are described by nontheoretical functions only) and secondary applications (which involve values of theoretical terms) makes it possible both to present a more realistic picture of the sets of applications of empirical theories and to retain the theoretical/nontheoretical dichotomy at the same time.

It is obtained from the standard version by 'blurring' the content $Cn(T_0)$. A precise formal explication of the notion of (admissible) blurs - and thus of the approximative empirical claim - can be obtained by utilizing standard topological concepts.[9]

These three remarks show that the strict version of the empirical claim stated above has to be treated with caution. The point we want to make in this article, however, which refers to a major deficit in the established view of the Ramsey-sentence of theory-nets, can be explained even with the help of the simple version of the empirical claim. Therefore, we will use this formulation throughout the following considerations.

3. Empirical Claim of Theory-Nets: The Standard View

Having discussed the empirical claim of theory-elements regarded in isolation, we now turn to the problem of how these Ramsey-sentences can be combined so as to provide the empirical claim of a theory-net. In this section we present the standard view on that issue. In the next section, we will argue that and why this view is misleading and how an adequate formulation can be obtained.

Def.: Let $N = \langle \bar{T}, \sigma \rangle$ be a treelike theory-net. $\bar{T} = \langle T_0, \ldots, T_n \rangle$ $(n \in \mathbb{N})$ is the set of theory-elements. It consists of the net's basic element T_0 and further theory-elements $T_1, \ldots, T_n$ obtained from T_0 with the help of the specialization relation σ. The theory-elements are ordered pairs $T_i = \langle K_i, I_i \rangle$, $0 \leq i \leq n$, with $K_i = \langle M_p^i, M^i, M_{pp}^i, GC^i, GL^i \rangle$. For each $T_i \in \bar{T}$, its theoretical content $Cn_{th}(T_i)$ and content $Cn(T_i)$ are defined as follows[10]:

$$(1) \qquad Cn_{th}(T_i) = Po(M^i) \cap GC^i \cap Po(GL^i),$$

$$(2) \qquad Cn(T_i) = \bar{\bar{r}}[Cn_{th}(T_i)].$$

At first glance, one might try to identify the empirical claim of a theory-net simply with the conjunction of the empirical claims associated with the theory-elements constituting this net:

$$\forall T_i \; [T_i = \langle K_i, I_i \rangle \in \bar{T} \rightarrow I_i \in Cn(T_i)].$$

[9] For a detailed description of how this can be achieved by means of uniformities, cf. *Architectonic*, p. 353-361.

[10] Since for all theory-elements belonging to the same specialization net, the corresponding sets of potential models are identical, we will use the same restriction functions $r, \bar{r}, \bar{\bar{r}}$ throughout the net.

In fact, this was one of the ideas underlying a proposal made in the first publication addressing the empirical claim of theory-nets.[11] However, this view was rightly criticized by Zandvoort (1993). The reason for its inadequacy can be illustrated with the help of a theory-net that consists of two theory-elements

$$T_0 = \langle K_0, I_0 \rangle, T_1 = \langle K_1, I_1 \rangle$$

such that T_1 is a specialization of T_0 ($T_1 \sigma T_0$)

$$T_0 = \langle K_0, I_0 \rangle \xrightarrow{\sigma} T_1 = \langle K_1, I_1 \rangle$$

Then, according to the definition of the specialization relation σ, the set I_1 of intended applications of T_1 is a subset of the set I_0 of T_0. The partial models of I_0 thus have to be extended both into the set of models M_1 of the specialized theory-element T_1 and into the set of models M_0 of the net's basic element T_0. One would rightly expect that for both purposes, the same theoretical functions are used so that, in both cases, identical theoretical descriptions are obtained for one and the same intended application. The fulfilment of this basic intuitive requirement, however, is not guaranteed when the above-stated version of the empirical claim is used.

In *Architectonic* the authors presented a version of the empirical claim of theory-nets which is not exposed to this objection. The basic idea underlying this proposal can be summarized as follows: if a theory-element T_j is obtained from another theory-element T_k by specialization ($T_j \sigma T_k$), then an inclusion relation has to hold between the sets $X_j \in Cn_{th}(T_j)$ and $X_k \in Cn_{th}(T_k)$ to which the corresponding sets of intended applications I_j, and I_k, respectively, are extended. This idea can be expressed by the following formulation[12] of the empirical claim associated with the specialization net N:

$$\exists X_0 \ldots \exists X_n [\forall i_{i \in \{0,1,\ldots,n\}} [X_i \in Cn_{th}(T_i) \wedge \bar{r}(X_i) = I_i]$$
$$\wedge \forall j, k_{j,k \in \{0,1,\ldots,n\}} [T_j \sigma T_k \rightarrow X_j \subseteq X_k]].$$

In the following sections, we will refer to this version as the *standard version* or *established view* of the empirical claim of theory-nets. We will demonstrate that this version, too, shows a major deficiency, and how this deficiency can be eliminated.

Before addressing that issue, however, we will point out a general problem connected with the structuralist view concerning the relation between theory-nets and the corresponding empirical claims. This problem

[11] Cf. Balzer & Sneed (1977/78).
[12] Cf. Gähde (1989a), p. 119.

can be summarized as follows: Many proponents of the structuralist approach have put special emphasis on the fact that this approach sharply differentiates between theory-nets as tools for the formulation of empirical claims and the claims themselves.[13] This supposedly strict separation, however, is watered down by the fact that already in the formulation of the theory-net (specialized) cores are assigned to certain sets of intended applications. Thus, the decisive information that should be reserved for the empirical claim, is already anticipated in the formulation of the net itself. If one wants to stick to the seperation mentioned above, the following path may be pursued: Instead of using theory-nets $N = \langle \bar{T}, \sigma \rangle$, one starts from the corresponding core nets $^{*}N = \langle \bar{K}, ^{*}\sigma \rangle$, which consist of the set of theory-cores $\bar{K}$ of N and a specialization relation $^{*}\sigma$ defined on that set. All intended applications are lumped together in one comprehensive set (which is identical with the set I_0 of intended applications of $N's$ basic theory-element). The ordered pair $\langle ^{*}N, I_0 \rangle$ may then be regarded as a weak version of a theory-net, and this version does not already anticipate the information expressed by the empirical claims that can be associated with it.[14]

4. Empirical Claim of Theory-Nets: A Refined Version (Including a Consistency Requirement)

The aim of this section is to show that the established view of the empirical claim of theory-nets presented in the last section is inadequate: in one crucial point it is too weak as the following may occur:

(1) The empirical claim is true, i.e., the corresponding existential claims can be fulfilled.

(2) However, they can be fulfilled only if at least one application of the theory in question is described in different ways which contradict each other.

This may occur due to the following fact: the definition of a theory-net allows for an overlap of the sets of intended applications of different specialized theory-elements. Be z a partial model that belongs to both sets of intended applications of theory-elements T_i and T_i, $(i, j \in \mathbb{N}, i \neq j)$.

[13] E.g. Stegmüller (1985).

[14] Note, that if this way is pursued, numerous different empirical claims may be associated with one and the same theory-net (in the weak sense as explained above). This might be regarded as an attractive consequence with respect to the reconstruction and analysis of certain developments in the history of science (e.g., various attempts to eliminate an anomaly; cf. section 4).

A situation may then arise where the empirical claim as formulated in section 3 can be fulfilled only in such a way that

(1) in T_1, z is extended to a model x_1 of the corresponding set of models M_1;

(2) in T_2, z is extended to a model x_2 of the corresponding set of models M_2;

(3) $x_1 \neq x_2$.

As x_1 and x_2 are identical with respect to all nontheoretical terms, they have to differ with respect to at least one theoretical function.

The meaning of this can be readily explained by the example of requirements concerning the force function in classical physics. *It does not mean* that in T_1 and T_2 requirements concerning different components of the force function are specified, e.g., that T_1 postulates that the first component of the force function satisfies Newton's law of universal gravitation, while T_2 postulates that the second component of the force function fulfills a special law describing velocity-dependent forces. These two postulates could well be fulfilled by one and the same model. (An illustrating example would be a model of classical mechanics that describes a free fall where frictional phenomena are taken into account.) By contrast, our considerations relate to the problematic case in which in T_1 and T_2 requirements concerning the force function are specified which cannot simultaneously be fulfilled by one and the same model. In our example, this would be the case if T_1 postulated that the first component of the force function follows Newton's law of gravitation, whereas T_2 required for the same force component that it violated this law. In this case, the requirements formulated in T_2 and T_2 would directly contradict each other. Nevertheless, in the formulation given above, the corresponding empirical claim could be true - in spite of an inconsistent treatment of at least one application.[15]

Even a brief look at the history of science shows that this situation would not be tolerated - at least as long as phenomena related to the approximative application of theories are not taken into account.[16] By contrast, the inconsistent treatment of some system by one and the same theory would be regarded as an argument against that theory - or, to be

[15] Whether this case might occur, depends crucially on the underdetermination of the theoretical functions by the theory's fundamental laws. As a consequence, one and the same partial model (intended application) might be extendable to different models of the theory, in which the theoretical functions fulfill different special laws in addition to the theory's axioms.

[16] In this section, we focus on the strict - non-approximative - version of the empirical claim; cf. section 7.

more precise, at least against its successful applicability to that special system. One rightly expects from an acceptable empirical theory that it provides more than theoretical descriptions of certain applications which are in accordance with the (nontheoretical) data base: a description of the whole set of applications that avoids any internal contradiction. This includes the basic requirement that one and the same application may not be described in different ways which are inconsistent which each other. This fundamental postulate, that might be called a *consistency requirement*, is not adequately expressed by any version of the Ramsey-sentence proposed so far.

In order to eliminate this deficiency, the formulation of the empirical claim needs to be essentially modified: an additional requirement has to be incorporated. According to this additional postulate, the Ramsey-sentence has to be fulfilled by a set X_0 which for each intended application of I_0 contains exactly one model. The modified empirical claim associated with an empirical theory can thus be formulated as follows:

$$\exists X_0 \ldots \exists X_n [\forall i_{i \in \{0,1,\ldots,n\}} [X_i \in Cn_{th}(T_i) \wedge \bar{r}(X_i) = I_i] \\ \wedge \neg \exists x, x' [x, x' \in X_0 \wedge x \neq x' \wedge r(x) = r(x')] \\ \wedge \forall j, k_{j,k \in \{0,1,\ldots,n\}} [T_j \sigma T_k \rightarrow X_j \subseteq X_k]].$$

This version of the empirical claim of theory-nets (which will be called the revised version here and after) is less strong than may appear at first glance: One has to consider carefully what is and what is not required:

1. *It does not claim* that each application $z \in I_i$ can be extended to exactly one model of T_i's set of models M^i, $(i = 0, 1, 2, \ldots)$. In many cases, some or all applications might even be extended to an infinite number of models.

2. *It does not claim* that each set I_i of intended applications can be extended to exactly one set of models $X_i \in M^i, (i = 0, 1, 2, \ldots)$ such that all constraints and links belonging to that theory-element are fulfilled. Again, there may be an infinite number of sets of models which fulfill these requirements and to which I_i can be extended.

3. *It does not claim* that the total set of intended applications I_0 of the theory-net can be extended to exactly one set X_0 of models of the net's basic element such that all requirements included in the standard version of the empirical claim connected with this net are met.

4. *By contrast, the revised Ramsey-sentence for theory-nets claims* that among the (possibly infinite number of) sets of models to which I_0 can be extended and for which all requirements (as specified in the established version of the Ramsey-sentence) are fulfilled, there is at least one set such that the following statement holds: For each application $z \in I_0$, it contains

exactly one model to which z has been extended. Intuitively speaking, this set of models thus supplies a theoretical description of the whole set of applications of this net that fulfills all requirements formulated in the standard version of the empirical claim as stated in section 3 and, in addition, is *consistent*.[17]

Against these considerations, the following objections may be raised: 'The history of science provides ample examples for situations in which one and the same (physical) system was described in different ways that were inconsistent which each other. In some cases, these alternative descriptions were obtained within the framework of one and the same theory, i.e., by making use of the same fundamental laws, but differed with respect to the special laws used in them.[18] In many cases, these alternative theoretical descriptions coexisted and concurred for several years or even decades. How does this undeniable historical fact fit in with the argumentation presented above? How can the occurrence of alternative inconsistent descriptions of one and the same phenomenon be represented within the framework of theory-nets?' Two possible answers to these questions are outlined below.

The first answer consists in the formulation of but one comprehensive

[17] Let us briefly discuss two trival formal objections against the consistency requirement as formulated above.The first objection refers to the following point: sometimes it might be useful to describe one and the same application by making use of different systems of units for different practical purposes. Strictly speaking, the theoretical functions used for the extension of this partial model are then different. This objection, however, can be trivially rejected: one can always find a standard system of units which can be used for the description of all applications of the theory-net in question. It goes without saying that, for the logical reconstruction of an empirical theory, such a standard system should in fact be chosen. - The second objection refers to the choice of a frame of reference: if one and the same 'segment of reality' is described as seen from different frames of reference, the theoretical functions used for these descriptions will differ as well. Thus, according to this argument, different theoretical functions are used for the description of one and the same intended application. This objection, however, is based on a misinterpretation of the structuralist conception of application. Let us illustrate that point by means of the term 'application' as used in the structuralist treatment of classical mechanics. Here, intended applications are triples that contain the position function s. The specification of this function for all objects involved, however, already relates to a certain frame of reference. In other words: within the structuralist framework, the term application already denotes a certain (e.g., physical) system, as described with the help of certain nontheoretical functions, and as seen from a certain frame of reference. If one and the same system is viewed from different frames of reference, this corresponds to different intended applications.

[18] An example is provided by the numerous attempts to eliminate the anomaly of Mercury's perihelion without leaving the framework of classical mechanics. In order to eliminate this anomaly (which was discovered by Le Verrier in 1859) in this way, different alternative special laws for the force function were developed (among them laws describing velocity-dependent gravitational forces) and used in addition to Newton's second axiom.

empirical claim. In this claim, all different approaches (all *schools*) with their varying descriptions of the system in question are represented. Thus, the problematic intended application will be assigned to different theory-elements simultaneously. What is more important: in each of these theory-elements, it will be theoretically described by a *different* model, i.e., with the help of different theoretical functions (which may obey different special laws, auxiliary hyptheses etc). Nevertheless, the Ramsey-sentence in its standard version may be true, as this version does not require that one and the same intended application be described consistently throughout the net.[19] If, however, but one of these approaches fails, this compehensive empirical claim will have failed as a whole: its existential claims cannot be fulfilled. This fact underscores the crucial deficit of the first answer to the above question: it cannot adequately distinguish between successful and unsuccessful attempts to deal with problematic intended applications.

When the second answer is chosen, this problem does not occur: here, each of the different alternative approaches is represented by a theory-net and a corresponding empirical claim of its own. These nets differ with respect to the theory-elements to which the problematic intended application is assigned. Internally, however, each of these approaches will claim to describe this application in a consistent way: it is extended to exactly one model.[20] This fact is mirrored by the consistency requirement included in the revised version of the empirical claim of theory-nets.

These nets and the corresponding empirical claims - each of which represents a specific approach (or *school*) - can coexist and concur during extended phases of scientific development. The empirical claims can be simultaneously true - although their theoretical descriptions of the system in question may contradict each other. This situation may suddenly change, however, if, due to the progress of scientific techniques, new nontheoreti-

[19] Another necessary condition for this case to occur is that the theoretical functions used to extend this partial model into the set of models are underdetermined by the theory's axioms; cf. section 5.

[20] Note, however, that this particular model may belong to the sets of models of different theory- elements. Furthermore, it should be kept in mind that, in this article, we focus on the strict (non-approximative) version of the empirical claim. If phenomena of approximation are considered, the situation may look somewhat different. For certain practical purposes, one might be satisfied with an approximative theoretical description of some intended application, although 'one knows better', i.e.,in principle a more precise and detailed description could be given. An example is provided by the free fall of bodies. For a number of practical purposes, it will be admissible to ignore friction, although it is known that these forces do occur, and although a precise description of this type of motion (which includes velocity-dependent frictional forces) could be given. In order to cope with these kinds of phenomena, the consistency requirement will have to be liberalized, if an approximative version of the empirical claim of theory-nets is going to be used (cf. the remark in section 7).

cal data become available.[21] Some of the concurring empirical claims may then turn into falsities: the existential claims contained in them cannot be fulfilled any longer. This may be due to different reasons: first, it may be that the (improved) nontheoretical data do no longer allow for a theoretical description of the isolated system in question which would be in accordance with the axioms and special laws required in the appropriate theory-element. Second, it may be that constraints or links connecting the theoretical description of this system with other applications cannot be fulfilled any longer. The failure of certain empirical claims thus mirrors the fact that the corresponding approach has dropped out of the race.

The second method of describing concurring approaches thus shows a significant advantage over the first: it enables a more detailed picture of the successes and failures of the different *schools* with respect to a changing data base, and it mirrors the claim of these schools to provide a consistent description of the set of intended applications. This claim is not articulated by the standard version of the Ramsey-sentence, but is expressed by the revised version presented above.

5. Empirical Claim, Holism, and Formal Conditions for the Distinction of Theoretical Terms

In the previous sections, we focused on two aspects of holism and showed how they are expressed by the empirical claim of theory- nets. The aim of this section is to demonstrate that there is a close connection between these holistic features of empirical theories, on the one hand, and the occurence of theoretical terms within them, on the other. We will concentrate on two formal properties of theoretical terms, which have a decisive impact on the holistic aspects described above. These properties can be articulated by two necessary - though not sufficient - conditions for theoreticity. As these have already been discussed in some detail in Gähde (1990), we will here confine ourselves to a short sketch of both conditions, placing special emphasis on their relation with the occurrence of holistic phenomena.

The task of the two requirements is to provide a rather complex answer to the question in which sense *theoretical terms derive from some empirical theory*. The first necessary condition for theoreticity starts by giving a negative answer: a function which is theoretical with respect to T cannot be uniquely determined by T-non-theoretical functions with the help of

[21] As we discussed in section 2, normally only a finite data base is available. If the experimental methods are improved, this data base may become more extensive. Furthermore, the precision of measurement processes may increase. This will be an important factor, if approximation is taken into account.

T's axioms alone. Theoretical terms are left fundamentally underdetermined by the theory's basic laws.[22] Obviously, this condition is neatly connected to Padoa's criterion of definability: it excludes that theoretical functions can be defined by the nontheoretical functions with the help of the theory's fundamental laws alone. Condition 1 thus takes up a basic postulate imposed on theoretical terms in informal metatheoretical concepts like Carnap's *two-level theory of scientific language*: the requirement that theoretical terms should neither be observational terms themselves (which is trivial), nor definable by observational terms.[23] As we will see below, it is exactly this specific underdetermination of the theoretical terms by the theory's fundamental laws which makes room for the two aspects of holism described above.

Before we turn to the second condition, we have to point out a crucial methodological feature of condition 1. At first glance, any attempt to apply this condition to concrete empirical theories seems to run into a vicious circle: on the one hand, it seems already to presuppose the theoretical/nontheoretical distinction (as the formulation of condition 1 explicitly refers to the set of nontheoretical terms). On the other hand, it is exactly with the help of this condition that the borderline between theoretical and nontheoretical terms is to be drawn. This impression of circularity is misleading, however. It is caused by a crucial methodological feature of both the first and the second condition: these conditions do not single out theoretical terms in succession, one after the other. They rather have to be interpreted as requirements imposed on all theoretical terms simultaneously. In order to apply them correctly, one first has to form all possible bipartitions of the set of T's nonlogical terms into two disjoint subsets. Then, for each of these partitions, one has to analyze whether or not, under the assumption that the terms of the first subset are given, the terms of the second subset are underdetermined by the theory's fundamental axiom(s). If the answer is 'yes', then the first condition is fulfilled, and the terms belonging to the second subset remain possible candidates for theoretical functions. This important methodological feature is shared by condition 2, to which we turn now.

Whereas condition 1 provides but a negative answer to the question in which sense theoretical terms *come from an empirical theory T*, condition 2 supplies positive information on that issue: according to this condition, a theoretical term - though underdetermined within the theory-net's basic element - can be uniquely determined (up to transformations of scale) in some models of a specialized theory-element. Again, several alterna-

[22] Different alternative formulations of condition 1 are discussed in Gähde (1990).

[23] See Carnap (1956).

tive formulations of this condition have to be distinguished.[24] Each of these possible reformulations, however, preserves the basic idea behind the second condition: theoretical terms can be determined by a specific cooperation between the fundamental axioms of *their* theory and suitable special laws, auxiliary hypotheses etc.[25]

We will now turn to the relation between these formal (necessary) conditions for theoreticity and the occurrence of the two aspects of holism discussed above. A first hint that there is such a correlation is given by the fact that the two conditions refer to exactly the same situation as the empirical claim of a theory-net (which, as we have seen, mirrors these holistic aspects): the situation, in which all nontheoretical terms are assumed as given and where one searches for proper theoretical terms that can be used for the extension of these partial structures into the theory's set of models.[26] Let us now assume that condition 1 is violated. In that case, the theoretical functions used for the extension of the theory's applications would be uniquely determined by the theory's fundamental law(s). This would have several major consequences. First, each application could be extended to one model of the theory at the most. No special laws, auxiliary hypotheses etc. would be necessary for that purpose. This does not

[24] These alternative formulations range from the rather weak postulate which only guarantees unique determinability of some theoretical functions in some partial models, to the exceptionally strong version which requires that, by a cooperation between axioms and special laws, for all partial models all theoretical functions can be uniquely determined (up to transformations of scale); cf. Gähde (1990).

[25] It is obvious that, if arbitrary requirements were admitted in addition to the theory's fundamental laws, any term could be made uniquely determinable, and so condition 2 would be trivialized. Thus, it is essential for any adequate reformulation of condition 2 that the class of additional special laws, auxiliary hypotheses etc. is appropriately narrowed down. One way to achieve this purpose has been described in Gähde (1983): There, only those specialized hypotheses were admitted that are in accordance with the invariance principles characteristic of the theory-net's basic element. An alternative way to reach this goal has been described in Gähde (1990): According to this approach, in order to single out the permissible special laws etc., one has explicitly to refer to the theory-elements that factually occur in the theory-net at a certain stage of its development. When the theory-net changes, in particular, when new specialized theory-elements are incorporated, some terms may become determinable that have been undeterminable before, which, in turn, may have a significant impact on where to draw the line between theoretical and nontheoretical terms. This seems only natural: *What one can do* with an empirical theory, which terms, in particular, can be determined with the help of this theory, crucially depends on the class of special laws, auxiliary hypotheses etc. as formulated in the framework of this theory. Note that, in general, these specializations will fulfill the same invariance principles as the theory's fundamental laws. Thus, the above condition concerning invariance principles will automatically be fulfilled in most cases. For an exception, however, compare the analysis of a Galilei-covariant subnet within the (Lorentz-invariant) net of relativistic electrodynamics in Bartelborth (1988, p. 102 ff.)

[26] We refer to the empirical claim of theory-nets as stated in sections 3, 4.

mean that no special laws could be found. Their role within the theory in question, however, would be completely different: they could not influence the theoretical description of the theory's applications (which, according to our assumption, is already uniquely fixed by the fundamental laws). Thus, the employment of a vast number of laws, auxiliary hyotheses, etc. in the description of the theory's intended applications, would be superfluous and the first aspect of holism would not occur.[27]

Second, the same holds with respect to the possibility of any correlation between the theoretical description of different applications - and thus to the second aspect of holism mentioned before. If the theoretical terms were already uniquely determined by the theory's axiom(s), bridge structures could have no impact whatsoever on the theoretical description of different systems. If, by chance, constraints or links are fulfilled, they would still not influence the values of the theoretical functions. If, by contrast, they are violated, there is no way to resolve the conflict without giving up either bridge structures or at least one intended application. In any case, constraints and links would have lost their power to induce any correlation between the theoretical descriptions of different applications; the second aspect of holism could not occur.

The situation is completely different, if condition 1 is fulfilled. Then, at least for some applications, special laws etc. are necessary to determine the values of the theoretical functions. The typical holistic complexes mentioned in section 1- consisting of the theory's fundamental laws, additional special laws, and eventually further auxiliary hypotheses - are then involved in the theoretical description of these applications. These holistic complexes are confronted with the (nontheoretical) data as a whole. If the corresponding empirical claim fails, one may then try to immunize the theory's axioms and modify more specialized hypotheses instead: the specific underdetermination of the theoretical functions by the theory's axioms - articulated by condition 1 - allows for the combination of a wide variety of special laws with the theory's fundamental principles. Thus, it accounts for the remarkable flexibility empirical theories show with respect to the task of adapting to new or modified sets of data. Due to this underdetermination, in most cases several alternative options will exist how an empirical theory can react to conflicts between *theory and experience.*[28]

Furthermore, we can now understand how the theoretical description of one application might be correlated with the theoretical description of another application - and thus the occurrence of the second aspect of

---

[27] Note, however, that still several fundamental laws (that are used to single out the set of models of the net's basic element) might be involved.

[28] For a case study of that issue cf. Gähde & Stegmüller (1986).

holism mentioned above: according to condition 2, the values of the theoretical functions can be uniquely determined (up to scale transformations) by a cooperation of fundamental laws and more specialized statements for at least some intended applications of the theory in question. By means of constraints and links, these theoretical data can be transfered to other intended applications. There, they may well influence the theoretical description of these applications - due to the underdetermination of the theoretical functions used to extend this partial model to a model of the theory.

Thus, the fulfillment of the two formal conditions for theoretical terms accounts for two facts: first, for the flexibility empirical theories show in dealing with intended applications regarded in isolation - due to the cooperation of numerous (lawlike or non-lawlike) statements and thus due to the first aspect of holism. Second, for the emergence of a network of systems whose theoretical descriptions are correlated by means of bridge structures - and thus for the second aspect of holism. Both facts are mirrored in the empirical claims of theory-nets.

6. A Formal Result: Substituting a Theory-Element for a Theory-Net

In this section we show that the revision of the empirical claim described in section 4 has an attractive formal consequence: For each treelike theory-net N, one can define in a straightforward manner a theory-element $T^\star$ such that the empirical claims associated with N and $T^\star$ are mathematically equivalent.

Theorem: Let $N = \langle \bar{T}, \sigma \rangle$ be a treelike theory-net, where $\bar{T} = \langle T_0, \ldots, T_n \rangle$ consists of ($n \in \mathbb{N}$) theory-elements

$$T_i = \langle K_i, I_i \rangle = \langle \langle M_p^i, M^i, M_{pp}^i, GC^i, GL^i \rangle, I_i \rangle$$

and σ is the specialization relation. Furthermore, let

$$T^\star = \langle K^\star, I^\star \rangle = \langle \langle M_p^\star, M^\star, M_{pp}^\star, GC^\star, GL^\star \rangle, I^\star \rangle$$

be a theory-element such that $M_p^\star = M_p^0, M_{pp}^\star = M_{pp}^0$, and $I^\star = I_0$, while $M^\star, GC^\star$, and $GL^\star$ are defined as follows:

(1) $M^\star = \{x \in M_0 \mid \forall i_{i=1,\ldots,n}[r(x) \in I_i \rightarrow x \in M^i]\}$,

(2) $GC^\star = \{X \in GC^0 \mid \neg\exists x, x'[x, x' \in X \wedge x \neq x' \wedge r(x) = r(x')] \wedge$
$\wedge \forall I_i \forall X_i[I_i \in \{I_1, \ldots, I_n\} \wedge \bar{r}(X_i) = I_i \wedge X_i \subseteq X \rightarrow X_i \in GC^i]\}$,

(3) $GL^\star = \{x \in GL^0 \mid \forall i_{i=1,\ldots,n}[r(x) \in I_i \rightarrow x \in GL^i]\}$.

Then the (revised) empirical claims (as stated in section 4.) of N and $T^\star$ are mathematically equivalent.
Comments

1. Obviously, $M^\star, GC^\star$, and $GL^\star$ have the formal properties of a set of models, a constraint, and a link, respectively, so that $T^\star$ is in fact a theory-element.

2. Note, however, that the consistency requirement is crucial for the validity of the theorem stated above. If this requirement is omitted from the empirical claim associated with N as well as from the claim associated with $T^\star$, the above theorem does not hold.[29]

3. The definitions of $M^\star, GC^\star$, and $GL^\star$ explicitly refer to the sets I_i $(i = 1, \ldots, n)$ of intended applications which, in general, cannot be characterized axiomatically. Thus, the same holds for $M^\star, GC^\star$, and $GL^\star$. If, as the theory develops, the sets of intended applications change, $M^\star, GC^\star$, and $GL^\star$ will change as well. As a consequence, the possibility of replacing a theory-net N with a theory-element $T^\star$ is not a suitable tool for the reconstruction of concrete empirical theories, and is of theoretical interest only: it shows that the total structure of a theory-net can be projected onto one single complex theory-element, if the revised version of the empirical claim is used.

Proof: For each theory-element

$$T_i = \langle K_i, I_i\rangle = \langle\langle M_p^i, M^i, M_{pp}^i, GC^i, GL^i\rangle, I_i\rangle,\ \ 0 \leq i \leq n,$$

its content is given by $Cn(T_i) = \bar{r}[Po(M^i) \cap GC^i \cap Po(GL^i)]$ (cf. sec. 3). We have to show that

$\exists X_0 \ldots \exists X_n [\forall i_{i=0,1,\ldots,n}[X_i \in Cn_{th}(T) \wedge \bar{r}(X) = I_i] \wedge$
$\wedge \neg\exists x, x'[x, x' \in X_0 \wedge x \neq x' \wedge r(x) = r(x')] \wedge$
$\wedge \forall j, k_{j,k=0,1,\ldots,n}[T_j \sigma T_k \rightarrow X_j \subseteq X_k]]$
$\longleftrightarrow\ I^\star \in Cn(T^\star)$.

'$\Rightarrow$:' Let $\tilde{X}_0, \ldots, \tilde{X}_n$ be sets that satisfy the left side of the equivalence above.
(α) $\tilde{X}_0 \in Po(M^\star)$:
Trivially, $\tilde{X}_0 \in Po(M^0)$. According to our assumptions, for every element of I_0 there is exactly one model in $\tilde{X}_0$ to which this application is extended, and for each set I_i of intended applications there is exactly one set of models $\tilde{X}_i \in \{\tilde{X}_1 \ldots, \tilde{X}_n\}$, such that $\bar{r}(\tilde{X}_i) = I_i$. Let $\tilde{x}$ be an element

[29] Cf. the remark at the end of this section.

of $\tilde{X}_0$ with $r(\tilde{x}) \in I_i$. Then, $\tilde{x} \in \tilde{X}_i$. According to our assumptions, $\tilde{X}_i \in [Po(M^i) \cap GC^i \cap Po(GL^i)]$, thus $\tilde{x} \in M^i$. The same can be shown for all elements of $\tilde{X}_0$. It follows that $\tilde{X}_0 \in Po(M^\star)$.
(β) $\tilde{X}_0 \in GC^\star$:
Trivially, $\tilde{X}_0 \in GC_0$. According to our premises, for each intended application of $I_0, \tilde{X}_0$ contains exactly one extension into the set of models. Thus, for any $I_i \in \{I_1, \ldots, I_n\}$ the subset $\tilde{X}_i \subseteq \tilde{X}_0$ with $\bar{r}(\tilde{X}_i) = I_i$ is uniquely determined. According to our premises, $\tilde{X}_i \in [Po(M^i) \cap GC^i \cap Po(GL^i)]$, thus $\tilde{X}_i \in GC^i$. This can be shown for all elements of $\{I_1, \ldots, I_n\}$ and the corresponding sets of models. It follows that $\tilde{X}_0 \in GC^\star$
(γ) $\tilde{X}_0 \in Po(GL^\star)$
Analogous to (α).
From (α) – (γ) we obtain $X_0 \in [Po(M^\star) \cap GC^\star \cap Po(GL^\star)]$. Because of $\bar{r}(\tilde{X}_0) = I_0$ it follows that $I_0 \in \bar{\bar{r}}[Po(M^\star) \cap GC^\star \cap Po(GL^\star)\}$.
'$\Leftarrow$:' $I_0 \in \bar{\bar{r}}[Po(M^\star) \cap GC^\star \cap Po(GL^\star)]$, thus

$$\exists X_0[X_0 \in [Po(M^\star) \cap GC^\star \cap Po(GL^\star)] \wedge \bar{r}(X_0) = I_0].$$

Let $\tilde{X}_0$ be a set of models which satisfies this existential sentence. According to the definition of $GC^\star$, for each intended application, $\tilde{X}_0$ contains exactly one model to which this partial model can be extended. Thus, for each $I_j \in \{, \ldots 1_n\}, \tilde{X}_0$ contains exactly one corresponding subset $\tilde{X}_j$ (i.e., $\tilde{X}_j \subseteq \tilde{X}_0$ and $\bar{r}(\tilde{X}_j) = I_j, j = 1, \ldots, n$). For $\tilde{X}_j$, the following statements hold:

(α) $\tilde{X}_j \in Po(M^j)$ (because $\tilde{X}_0 \in Po(M^\star)$);

(β) $\tilde{X}_j \in GC^j$ (because $\tilde{X}_0 \in GC^\star$);

(γ) $\tilde{X}_j \in Po(GL^j)$ (because $\tilde{X}_0 \in Po(GL^\star)$).

Thus, $\tilde{X}_j \in [Po(M^j) \cap GC^j \cap Po(GL^j)]$. Let T_k, T_l be two theory-elements of net N with $T_k \sigma T_l$. Thus, $I_k \subseteq I_l$. Let $\tilde{X}_k, \tilde{X}_l$ be the corresponding sets of models to which I_k, I_l can be extended such that $\tilde{X}_k \subseteq \tilde{X}_0, \tilde{X}_l \subseteq \tilde{X}_0$. Because (1) $\tilde{X}_0$ contains exactly one model for each element of I_0, and (2) $\bar{r}(\tilde{X}_k) \subseteq \bar{r}(\tilde{X}_l)$, it follows that $\tilde{X}_k \subseteq \tilde{X}_l$. Q.E.D.
Remark: As has been stated above, it is crucial for the above theorem that the revised version of the empirical claim is used. One easily proves that a theory-net N can be found, such that the following statement holds: if the consistency requirement is omitted from the empirical claim associated with N as well as from the empirical claim associated with the corresponding theory-element $T^\star$, the remaining Ramsey-sentences are not equivalent. This can be shown as follows: Let N be a theory-net which

contains two theory-elements T_i, T_j such that $I_i \cap I_j \neq \emptyset$ and $M^i \cap M^j = \emptyset$. Let us assume that $z \in I_j \cap I_j$, and that the standard version of the empirical claim (without consistency requirement) of N can be satisfied by extending z to models $x \in M^i, x' \in M^j$. In that case, the empirical claim of the corresponding theory-element $T^\star$ can not be satisfied: since $z \in I_i$, it has to be extended to a model $x \in M^i$ (according to the definition of $M^\star$). Then, however, $x \notin M^j$ (since $M^i \cap M^j = \emptyset$), although $r(x) \in I_j$. Thus, $\exists i_{i \in \{1,\ldots,n\}}[r(x) \in I_i \wedge x \notin M^i]$. It follows, that $x \notin M^\star$.[30]

7. *Further Questions*

In the previous sections, we have shown how the empirical claim of theory-nets mirrors two crucial aspects of the holism of empirical theories: the involvement of a vast number of hypotheses in the theoretical description of the theories' intended applications regarded in isolation, and the correlations that might be established between these descriptions by means of bridge structures. By providing both a detailed and precise picture of these two features - which can be tested by an analysis of concrete empirical theories - the structuralist approach surpasses most informal metatheoretical approaches to the holistic nature of science.

Several important problems have not been addressed in this article so far. One problem concerns how the consistency requirement has to be modified to fit in with an approximative version of the empirical claim of theory-nets. We leave this question open for future research.

Another important problem concerns the question of how far the holistic features of empirical theories extend. Quine's holism, as is well known, is unlimited in the following sense: at least in principle, it is always the totality of our knowledge that is confronted with experience; no part of it is completely immune to possible revision. By contrast, for the description of the real- life activity of scientists, a moderate form of holism seems to be far more adequate: in general, scientists will be aware that the theoretical description of any experiment or observation involves a bundle of hypotheses, but they will nevertheless deny that the totality of our knowledge is always at stake.

Within the structuralist view of empirical theories, the development of a *moderate holistic position* has been obstructed by a dogma that can

[30] If the consistency requirement is not included, a similar problem arises with respect to the definition of the constraint $GC^\star$. In that case, $GC^\star$ could not be defined by simply referring to the sets $I_1, \ldots, I_n$ of intended applications, as one and the same set of partial models may have to be extended to different sets of models, which, in turn, have to satisfy different constraints. Instead, one would have to refer explicitly to these sets of models (or the corresponding theory-elements). Thus, the definitions of $M^\star$ and $GC^\star$ could not be decorrelated in the way shown in the above theorem.

already be found in Sneed's classical work,[31] and has since been echoed in many publications:[32] the dogma that the Ramsey-sentence of an empirical theory forms one inseparable unit that has to be either accepted or rejected in toto. This view has been substantiated by pointing out the role that bridge structures play whithin the empirical claim: they facilitate a tight net of correlations between the theoretical descriptions of the theory's applications, and thus prohibit the corresponding empirical claim from splitting into several smaller Ramsey-sentences which can be tested independently of each other.

As has been shown in Gähde (1989a), however, this view is misleading. It may be true that in some (ontologically dubious) sense 'everything is connected with everything'. What matters in the structuralist analysis of the holistic features of concrete theories, however, are those correlations that are *de facto* established by these empirical theories at a certain stage of their development. The analysis in Gähde (1989a) shows that, under certain circumstances, an empirical theory's set of intended applications can be split into subsets such that the following statements hold: First, the theoretical descriptions of all applications that belong to one of these subsets are effectively interrelated by bridge structures; they form holistic complexes which have to be described by a Ramsey-sentence that cannot be split up further. Second, however, constraints and links do not facilitate an interrelation between the theoretical descriptions of intended applications belonging to different subsets. Thus, the total empirical claim of that theory may well be dissected into a conjunction of claims, each of which refers to one of the holistic complexes mentioned above (and thus to one of the subsets of the theory's set of intended applications).

In the initial stage of their development, most empirical theories will concentrate on the theoretical descriptions of their intended applications in isolation. Only few correlations between these systems will be considered. As the theories are developed further and become more refined, however, the net of correlations established between the theoretical descriptions of their intended applications will become increasingly tight: new constraints and links will be added, while other correlations may break down. Both types of processes can lead to changes in the holistic features shown by that theory and mirrored in the corresponding empirical claim: some holistic complexes may unite, while others disintegrate. The structuralist reconstruction of these processes thus enables first insights into the dynamics of holistic phenomena.[33]

---

[31] See Sneed (1971), S. 70 f.

[32] E.g. in Stegmüller (1986).

[33] A case study of the dynamics of holistic phenomena has been presented in Gähde (1989a). A more detailed discussion is given in Gähde (1989b).

The analysis of whether and how the empirical claim of theory-nets can be split up into subclaims presupposes an adequate formal explication of this claim. The consistency requirement discussed in the previous sections contributes to that aim. As has been shown in more detail in Gähde (1989b), it thus plays an important role in the development of a moderate holistic position for the philosophy of science.

References

Balzer, W. & Lauth, B. & Zoubek, G., 1993, 'A Model for Science Kinematics', *Studia Logica* 52, 519-448 (1993).

Balzer, W. & Moulines, C.U. & Sneed, J.D., 1987, *An Architectonic for Science*, Dordrecht.

Balzer, W. & Sneed, J.D., 1977/78, 'Generalized Net-Structures of Empirical Theories', part I, *Studia Logica* 36 (1977), 195-211, part II, *Studia Logica* 37 (1978), 167-194.

Bartelborth, T., 1988, *Eine logische Rekonstruktion der klassischen Elektrodynamik* , Frankfurt am Main.

Carnap, R., 1956, 'The Methodological Character of Theoretical Concepts'. In: Feigl, H. & Sriven, M. (eds.), *Minnesota Studies in the Philosophy of Science* , vol. I, Minneapolis.

Duhem, P., 1906, *La thorie physique, son objet et sa structure* , Paris.

Gähde, U., 1983, *T-Theoretizität und Holismus* , Frankfurt am Main.

Gähde, U., 1989, *Theorie und Hypothese. Zur Eingrenzung von Konflikten zwischen Theorie und Erfahrung* , Habilitationsschrift, Bielefeld, (cited as Gähde 1989a).

Gähde, U., 1989, 'Bridge Structures and the Borderline Between the Internal and External History of Science'. In: Gavroglu, K., et. al., *Imre Lakatos and Theories of Scientific Change* , Boston Studies in the Philosophy of Science, Dordrecht: Kluwer.

Gähde, U., 1990, 'On Innertheoretical Conditions for Theoretical Terms', *Erkenntnis* 32, 215-233.

Gähde, U. & Stegmüller, W., 1986, 'An Argument in Favor of the Duhem-Quine-Thesis: From the Structuralist Point of View'. In: Hahn, L.E. & Schilpp, P.H. (eds.), *The Philosophy of W. V. Quine* , The Library of Living Philosophers, vol. 15, La Salle, Il.

Quine, W.V.O. & Ullian, J.S., 1978; *The Web of Belief* , New York.

Roseveare, N.T., 1992, *Mercury's Perihelion. From Le Verrier to Einstein,* Oxford.

Sneed, J.D., 1979, *The Logical Structure of Mathematical Physics* , second revised edition, Dordrecht, Reidel.

Stegmüller, W., 1986, *Theorie und Erfahrung*, zweiter Teilband, *Theorienstrukturen und Theoriendynamik*, 1973, second revised edition 1985, dritter Teilband, *Die Entwicklung des neuen Strukturalismus seit 1973*, Berlin.

Zandvoort, H., 1982, 'Comments on the Notion 'Empirical Claim of a Specialization Net' within the Structuralist Conception of Theories', *Erkenntnis* 18, 25-38.

Chapter 10: Symmetry and Invariance

Felix Mühlhölzer

The aim of this paper is to introduce symmetry groups, and with them the notion of invariance, into the structuralist machinery. Nothing new will be said about symmetries and invariance as such, and in comparison to what is the daily bread of contemporary theoretical physicists[1] the level of sophistication will remain rather low. I am only concerned with the first beginnings of an expansion of the structuralist framework which proves to be necessary if symmetries are taken seriously. Structuralist philosophers of science have felt for long that the topic 'symmetry and invariance' has been unduly neglected in their writings.[2] One reason for this neglect may be the empiricist orientation of structuralism. In fact, it is not an easy matter to understand the contribution of symmetries to the empirical content and especially to the Ramsey sentence of a theory, and, unfortunately, even the present study will remain more or less silent on this point. It will have to be investigated on a later occasion. In Architectonic, which is by now the standard textbook of structuralism, symmetries and invariances have been explicitly left out of consideration on the grounds that they should be dealt with in a category-theoretic setting which would exceed the Bourbakian set-theoretic framework chosen in Architectonic.[3] However, it is by no means necessary to make use of category theory here. In what follows I will remain entirely within the Bourbakian framework of Architectonic, simply proposing an addition to, and not an alteration of, the explication of the concept of a scientific theory as presented in Architectonic. The main idea will be to include symmetry groups, together with certain 'actions' of these groups, within the cores of the structuralistically reconstructed theories.

---

[1] See, e.g., Manin (1981), pp. 95-99.

[2] See Diederich (1989), p. 368: '(Invariance principles) should be a major topic of structuralist analysis, but unfortunately never (have) been the subject of sufficiently detailed studies.' At this point it might be appropriate, however, to mention some of those structuralist studies which at least consider invariance: Sneed (1971) (on pp. 149f.); Stegmüller (1973) (on p. 119); Sneed (1979); Balzer (1982); Balzer (1983); Gähde (1983); Kamlah (1983); Sneed (1984); Bartelborth (1988); Mühlhölzer (1988b); Mühlhölzer (1989) (in Chapter 3); Gähde (1990). And there is, of course, the pre-structuralist McKinsey & Suppes (1955) which, together with McKinsey & Sugar & Suppes (1953), was crucial to Sneed's approach.

[3] See Architectonic p. xxii.

1. General Analysis of the Notions 'Symmetry' and 'Invariance'

The eminent importance of the notion of symmetry within an architectonic for science can be demonstrated on countless examples. Let us consider a very simple and paradigmatic one, which has been used in a particularly illuminating way by Ludwig Wittgenstein.[4] It concerns the child's question why, if the earth is a sphere, people on the bottom don't fall off. A supposedly scientific answer to this question would be that the people on the bottom are kept attached to the earth by gravity. This answer, however, is misleading since it suggests that only they are in need of gravity, and not the people on the top as well. Wittgenstein, instead, would draw a circle with a stick figure atop, turn it upside down, and say 'Now *he* falls into space'. With this move, Wittgenstein all of a sudden uncovers the symmetry of the situation and leads the child to a change of its question. An appropriate question, then, might be, 'Why do all objects fall towards the earth's center?', and only at that point, when the child is prepared to look at the situation differently, gravity should be brought into play. Thus symmetry considerations help to clarify scientific problems. And in innumerable cases, of course, they immediately help to solve them. A classical example is Archimedes' derivation of the principles of the lever. It essentially relies on the assumption that bodies of equal weight positioned at equal distances from the pivot of a beam-balance are in equilibrium, which in turn is justified by nothing else but the symmetry of the situation.[5]

From examples like these, the general idea of symmetry can be easily drawn. The Archimedean case concerns bilateral symmetry which consists in the fact that a transformation that switches the left- and right-hand parts of the beam-balance leaves the situation unchanged or, to be more precise, unchanged in relevant respects. In the example of the globe, what we meet is not only the symmetry of up and down, but (at least) the whole rotational symmetry of a sphere. If we rotate the globe about its center, the situation after the rotation is, in all relevant respects, exactly like the original one. A transformation of this kind is called a *symmetry (transformation)*, or an *automorphism*. That is, a symmetry (= automorphism) is a transformation which leaves all relevant structure unchanged. In all relevant respects, its result is always exactly like the original.

With respect to a given situation (beam-balance; globe; etc.) we can gather all these transformations into a set G. Any transformations $\sigma_1, \sigma_2 \in G$[6] can be combined by first applying σ_1 followed by an application of σ_2.

[4] Here I draw upon the sympathetic description in Goldfarb (1992), p. 111, which in turn relies on the report given in Kreisel (1960), p. 239.

[5] See the first chapter of Mach (1933).

[6] 'σ' like 'symmetry', or 'substitution', or 'Stellvertreterfunktion' ('proxy function').

The result obviously is a symmetry transformation again, which is called the *product* of σ_1 and σ_2 and denoted by $\sigma_2\sigma_1$ (to be read from right to left). This product obviously is associative: $\sigma_3(\sigma_2\sigma_1) = (\sigma_3\sigma_2)\sigma_1$, $\forall\sigma_1, \sigma_2, \sigma_3 \in G$. Furthermore, the transformation which doesn't change anything – the identity ϵ[7] – belongs to G. And for any $\sigma \in G$ there is exactly one transformation which reverses σ. It is called the *inverse* of σ and denoted by σ^{-1}. Thus: $\sigma^{-1}\sigma = \sigma\sigma^{-1} = \epsilon$, $\forall\sigma \in G$. Consequently, G, together with the product mapping $(\sigma_2, \sigma_1) \mapsto \sigma_2\sigma_1$ just described, is a group.[8] It is called *symmetry group* or *automorphism group*.

That they form a group is certainly the basic fact about symmetries[9]. Such a group G acts on a certain set X and thereby 'measures' the symmetry of the situation represented by X. In the Archimedean case, X may be a set of beam-balances in equilibrium and G the two-element group consisting of the identity transformation and the transformation switching the left- and right-hand parts of the balances. In the case of the globe, X may be the set of points which the globe occupies in space and G the group of all rotations about the center (and, possibly, all reflections at planes through the center).

The abstract mathematical description is as follows. Let G be a group and X a set.

Definition. An *action* (also called *operation*) of G on X is a mapping $G \times X \to X, (\sigma, x) \mapsto \sigma x, \forall\sigma \in G, x \in X$, with the properties

(1) $(\tau\sigma)x = \tau(\sigma x)$, $\forall\sigma, \tau \in G, x \in X$;

(2) $\epsilon x = x, \forall x \in X$, where ϵ is the identity of G.

If G acts on X, then any $\sigma \in G$ defines a mapping $x \mapsto \sigma x, \forall x \in X$, which, because of (1) and (2), is a bijection of X and which will be denoted by σ as well. In other words, any $\sigma \in G$ will be identified with this bijection.

To any action of G on a set X corresponds a canonical action of G on the power set of $X, Po(X)$, defined by:
(*a*) $\sigma Y := \sigma[Y], \forall Y \in Po(X)$, where $\sigma[Y] := \{\sigma y \mid y \in Y\}$.
And to any actions of G on sets X and Y corresponds a canonical action of G on the cartesian product of X and Y, $X \times Y$, defined by:
(*b*) $\sigma(x, y) := (\sigma x, \sigma y)$, $\forall(x, y) \in X \times Y$.
Thus, if G acts on the base sets of an echelon (in the sense of Bourbaki[10]), then via iterative application of (a) and (b) there is a canonical action

[7] Like 'Eins(element)'.

[8] In what follows this product mapping will never be mentioned expressly and G alone will be called a group.

[9] See Weyl (1949), pp. 72f.

[10] See Bourbaki (1968), pp. 259f.

of G on the echelon as well, which I will call the *canonical extension* of the actions of G on the base sets. In what follows, canonical extensions of group actions will be used without further mention. Furthermore, in accordance with the spirit of the Bourbakian distinction between principal and auxiliary base sets,[11] the action of G on the auxiliary base sets will normally be presupposed to be the trivial one; i.e., if A is an auxiliary base set, then (normally), for any $\sigma \in G, \sigma a = a, \forall a \in A$.

An action of G on a set X leads to the following relation $\sim_G$ on X, which I will call the relation of *G-indistinguishability*: $x \sim_G y$ iff $\exists \sigma \in G : y = \sigma x$. One easily verifies that G-indistinguishability is an equivalence relation on X. An equivalence class of $\sim_G$ containing $x \in X$ –i.e., the class $Gx := \{\sigma x \mid \sigma \in G\}$– is called the *orbit* of x. The orbits, being equivalence classes, form a partition of X: Their union is the whole of X, and they are mutually disjoint (i.e., $Gx \cap Gy \neq \phi \Rightarrow Gx = Gy$).[12]

If G acts on X, then the notion of an invariant is defined as follows:

Definition. $x \in X$ is a *(G-)invariant* iff $\sigma x = x,\ \forall \sigma \in G$.

In the special case where $X = Po(Z)$ and the action of G on X is the canonical extension of an action of G on Z, it follows from what has been said earlier that $Y \subseteq Z$ is a G-invariant iff $\sigma[Y] = Y, \forall \sigma \in G$.[13]

What is the significance of this notion of an invariant? The most comprehensive answer to this question can be found in the work of Hermann Weyl. Invariance is intimately connected with objectivity, where 'objectivity' is understood in the sense of independence with respect to arbitrary or contingent factors (especially if these factors have not been made explicit).[14] Think of our set X as equipped with an objective structure

[11] See Bourbaki (1968), pp. 261f.

[12] So any action of G on X in a natural way induces an equivalence relation on X and the corresponding partition of X by equivalence classes. On p. 246 of his (1989), van Fraassen claims something much stronger: 'the three concepts: *equivalence relation*, *partition*, and *group of transformations*, amount really to the same concepts.' This, however, cannot be the case since different groups, acting on the same X, may lead to the same equivalence relation on X. Simple example: X = Euclidean plane; G = group of all translations of the plane; G' = group of all continuous transformations of the plane; obvious actions of G and G' on X ; in both cases the corresponding equivalence relation is the same (and a trivial one):$Gx = G'x = X, \forall x \in X$. Obviously, the concept of a group action on a set X contains much more information than the simple concept of an equivalence relation on X.

[13] There is also the natural, but weaker notion of *G-stability*: $Y \subseteq Z$ is *(G-)stable* iff $\sigma[Y] \subseteq Y, \forall \sigma \in G$; see Dieudonné & Carrell (1971), p. 1. I will not make use of this notion.

[14] See Weyl (1946), §§4f., and Weyl (1949), §§13-17. See my (1988a) for a detailed analysis of this notion of objectivity and for an attempt to clarify Weyl's expositions.

in this sense. X may be, e.g., Euclidean space. If we state an axiomatic theory for Euclidean space, like the Hilbertian one, say, we will certainly make use of structures which we deem to be objective. A paradigmatic example of a structure of this kind is the congruence relation, i.e. the 4-place relation 'the distance between x and y is the same as the distance between z and w'. We will not make use, however, of non-objective structures as, e.g., the 2-place relation 'x and y have the same distance from a', where a is an arbitrarily chosen point. If we presuppose that our axioms contain everything that can be said about Euclidean space, then the symmetry group G of Euclidean space is nothing else but the group of all bijections of X that leave invariant all the structures referred to in the axioms.[15]
With the help of G, we can state the following
Necessary Condition for Objectivity: If a structure S on X is objective, then it is a G-invariant.[16]
This condition can be justified as follows: Suppose S is not a G-invariant. Then there is a $\sigma \in G$ such that $\sigma S \neq S$. σS and S, however, are objectively indistinguishable, i.e., they are indistinguishable with respect to the objective background consisting of the structures referred to by our theory. Thus, this background is not sufficient for distinguishing between σS and S and thus not sufficient for determining S. So, on the presupposition that our theory says everything that can be objectively said about the structure which X is equipped with, S cannot be objective.

This justification is of a very general form and applies to every situation where G is the group of automorphisms of structures on X referred to by a theory which says everything that can be objectively said about a certain domain. In such a case, G may be called the ***objectivity group*** of the domain, and two G-indistinguishable structures S and S' on X, i.e. two structures S and S' such that there is a $\sigma \in G$ with $S' = \sigma S$, will be called ***objectively indistinguishable***. From now on I will generally presuppose that G in fact plays this role of an objectivity group and, consequently, that our Necessary Condition for Objectivity holds.

The justification of this condition contains the essence of what is called *relativity*. Since the structures σS and S cannot be distinguished on the basis of the objective background structure alone, their distinction must

[15] I.e., the group of automorphisms of this structure. In cases like this, where G is given as a group of bijections of X, it goes without saying that the action of G on X is simply the application of these bijections to the elements of X. This action will always be presupposed.

[16] For this and the following see the much more detailed exposition in my (1988a), especially in Sections 5-8. By a 'structure on X' I simply mean (in the spirit of Bourbaki) an element of an echelon on the base sets $E_1, \ldots, E_n$, where $E_1 = X$ and $E_2, \ldots, E_n$ are auxiliary base sets.

depend on some further, non-objective structural elements; in other words, σS and S can only be distinguished *relative* to these further elements. Think of our example of the globe. There is no objective distinction between an 'upper' and a 'lower' hemisphere since any two complementary hemispheres are related by a rotation transforming the supposed upper one into the supposed lower one and vice versa. Consequently, one needs some further structural elements – e.g. an arrow through an axis of the globe – in order to save the distinction.[17]

What about the following counterpart of the objectivity condition just stated?

Sufficient Condition for Objectivity: If a structure S on X is a G-invariant, then it is objective.

Unfortunately, there seem to be clear-cut cases where G-invariance is not sufficient for objectivity; e.g., if the automorphism group G acting on X is the trivial group (i.e., the group consisting of the identity only), as in all realistic spacetime models of general relativity theory, where the irregularities of the distribution of energy and momentum are mirrored in the irregularities of the spacetime structure itself. In such a case, every structure on X is a G-invariant, but we certainly do not wish to count every structure as objective. On the other hand, there also seem to be clear-cut cases where the Sufficient Condition for Objectivity holds. Such cases are: Euclidean geometry, Newtonian, Neo-Newtonian[18] and special-relativistic spacetime theory. It is a difficult problem to specify in a general way just under which circumstances G-invariance is sufficient for objectivity, and I cannot go into that here.[19] However, when in what follows Euclidean geometry, Newtonian, Neo-Newtonian or special-relativistic spacetime theory are under consideration, I will always presuppose that the Sufficient Condition for Objectivity may be accepted, and I will make rather frequent use of it.

The foregoing are the most important conceptual tools for dealing with symmetry and invariance, which are to be applied now within the structuralist framework.

2. Introduction of Symmetry Groups into Theory-Cores

Let us start with the paradigm of a structuralistically reconstructed theory: classical particle mechanics (*CPM*), as it is formulated in Section I.7 of

[17] It makes sense, however, to call the distinction, then, objective *relative* to these further structural elements; see again my (1988a).

[18] In Neo-Newtonian spacetime theory, Newton's absolute space is replaced by the whole system of inertial frames; see below.

[19] Attempts at a solution of this problem are to be found in Winnie (1986), Chapter V.

Architectonic.[20]

Definition. x is a model of classical particle mechanics ($x \in M(CPM)$) iff there exist $P, T, S, c_1, c_2, s, m, f$ such that

(1) $x = \langle P, T, S, c_1, c_2, s, m, f \rangle$;

(2) P, T, S are non-empty sets, P is finite;

(3) $c_1 : T \to \mathbf{R}, c_2 : S \to \mathbf{R}^3$ and c_1, c_2 are bijective;

(4) $s : P \times T \to S$ and $c_2 \circ s_p \circ c_1^{-1} =: r_p$ is smooth $\forall p \in P$;

(5) $m : P \to \mathbf{R}^+$;

(6) $f : P \times T \times \mathbf{N} \to \mathbf{R}^3$;

(7) $m(p)D^2 r_p(\alpha) = \sum_{i \in \mathbf{N}} f(p, c_1^{-1}(\alpha), i), \forall p \in P, \alpha \in \mathbf{R}$.

The occurrence of specific coordinate systems c_1 and c_2 in the models – a characteristic feature of many structuralist reconstructions – is commented on in Architectonic as follows: "c_1 and c_2 are bijective and therefore induce the structures of $\mathbf{R}$ and $\mathbf{R}^3$ on T and S. In this way, time and space are supplied with their usual, classical structures".[21] The idea obviously is to simply transport the structures of $\mathbf{R}$ and $\mathbf{R}^3$ via c_1^{-1} and c_2^{-1} to T and S. In this manner we can avoid the explicit formulation of a theory of time and space within CPM. In the words of Architectonic: "it does not seem adequate to make explicit in a reconstruction of classical mechanics all the structure and axioms for space and time which are used implicitly. In these circumstances the simplest way to get the desired structures is to put them on T and S 'from the outside', namely via c_1 and c_2".[22]

This method, however, has the disadvantage of inducing too much structure on T and S. The coordinate systems c_1 and c_2 induce the full structure of $\mathbf{R}^n$ on T and S, and this structure contains elements – a distinguished origin, distinguished vector bases and (perhaps) distinguished orientations of these bases[23] – which certainly are foreign to physical time and physical space (as seen from the point of view of Newtonian physics

[20] With some minor terminological and notational changes and without explicitly listing the auxiliary base sets $\mathbf{N}$ and $\mathbf{R}$ in the models.

[21] *Architectonic*, p. 30.

[22] *Architectonic* p. 31.

[23] The distinguished unit of length should be mentioned, too, but in what follows I will ignore issues concerning the choice of units of measurement. Their consideration would only lead to fruitless complications of exposition.

at least). They are, in the terminology developed above, non-objective structures and should be eliminated from our representation of reality.

This can be done quite easily by means of symmetry groups. Consider for example the set S, introduced to represent physical space, which, by means of the bijection $c_2 : S \rightarrow \mathbf{R}^3$, receives a certain structure, but too much structure. There is an amount of non-objective surplus structure. Let us suppose that we have got a theory of the objective structure of space S. It is, of course, the structure of Euclidean geometry. Then we have also got the group G of automorphisms of the structural elements referred to in this theory and the natural action of G on S. G contains all translations and all rotations of space S (and, possibly, all reflections as well). As our Conditions for Objectivity show, G then allows us to distinguish between the objective and non-objective structures on S, in particular in the case of the structures induced by c_2. The non-objective structures are just those which are not invariant under the action of G.

By using G, we avoid what Architectonic wanted to avoid, namely having to make explicit all the structure of and axioms for space and time which are used implicitly in *CPM*. At the same time we have got the means to eliminate the non-objective structures which had bothered us in the coordinate-dependent formulation of Architectonic. Of course, in most cases G is given to us by means of an axiomatic theory, namely, as the group of automorphisms of just those structures dealt with in the axioms of the theory. But we need not refer to these axioms in our formulation of *CPM*; it is enough to use the group G.

Of course, the structure c_2 itself[24] is not an objective structure on S; it is non-objective to the extent of not being invariant under any non-trivial transformation of G. From c_2, however, we can get something objective by simply adding further coordinate systems until we reach the class $K[c_2]$ defined as follows:

Definition. $K[c_2] := \{\sigma c_2 \mid \sigma \in G\}$.

Note that, if $c : S \rightarrow \mathbf{R}^3$, then, according to our definition of a canonical extension of a group action, $\sigma c = \{(\sigma x, r) \mid x \in S, r \in \mathbf{R}^3, (x, r) \in c\}$, $\forall \sigma \in G$. This means that $\sigma c = c \circ \sigma^{-1}$; i.e., the following diagram is commutative:

[24] $c_2 \in Po(S \times \mathbf{R}^3)$. In Bourbakian terminology, $Po(S \times \mathbf{R}^3)$ is an echelon on the (principal) base set S and the (auxiliary) base set $\mathbf{R}$.

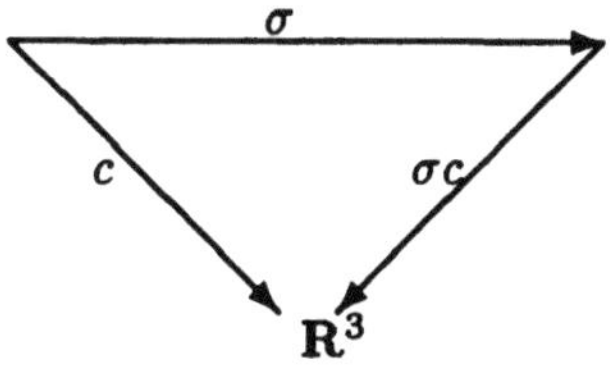

One immediately verifies that, since G is a group, $K[c_2]$ is the smallest class of bijections $c : S \to \mathbf{R}^3$ which contains c_2 and is closed under the action of G, and that it is a G-invariant. In other words: $K[c_2]$ is the class of just those coordinate systems $c : S \to \mathbf{R}^3$ which are objectively indistinguishable from c_2 (and thus objectively indistinguishable from one another), and as such $K[c_2]$ is objective.

It is an important fact that G not only leads naturally to $K[c_2]$ if c_2 is given, but that, conversely, from $K[c_2]$ we can recover G by the following natural constructions:

Definition. $\Gamma := \{c' \circ c^{-1} \mid c, c' \in K[c_2]\}$.

Γ is a set of bijections of $\mathbf{R}^3$ – to be interpreted as the set of 'coordinate transformations' between coordinate systems of $K[c_2]$, and one easily verifies that Γ is a group. Furthermore, for any $c \in K[c_2]$ we define:

Definition. $c_\star : G \to \Gamma, c_\star : \sigma \mapsto c \circ \sigma \circ c^{-1}$, $\forall \sigma \in G$.

A simple calculation shows that $c_\star$ is a group isomorphism. Consequently, $G = c^{-1} \circ \Gamma \circ c$ $(:= \{c^{-1} \circ \kappa \circ c \mid \kappa \in \Gamma\})$ and we have got back our group G.

Note that this construction is possible for any $c \in [c_2]$ and that it does not depend on any specific property of the coordinate system c_2. Any bijection $c_2 : S \to \mathbf{R}^3$ whatsoever will lead us to the group G, if we have available the class $K[c_2]$ of all coordinate systems which are objectively indistinguishable from c_2.

There are in fact important cases where $K[c_2]$ is given first, and G and the invariants of G are constructed only afterwards. Historically this was the case with the development of special relativity theory, where S now has to be taken to be the set of events (= space-time points) of the physical world. Einstein first was quite content with a class of objectively indistinguishable coordinate systems $c : S \to \mathbf{R}^4$ –namely, the class of inertial coordinate systems – and its corresponding class Γ of coordinate transformations, and only afterwards, essentially through Minkowski, Γ was identified with the objectivity group of the spacetime structure, and only then the most important invariant of this group, the Minkowski metric, became the focus of attention. Another important example of this kind will be discussed in Section 6.

All the foregoing considerations concerning the coordinatization of Newtonian physical space S also apply, of course, to Newtonian physical time T and to pre-general-relativistic spacetimes. So, in *CPM*, as it is formulated above, not only the automorphism group of physical space S, but also the automorphism group of physical time T has to be taken into account. The direct product of these two groups, then, is the automorphism group of the spacetime $S \times T$ of *CPM*. This, however, is not the correct automorphism group of pre-relativistic physics. The correct group is the (inhomogeneous)[25] Galilei group, and we have to reformulate *CPM* in order to let this group act on spacetime. The decisive step in the following reformulation is the replacement of $S \times T$ by a set M, the set of events (= space-time points), the objective structure of which will not be the structure of a product of space and time.[26]

Definition x is a model of *Neo-Newtonian particle mechanics* ($x \in M(NNPM)$) iff there exist M, c, P, s, m, f such that

(1) $x = \langle M, c, P, s, m, f \rangle$;

(2) M is a non-empty set;

(3) $c : M \to \mathbf{R}^4$ and c is bijective;

(4) P is a non-empty, finite subset of $Po(M)$;

(5) $s : P \times \mathbf{R} \to M$ and $s_p[\mathbf{R}] = p$ and $c \circ s_p =: r_p$ has the following properties:

  (a) it is smooth;

  (b) $\pi_4 \circ r_p =$ identity, where $\pi_4 =$ projection of $\mathbf{R}^4$ onto its fourth component (the 'time component');

(6) $m : P \to \mathbf{R}^+$;

(7) $f : P \times \mathbf{R} \times \mathbf{N} \to \mathbf{R}^4$, where $\pi_4 \circ f = O$;

(8) $m(p)D^2 r_p(\alpha) = \sum_{i \in \mathbf{N}} f(p, \alpha, i), \ \forall p \in P, \ \alpha \in \mathbf{R}$

[25] 'Inhomogeneous' means: including the (non-linear!) spacetime translations. The so-called homogeneous Galilei group consists of linear transformations only. In this paper 'Galilei group' will always refer to the inhomogeneous group since this is the objectivity group appropriate to Neo-Newtonian spacetime.

[26] See Ehlers (1973b) for more details about this transition from $S \times T$ to M.

In this definition, we have identified every $p \in P$ with its worldline, i.e., with a subset of M,[27] and the coordinate system $c : M \rightarrow \mathbf{R}^4$ is the usual 'Galilean' space-time coordinate system with the first three components as space-components and the fourth component as time-component. Everything else in this definition, then, is obvious.

As in the case of *CPM*, the coordinatization $c : M \rightarrow \mathbf{R}^4$ induces too much structure on M, and so we now use the Galilei group G and its action on M —which we may have got from an explicit theory about the geometric structure of M, or, alternatively, from a whole class of objectively indistinguishable coordinate systems on M— in order to eliminate the non-objective structural elements induced by c.

In cases like this, where the models contain non-objective components, we should enrich the framework of structuralism by including the respective objectivity groups into the theory-cores. More precisely, we should include into a theory-core a function which assigns to every model x certain group actions on certain components of x. To an $x = \langle P, T, S, c_1, c_2, s, m, f\rangle \in M(CPM)$, e.g., we would assign an action of the objectivity group of physical time on T and an action of the objectivity group of physical space on S; to an $x = \langle M, c, P, s, m, f\rangle \in M(NNPM)$, we would assign an action of the objectivity group of Neo-Newtonian spacetime, i.e. the Galilei group, on M. Furthermore, we could then envisage actions of these groups not only on space, time and spacetime, but also on the other structural components of the models, or even on the class of models as a whole.[28]

3. Structural Division of Labor and General Covariance

The method of objectivization described above is not fully satisfactory since there still appear non-objective structures in the models, even if one succeeds in eliminating them —or better: singling them out— afterwards. It would certainly be desirable to have a theory formulation that does without these structures from the outset. This desire has been expressed, but not answered, by McKinsey and Suppes in the very last sentence of their paper 'On the Notion of Invariance in Classical Mechanics'.[29] It can be answered, however, quite easily by means of notions from differential geometry. The key to the answer lies in the role of the coordinate systems used in our theories. Apart from containing non-objective structural elements, the coordinate system c occurring in a model x of *NNPM* has the further disadvantage of not distinguishing between structures —even among the

[27] Such that it would be possible in principle to formulate it without explicit mention of s and without using the auxiliary set $\mathbf{R}$ in (5) and (7).

[28] See, e.g., McKinsey & Suppes (1955).

[29] McKinsey & Suppes (1955), p. 302.

objective ones– that should be regarded as very different, for example between the differentiable structure, the metric structure of space and of time (and, possibly, the orientations of space and time). These different kinds of structures enter in our theory formulation as a mixture that is not easy to see through. The replacement of c by the objective class $K[c]$, determined by the objectivity group G, does not overcome this defect.

It can be overcome by replacing c not by $K[c]$ but by a whole differentiable atlas $\mathcal{A}$, and by charging our coordinate systems with the sole task of representing the differentiable structure. The other structures can be represented then, and also neatly distinguished, by coordinate-independent ('generally covariant') means. In the 'Autobiographical Notes' of the Schilpp volume dedicated to him, Albert Einstein admitted that this step caused him great trouble in the development of general relativity theory, since, as he said, "It is not so easy to free oneself from the idea that coordinates must have an immediate metrical meaning".[30] Taking this step belongs to what could be called a *structural division of labor*: Different structural components of a theory should be carefully distinguished.[31]

In the case of our exemplary theory *NNPM*, then, we should give, without making use of specific coordinate systems, an explicit account of the spacetime structure. This, of course, is a disadvantage from the point of view of Architectonic, p. 31, but the advantages outweigh the disadvantages.

Definition. x is a model of *Cartanian*[32] *classical spacetime theory* ($x \in M(CCST)$) iff there exist $M, \mathcal{A}, t_a, h^{ab}, D$ such that

(1) $x = \langle M, \mathcal{A}, t_a, h^{ab}, D\rangle$;

(2) M is a non-empty set;

(3) $\mathcal{A}$ is a differentiable atlas on M providing M with the structure of a 4-dimensional differentiable manifold;

(4) t_a is a nowhere vanishing 1-form on M;

(5) h^{ab} is a symmetric tensorfield on M of type (2,0) and signature (3,0,1);

---

[30] Einstein (1949), p. 67.

[31] See also Mühlhölzer (1994). What are the criteria of distinctness of structural components? Why, e.g., should we distinguish between differentiable structure and metric structure? The criteria, I think, are mainly of a systematic nature in the spirit of Bourbaki: A certain structure S has to be considered as autonomous and to be distinguished from a structure S', if S comes up in very diverse contexts and if the S-contexts do not coincide with the S'–contexts.

[32] The following formulation essentially goes back to Elie Cartan. See Havas (1964), p. 939.

(6) D is a torsion-free linear connection on M;

(7) $h^{ab}t_a = O$;

(8) $D_a t_b = O$;

(9) $D_a h^{bc} = O$;

(10) $R^d_{abc} = O$, where R^d_{abc} is the curvature tensor determined by D.

I will give no detailed explication of this formulation here. I have used the so-called abstract index notation, which is perspicuously described in Wald (1984), Chapters 2 and 3; apart from that my formulation is more or less identical with Friedman's in Chapter III of his (1983), whose explications may be consulted. For our present purposes it is sufficient to know that: M is to represent the set of physical events; $\mathcal{A}$ is a set of 4–dimensional coordinate systems defined on subsets of M such that every transformation between two coordinate systems of $\mathcal{A}$ is differentiable ($\mathcal{A}$ then makes it possible to speak of the differentiability of functions defined on M); t_a is to represent the metric of time; h_{ab} is to represent the metric of space; D is to represent the inertial structure of spacetime (D makes it possible to speak of the straightness of worldlines); conditions (7) to (9) express the mutual compatibility of t_a, h^{ab} and D; condition (10) says that the inertial structure of spacetime is flat (i.e., with respect to the straightness of worldlines, spacetime is like 4-dimensional Euclidean space).

None of the entities and quantities occurring in this formulation depend on any specific coordinate system; in particular, the definitions of t_a, h^{ab} and D are based on the class $\mathcal{A}$ as a whole and not on any specific $c \in \mathcal{A}$. This formulation may rightly be considered an objective one, and the automorphism group of an $x = \langle M, \mathcal{A}, t_a, h^{ab}, D \rangle \in M(CCST))$ then is the objectivity group of spacetime. Note that our characterization of such an x is only a local one and the manifold M which it refers to need not be understood as representing the whole universe. Thus, the automorphism group can be different for different models. However, under appropriate global conditions –e.g. under conditions which guarantee that $\mathcal{A}$ contains a 'Galilean' coordinate system $c : M \rightarrow \mathbf{R}^4$– this objectivity group will be the Galilei group.

The sought-for objective counterpart of *NNPM* can now be defined as follows:

Definition. x is a model of *Cartanian classical particle mechanics* ($x \in M(CCPM)$) iff there exist $M, \mathcal{A}, t_a, h^{ab}, D, P, s, m, f$ such that

(1) $x = \langle M, \mathcal{A}, t_a, h^{ab}, D, P, s, m, f \rangle$;

(2) $\langle M, \mathcal{A}, t_a, h^{ab}, D\rangle \in M(CCST)$;

(3) P is a non-empty, finite subset of $Po(M)$;

(4) $s : P \times \mathbf{R} \to M$ and $s_p[\mathbf{R}] = p$ and s_p has the following properties:

  (a) it is smooth;

  (b) $t_a \dot{s}_p(t)^a = 1, \forall p \in P, t \in \mathbf{R}$, where $\dot{s}_p(t)^a$ is the tangent vector of s_p at t;

(5) $m : P \to \mathbf{R}^+$;

(6) $f : P \times \mathbf{R} \times \mathbf{N} \to TM$, where $f(p,t,i) \in SM_{s(p,t)}$, $\forall p \in P, t \in \mathbf{R}, i \in \mathbf{N}$ ($TM := \bigcup_{e \in M} TM_e$, where $TM_e :=$ tangent space of M at the point e; and $SM_{s(p,t)} := \{v^a \in TM_{s(p,t)} \mid t_a v^a = O\}$, i.e., the set of 'spacelike' tangent vectors in $s(p,t)$);

(7) $m(p)\dot{s}_p(t)^a D_a \dot{s}_p(t)^b = \sum_{i \in \mathbf{N}} f^b(p,t,i)$, $\forall p \in P$, $t \in \mathbf{R}$.

As in *NNPM*, every particle $p \in P$ is identified with its worldline in M. The function s essentially provides a parametrization of $p \in P$ with absolute time. m and f are the mass and force functions respectively, and (7) states Newton's Second Law in its coordinate-free version. In principle, an explicit reference to s and to the corresponding time parametrization could be avoided by stating appropriate conditions for any $p \in P$, implying the existence of s. The same applies to the force function f, whereby an explicit use of the auxiliary set $\mathbf{R}$ could be avoided. On simplicity grounds I refrain from giving this more adequate –the really objective– formulation here. One should bear in mind, however, that such a formulation is possible. With it, McKinsey's and Suppes' demand has been answered.

I would like to stress again that in this formulation, which accords with our demand for a structural division of labor, the coordinate systems have got a single task only: to represent the differentiable structure and nothing else. Further tasks are not required; they can be taken over by other structural elements formulated in a coordinate-independent way. The equations formulated with the help of these structural elements, then, are coordinate-independent, too. They meet Einstein's notorious *principle of general covariance*: 'The general laws of nature are to be expressed by equations which hold good for all systems of co-ordinates, that is, are covariant with respect to any substitution whatever (generally covariant).'[33] This principle can be justified in very different ways,[34] but to my mind

[33] See Einstein (1916); English translation from Norton (1993), p. 12.

[34] See Norton (1993).

the most convincing justification is that of Hilbert, who writes: "Just as in the theory of curves and surfaces an assertion, for which a parameter representation of the curve or surface has been chosen, has no meaning with regard to the curve or surface itself, if it does not remain invariant under an arbitrary transformation of the parameters or can be brought into an invariant form, so also in physics an assertion which does not remain invariant under any arbitrary transformation of the coordinate system is marked as physically meaningless".[35] Instead of 'physically meaningless' I would prefer the term 'not objective', but this is only a terminological point. The actual content of Hilbert's argument is such that we seem to be more or less forced to formulate any theory in a generally covariant way and, furthermore, to believe that such formulations must always be possible.

Against this it has been argued that, after all, specific coordinate systems are important in the applications of our theories and that we should, therefore, offer coordinate-dependent formulations of the theories to be applied.[36] *These* coordinate systems, however, can very well be represented by coordinate-independent means (i.e., by means independent of the coordinate systems of $\mathcal{A}$), e.g. by a family of worldlines O (representing observers) and certain basis vector fields associated with O (representing the reference frames used by the observers).[37] Of course, the intended applications, too, as they are ordinarily understood within the structuralist setting, can be represented by coordinate independent means. In *CCPM* they are simply structures $\langle M, \mathcal{A}, t_a, h^{ab}, D, P, s, m\rangle \in M_{pp}(CCPM)$. Hilbert's argument suggests that everything should be represented —and should be able to be represented— in this manner.

4. Active and Passive Transformations

I do not want to forbid the use of coordinate systems in different functional roles. But then in the spirit of the structural division of labor these differences should be made explicit. E.g., one should distinguish between the 'mathematical' coordinate systems used in the atlas $\mathcal{A}$ and the 'physical' coordinate systems used in the applications of a theory.[38] There is one particularly elementary and fundamental role of coordinate systems which

[35] Hilbert (1964), p. 63 (my translation).

[36] See, e.g., Bartelborth (1988), p. 25: 'erscheint es (...(naheliegend, daß jedes Modell nur mit einer bestimmten Koordinatisierung versehen wird, für die die Grössen des Modells dann definiert werden'. See also the reconstructions in Bartelborth (1993).

[37] See Wald (1984), pp. 56-59.

[38] See Coleman & Korté (1994); in the first paragraph they write: "In fact a great deal of structure has to be introduced before physical coordinate systems can be introduced; for example, the spacetime geometry is introduced as a cross section of a bundle of

I shortly want to discuss here because it is the basis of the important distinction between 'active' and 'passive' transformations. It is the simple role of naming –and thereby identifying– the physical events of M. In pre-general-relativistic physics the objective geometrical structure of M does not suffice to distinguish the elements of M.[39] In a sense, it is precisely this homogeneity of M which the symmetry group of the geometric structure of M measures. In order to identify the elements of M, we can use a coordinate system $c : M \rightarrow \mathbf{R}^4$ which gives every element $e \in M$ the unique 'name' $c(e) \in \mathbf{R}^4$. To serve this purpose it is quite inessential that the range of c is the set $\mathbf{R}^4$; any set N of 'names', which has the same cardinality as M, will do. I.e., any bijection $c : M \rightarrow N$ will do, if N is a set of well-distinguished (individual) entities. Such a bijection need not have anything to do with the coordinate systems of $\mathcal{A}$ or with any other structure on M.

Given the objectivity group G acting on M, we can recapitulate, then, the considerations of Section 2. Any coordinate system (i.e. bijection) $c : M \rightarrow N$ determines the objective class of coordinate systems $K[c] = \{\sigma c \mid \sigma \in G\}$, which in turn determines the group $\Gamma = \{c''c'^{-1} \mid c', c'' \in K[c]\}$, which is isomorphic to G via

$$c_\star : G \rightarrow \Gamma,\ c_\star : \sigma \mapsto c \circ \sigma \circ c^{-1}, \forall \sigma \in G.$$

Thus we have the following commutative diagram:

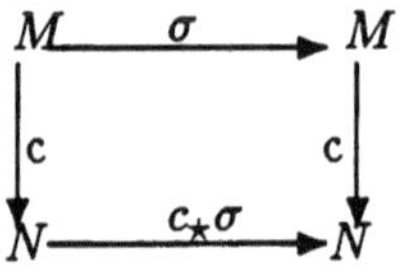

$c_\star\sigma$ is the way in which the transformation $\sigma \in G$ is given to us if we represent the set M by means of the coordinatization c. σ then appears as a so-called *active* transformation which describes real changes in the physical situation, as can be seen, e.g., with respect to the particles $p \in P$. Any particle $p \in P$, which is a subset of M, is described via c as the set $c[p] \subseteq N$, and σ transforms it into the set $\sigma[p]$, described via c as the set $(c \circ \sigma)[p]$. $c[p]$ and $(c \circ \sigma)[p]$ are different sets and, because c is a bijection, we may conclude that they describe different situations: $c[p]$

symmetric, second order tensors of appropriate signature and at least the basic facts of particle motion and of the propagation of electromagnetic fields need to be presented along with numerous other details. Physical radar coordinate systems are in fact rather complex products of modern technology. From a theoretical point of view, they are an advanced topic."

[39] In the general-relativistic case it can be otherwise; see Schmidt (1993).

describes p, $(c \circ \sigma)[p]$ describes $\sigma[p]$. Note that p and $\sigma[p]$ are objectively indistinguishable. They can be distinguished, however, relative to the fixed coordinate system c, and only by fixing such *a* c (or any other means of distinguishing p and $\sigma[p]$) are we allowed to interpret $c_\star\sigma$ as representing an active transformation.

$(c \circ \sigma)[p]$, however, not only is a description of $\sigma[p]$ by means of the coordinate system c, but at the same time also a description of p by means of the coordinate system $(c \circ \sigma)$, and one may say, therefore, that $c_\star\sigma$ does not represent a change of the physical situation but only a change of the coordinate description of one and the same physical situation. $c_\star\sigma$ itself then is called a *passive* transformation. Note that $c_\star\sigma$ per se is simply a bijection of N. It becomes a 'coordinate transformation' only by anchoring it to a coordinate system c. Then it leads from c to $c \circ \sigma$.

Now, the two cases just mentioned cannot be objectively distinguished. It is true that, relative to the fixed coordinate system c, σ is represented by $c_\star\sigma$ as an active transformation. But because of the objective indistinguishability of c and $c \circ \sigma$ we cannot tell whether c in fact remained fixed in the transition from, say, $\sigma[p]$ to $(c \circ \sigma)[p]$, or whether it was replaced by $c \circ \sigma$. This is the notorious *equivalence of active and passive transformations*. It is the equivalence of, on the one hand, the transformation $\sigma : M \to M$, represented as $c_\star\sigma$ by means of a fixed coordinate system c, and, on the other hand, the transformation $c_\star\sigma$ itself, now considered not as representing σ, but as a coordinate transformation leading from the coordinate system c to the coordinate system $c \circ \sigma$. This may be expressed more formally as follows: It is the coordinate system $c : M \to N$ which renders every $\sigma \in G$ into an active transformation and every $\kappa \in \Gamma$ into a passive transformation (= coordinate transformation). And it is the group isomorphism $c_\star : G \to \Gamma$ which constitutes the equivalence of active and passive transformation: $\sigma \in G$ is equivalent to $c_\star\sigma \in \Gamma$. Our foregoing diagram conveys just this equivalence, if one takes c as fixed and σ as variable. Note that the equivalence between active and passive transformations is objective in the sense that (as one easily verifies) the structure $c_\star$ itself[40] is a G-invariant.

[40] Obviously, if $\sigma \in G$ is considered as a bijection of M, $c_\star \subseteq Po((M \times M) \times (N \times N))$. The set N here functions as an auxiliary base set on which G acts trivially.

5. Spacetime Symmetries and Other Symmetries

So far we were mainly concerned with spacetime symmetries. In this Section, let us presuppose that all the theories T under consideration are formulated generally covariant and that the models $x \in M(T)$ are of the form $x = \langle M, S_1, \ldots, S_m, P_1, \ldots, P_n \rangle$, where $S_1, \ldots, S_m, P_1, \ldots, P_n$ are structures on a single set M and where the structures $S_1, \ldots, S_m$ represent the spacetime geometry and the structures $P_1, \ldots, P_n$ represent the physical structure. The spacetime symmetries, then, are defined as follows:

Definition. A *spacetime symmetry* of x is a bijection $\sigma : M \to M$ that leaves the spacetime geometry invariant.

Let $G_S(x)$ be the set of all spacetime symmetries of x. Then it immediately follows from the definition, that $G_S(x)$ is a group. In the foregoing Sections we were concerned with groups of this kind because we were interested in the objective geometric structure, and that is just the $G_S(x)$ -invariant structure, at least in the case of Newtonian, Neo-Newtonian and special-relativistic spacetime theory. What is important about the geometric structure of spacetime such that the distinction between the $S's$ and the $P's$ is a natural one? What is the basis of this distinction? This is a very difficult question which I cannot really go into here.[41] Some very rough suggestions as to where the answers may be sought must suffice. The most natural answer simply says that the geometric structure of spacetime is the arena in which the physical happenings take place. According to this view, physics is mainly concerned with the question which structures $P_1, \ldots, P_n$ are possible, given the fixed background structure $S_1, \ldots, S_m$. With the advent of general relativity theory the idea of a 'fixed' background structure has become obsolete, but to my mind the idea of a 'background structure', even if not a fixed one, remains important. The well-known attempts of John Wheeler and others to rescind the dichotomy between the arena and what is happening in the arena, by geometrizing the whole of physics, have been without success,[42] and it seems that we still have to live with the dichotomy for a while. Then, however, we should become clear about its real significance.

Part of this significance lies in the fact that the spacetime structure is, in a sense, the fundamental structure needed for the physical *interpretation* of the remaining structures $P_1, \ldots, P_n$.[43] In general, physical

[41] Kanitscheider (1987) is a useful survey concerning the answers that have been given to this question.

[42] See Kanitscheider (1987).

[43] See my (1988b), pp. 125-128, and (1989), Section 3.5. In pre-general-relativistic

quantities essentially depend on the metric of space and time, i.e., they are not determined –have no 'physical meaning', as it were– if the metric of space and time is not presupposed.[44] In its function as an interpretative background,[45] the metric represents the measurement of length and time, and without this measurement many important concepts of physics, e.g. the concept 'energy', have no content. This fact has sometimes been forgotten, as the debate about Mach's principle shows. In the context of general relativity theory, Mach's principle has often been formulated as follows: The spacetime metric g_{ab} should be completely determined by the energy-momentum tensorfield T_{ab}. One may tend to think, then, of T_{ab}, which represents energy and momentum, as given first and g_{ab}, which represents the metric, as determined afterwards ('first' and 'afterwards' in a logical sense, of course). This, however, hardly makes sense, since to talk about 'energy' already presupposes a metric background structure.[46]

If we accept the dichotomy between spacetime geometry S and physical structure P, then other symmetry concepts, apart from the concept of a spacetime symmetry, suggest themselves.

Definition *A physical symmetry* of x is a bijection $\sigma : M \to M$ that leaves the physical structure invariant.

The physical symmetries of x simply measure the symmetry of the physical world, as represented by x. Let $G_P(x)$ be the set of all physical symmetries of x. It immediately follows from the definition that $G_P(x)$ is a group. In all realistic cases, of course, there will be no (non-trivial) physical symmetries; i.e., $G_P(x) = \{\epsilon\}$ and the consideration of $G_P(x)$ is uninteresting. In order to assess a theory **T**, however, it may be important to see what **T** says about non-realistic, counterfactual cases, and then a consideration of $G_P(x)$, for $x \in \mathbf{M}(\mathbf{T})$, may be instructive.

Another important symmetry concept is the following:[47]

physics the geometric structure also plays the role of the background on which our *explanations* are based; see my (1989), Section 6.3.

[44]There are, however, important exceptions, for example in the theory of the electromagnetic field; see Meetz & Engel (1980).

[45]According to general relativity theory, the metric field at the same time has the character of a genuine physical field which is able to carry energy and momentum (as gravitational waves show). This Janus-faced appearance of the spacetime metric leads to strange complications which cannot be discussed here.

[46]See Albert Einstein, as quoted in Torretti (1983), p. 202.

[47]I have taken over this concept – with some modification, however – from Earman (1989), Chapter 3, Section 4. The following considerations are influenced by Earman's way of discussing symmetries.

Definition A *dynamical symmetry* of $x = \langle M, S_1, \ldots, S_m, P_1, \ldots, P_n\rangle$ with respect to T is a bijection $\sigma : M \to M$ such that for all structures $x' = \langle M, S_1, \ldots, S_m, P'_1, \ldots, P'_n\rangle$ (i.e. for all structures sharing with x the first $m+1$ components): $x' = \langle M, S_1, \ldots, S_m, P'_1, \ldots, P'_n\rangle \in M(T) \Rightarrow \langle M, S_1, \ldots, S_m, \sigma P'_1, \ldots, \sigma P'_n\rangle \in M(T)$.

A dynamical symmetry is a transformation of the physical structure before the background of a fixed geometric spacetime structure, which leads from structures satisfying the laws of **T** to structures which again satisfy the laws of **T**. Let $C_T(x)$ be the set of all dynamical symmetries of x with respect to **T**. Then, if $x \in \mathbf{M(T)}, C_T(x)$ measures the extent to which, given a fixed spacetime background, the physical structure of x may be transformed without violating the laws of **T**. From this definition it does not follow that $C_T(x)$ is a group, and therefore the label 'symmetry' may not seem appropriate. However, as we will see in a moment, there are good reasons to hold that $C_T(x)$ should be identical with $G_S(x)$, in which case $C_T(x)$ is a group and the label 'symmetry' is in order.

One can argue, in fact, for the following symmetry principles $S1$ to $S3$, which may be used, then, to assess the acceptability of our theories. (In what follows it is always presupposed that $x = \langle M, S_1, \ldots, S_m, P_1, \ldots, P_n\rangle \in M(T)$):

$$(S1) \qquad G_S(x) \subseteq C_T(x)$$

S1 is an immediate consequence of what Bourbaki calls 'transportability' and which I will presuppose throughout.[48] In the structuralist framework, transportability simply says that, if T is a theory, then for every $x \in M(T)$ all isomorphic images of x belong to $M(T)$ as well.[49] Therefore, if σ is a bijection on M and $x' = \langle M, \sigma S_1, \ldots, \sigma S_m P'_1, \ldots, P'_n\rangle \in M(T)$, then $\sigma x' = \langle M, \sigma S_1, \ldots, \sigma S_m, \sigma P'_1, \ldots, \sigma P'_n\rangle \in M(T)$. And if, in particular, $\sigma \in G_S(x)$, then $\sigma x' = \langle M, S_1, \ldots, S_m, \sigma P'_1, \ldots, \sigma P'_n\rangle$. Thus, $G_S(x) \subseteq$

[48] Bourbaki (1968), pp. 261f. Under the name of *canonical invariance*, transportability plays an essential role in Scheibe's analyses of invariance; see Scheibe (1981) and (1994).

[49] Auxiliary base sets remain unchanged under isomorphism; they are treated, so to speak, as logical constants; thus transportability may be considered to be simply a case of the more or less trivial model-indeterminacy occurring in usual model theoretic semantics. In Architectonic transportability has been dropped as an explicit metatheoretic demand on our theories; see Architectonic, pp. 11f., 37f. and 86f. It seems to me, however, that it should be uphold. Not only is it a condition very appropriate for a 'structuralist' account of theories, but it is also the actual basis of the distinction between principal and auxiliary base sets, which structuralism makes extensive use of. To my mind, even a purely philosophical justification of this demand might be given, relating to what Quine calls 'inscrutability of reference'; but I cannot go into that here.

$C_T(x)$.

Obviously, $S1$ does not have very much substance, and it should be accepted throughout. The next principle is more substantial:

$$(S2) \qquad C_T(x) \subseteq G_S(x).$$

$S2$ can be justified as follows. Suppose it is false. Then there exists a $\sigma \in C_T(x)$ and an $x' = \langle M, S_1, \ldots, S_m, P_1', \ldots, P_n' \rangle \in M(T)$ such that $\langle M, S_1, \ldots, S_m, \sigma P_1', \ldots, \sigma P_n' \rangle \in M(T)$ and $\langle S_1, \ldots, S_m \rangle \neq \langle \sigma S_1, \ldots, \sigma S_m \rangle$. Because of transportability we also have $\langle M, \sigma S_1, \ldots, \sigma S_m, \sigma P_1', \ldots, \sigma P_n' \rangle \in M(T)$. Thus, one and the same physical structure $-\langle M, \sigma P_1', \ldots, \sigma P_n' \rangle-$ is compatible with different spacetime geometries. This means that the laws which govern the physical structure, and which determine what is a model of T and what is not, do not determine the spacetime geometry. In other words: The spacetime structure does not supervene on the physical structure of the world. This may be considered an undesirable state of affairs. It contradicts our Machian intuitions which suggest that the ethereal spacetime structure should in no way have an existence independent of the physical structure.

This view about the structure of spacetime, underlying $S2$, is in a sense complementary to the view described above, when we were concerned with justifying the dichotomy between spacetime structure and physical structure. There, spacetime appeared as the necessary background for interpreting the quantities that make up the physical structure. Now, spacetime appears as heavily dependent on the physical structure. This is not a contradiction since the dependence of physics on spacetime appears as a 'semantical' dependence, whereas the dependence of spacetime on physics as an ontological one. What we have here, however, is an interdependence which is not very well understood and which is characteristic of the riddles connected with spacetime.

$S2$ is satisfied, e.g., by $CCPM$. Thus, in this case – which, of course, also satisfies $S1$ – $G_S(x) = C_T(x)$, and consequently $C_T(x)$ is a group. $S2$ is not satisfied in Newtonian mechanics with absolute space. In this theory absolute space has got an existence of its own, quite independent of the physical structure, and it is the violation of $S2$ which indicates this defect.

A considerably more uncomfortable and debatable demand than the foregoing ones is expressed in principle[50]

$$(S3) \qquad G_S(x) \subseteq G_P(x).$$

$S3$ says that there cannot be more indiscernibility (= symmetry) in the

[50] See also my (1988b) for a statement and discussion of this principle.

spacetime structure than there is indiscernibility (= symmetry) in the remaining physical structure. It postulates a dependence of spacetime on the physical structure which is much stronger than the dependence postulated in $S2$. Obviously, $S3$ is violated in Newtonian, Neo-Newtonian and special-relativistic theories where the spacetime structure remains completely untouched by the irregularities of the physical world.

Is $S3$ satisfied by general relativity theory? The models of general relativity theory are of the form $x = \langle M, \mathcal{A}, g_{ab}, T_{ab}, P_2, \ldots, P_n \rangle$, where $\langle M, \mathcal{A} \rangle$ = differentiable manifold, g_{ab} = metric tensor field, T_{ab} = energy-momentum tensor field (where, since T_{ab} belongs to the physical structure, we have to set $T_{ab} = P_1$), and $\langle P_2, \ldots, P_n \rangle$ = remaining physical structure. It immediately follows from Einstein's field equation that any $\sigma \in G_S(x)$, i.e. any symmetry of $\langle M, \mathcal{A}, g_{ab} \rangle$, is also a symmetry of $\langle M, T_{ab} \rangle$. But is it also a symmetry of the physical structure $\langle M, T_{ab}, P_2, \ldots, P_n \rangle$ as a whole? Since T_{ab} is in a sense 'constituted' by $\langle P_2, \ldots, P_n \rangle$, one might suspect that the answer must be yes. But this is wrong. There are cases where T_{ab} has more symmetries than the remaining $\langle P_2, \ldots, P_n \rangle$ which constitutes T_{ab}.[51] However, my suspicion is that all situations where symmetries of this sort turn up, are not realistic[52]. If this is the case, then, if restricted to the realistic situations, general relativity would satisfy S3.

Of course, the foregoing considerations concerning symmetries which are different from spacetime symmetries are very far from being complete. I have left out important kinds of symmetries – e.g., gauge symmetries – which are at the center of attention of contemporary physics and natural science in general. Their investigation and reconstruction within the structuralist setting is a task for the future.

6. Introduction of Symmetry Groups into the Cores of Generally Covariant Theories

In the case of the non-generally covariant theory *NNPM* it was necessary to explicitly mention the objectivity group G, i.e. the Galilei group, acting on the principal base sets M of the models. It was, therefore, necessary to explicitly integrate these group actions into (say) the theory-cores. In the case of *CCPM* – the generally covariant reformulation of *NNPM* – this is not necessary. To simplify matters, let us presuppose that the models $\langle M, \mathcal{A}, t_a, h^{ab}, D \rangle \in M(CCST)$, and the corresponding models of

[51] See Lukács & Perjés (1977) and Goenner (1984).

[52] E.g. the situation envisaged in note 4 of Norton (1989). If my suspicion is correct, one would have a good argument in defense of what John Earman and John Norton call *manifold plus further structure substantivalism*; see Earman & Norton (1987), Norton (1989) and Earman (1989), Chapter 9.

$M(CCPM)$, possess Galilean coordinate systems $c : M \rightarrow \mathbf{R}^4$. Then, the Galilei group G simply turns out to be the automorphism group of the spacetime structure of the models, i.e. the automorphism group of the structures $\langle M, \mathcal{A}, t_a, h^{ab}, D\rangle \in M(CCST)$ which is already given when the models are given. Thus it need not be mentioned in addition.

There are, however, cases even of generally covariant theories where the objectivity group cannot be identified with any automorphism group of the structures already occurring in the models. To see this, let us have a closer look at the models of $CCST$. Let us fix one model $x = \langle M, \mathcal{A}, t_a, h^{ab}, D\rangle$. Then a Galilean coordinate system $c : M \rightarrow \mathbf{R}^4$, the existence of which we presuppose, is nothing but a differentiable coordinate system with respect to which t_a is represented by $(0,0,0,1)$, h^{ab} is represented by a matrix with diagonal $(1,1,1,0)$ and vanishing components otherwise, and D is represented by components which vanish altogether. Let us fix a Galilean coordinate system $c : M \rightarrow \mathbf{R}^4$ and let us apply, then, a coordinate transformation $\kappa : \mathbf{R}^4 \rightarrow \mathbf{R}^4$ to c, which represents a (possibly) time dependent acceleration of the spatial part of c; in other words: $\kappa : (r^1, r^2, r^3, r^4) \mapsto (r'^1, r'^2, r'^3, r'^4)$ with

$$r'^{\alpha} = R^{\alpha}_{\beta} r^{\beta} + a^{\alpha}(r^4)$$

$$r'^4 = r^4 + constant,$$

where the Einstein summation convention on repeated indices is in effect, Greek indices are understood to run from 1 to 3, $(R^{\alpha}_{\beta})_{\alpha,\beta}$ is a constant orthogonal matrix, and $(a^1(r^4), a^2(r^4), a^3(r^4))$ is the linear acceleration at time r^4 just mentioned. Let Γ be the set of all coordinate transformations $\kappa : \mathbf{R}^4 \rightarrow \mathbf{R}^4$ of this kind. Then Γ obviously is a group. For any $\kappa \in \Gamma$ we get a coordinate system $\kappa \circ c : M \rightarrow \mathbf{R}^4$ which we call *non-rotating with respect to c*.

Let $K[c]$ be the set of coordinate systems $M \rightarrow \mathbf{R}^4$ which are non-rotating w.r.t. c; i.e., $K[c] := \Gamma \circ c := \{\kappa \circ c \mid \kappa \in \Gamma\}$. Then, from the point of view of Newton's theory of gravitation, all the coordinate systems of K[c] are indistinguishable from each other.[53] Furthermore, if we recognize that c represents the inertial structure of spacetime and if we accept Einstein's insight that inertia and gravity are essentially identical ('wesensgleich'[54]), then we must conclude that all the coordinate systems of $K[c]$ are objectively indistinguishable from each other.

---

[53] At least under the realistic global condition that our universe is not an 'island universe'. See Ehlers (1973a), Friedman (1983), Chapter III, and my (1988b), where the facts which I have just referred to are explained in more detail.

[54] See Einstein (1973), p. 60.

Consequently, as shown in Section 2 (and repeated in Section 4), the group G of bijections of M which is the image of the group Γ under the isomorphism $c_\star^{-1} : \Gamma \rightarrow G$ ($c_\star^{-1} : \kappa \mapsto c^{-1} \circ \kappa \circ c, \forall \kappa \in \Gamma$) has to be regarded as the real objectivity group of spacetime. Note that this group is[55] not an automorphism group of any of the structures occurring in $x = \langle M, \mathcal{A}, t_a, h^{ab}, D \rangle \in M(CCST)$. In fact, it contains the automorphism group of $\langle M, \mathcal{A}, t_a, h^{ab}, D \rangle$, i.e., the Galilei group, as a proper subgroup, and it is itself a proper subgroup of the automorphism group of $\langle M, \mathcal{A}, t_a, h^{ab} \rangle$, the so-called kinematical group. In particular, it has the structures $\mathcal{A}, t_a$ and h^{ab} as invariants, but not the structure D. D turns out to be non-objective.

Thus in *CCPM* we are in the same situation as in the case of the non-generally covariant theories *CPM* and *NNPM*. There is a non-objective structural element in our models which has to be singled out by means of actions of an objectivity group, which has to be introduced from the outside. The proper formal places of these actions are the cores of the theory elements of *CCPM*. Furthermore, in this case, too, there is the theoretically progressive step analogous to the step from *NNPM* to *CCPM* of replacing the non-objective structural element by an objective one; namely: by a non-flat connection on M, the geodesics of which represent free-falling test particles. This is the so-called *Newton-Cartan theory* of gravitation.[56]

An example like this shows that our initial maneuvre of introducing group actions into the cores of theory elements is of a very general nature, and that we can never be certain whether such an introduction may not prove to be necessary in order to single out non-objective structures and in order to direct our steps to a better theory. The corresponding general *pattern of progress*, then, is of the following form: theory without explicit actions of objectivity groups → introduction of actions of objectivity groups from the outside → better theory without explicit actions of objectivity groups (i.e., where mention of these actions is no longer necessary).

---

[55] Which is discussed in Rosen (1972) and Ehlers (1973b).

[56] See again Ehlers (1973a), Friedman (1983), Chapter III, and my (1988b).

7. *Prospects*

The foregoing considerations have been 'structuralist' only insofar as they presupposed a model-theoretic conception of scientific theories ('non-statement view') and as they started with the paradigm of structuralism, namely, classical particle mechanics as formulated in McKinsey & Sugar & Suppes (1953). The other most characteristic tenets of structuralism – constraints, links, theoreticity, intended applications and a sophisticated Ramsey sentence as the empirical claim – have been neglected. Of course, they all are intimately connected with the symmetries inherent in the respective theories and with the corresponding relation of objective indistinguishability and the corresponding invariants. In the literature there are a few reflections concerning these matters, e.g. with respect to theoreticity[57] and with respect to certain intertheoretical links,[58] but in general, symmetries are not really discussed. This is a situation that should be altered. To my mind, it is particularly important to take symmetries into account in the formulation of the Ramsey sentence of a theory, in order to get a clear understanding of their empirical significance. Furthermore, investigations are needed which clarify the role of symmetries with respect to a sufficiently substantive account of a scientific law[59] and in scientific explanations[60]. As already stressed above, there are very different kinds of symmetries which are important in contemporary science, and all of them should be considered in structuralist reconstructions.[61]

References

Agassi, J. & Cohen, R.S. (eds.), 1981, *Scientific Philosophy Today*. Dordrecht.

Balzer, W., 1982, *Empirische Theorien, Modelle, Strukturen, Beispiele*. Braunschweig.

Balzer, W., 'The origin and role of invariance in classical kinematics', In, Mayr, D. & Süßmann, G. (1983), 149-168.

[57] Gähde's criterion of theoreticity, as developed in Gähde (1983) and (1990), needs a sufficiently substantive notion of 'special laws', and Gähde's demand is that special laws should be invariant under certain symmetry transformations.

[58] See my (1988b).

[59] See Architectonic, p. 15, and the beginnings of such an account in Gähde (1983) and (1990).

[60] Without reference to the structuralist setting, I have, in my (1994), stated some conditions of adequacy for scientific explanations which essentially involve symmetries. These conditions should be translated, as it were, into the structuralist manner of speaking. (For a structuralist conception of explanation see Bartelborth's contribution to this volume.)

[61] I am grateful to Arthur Merin for correcting my English.

Balzer, W. & Moulines, C.U. & Sneed, J.D., 1987, *An Architectonic for Science*, Dordrecht.
Balzer, W. & Pearce, D. & Schmidt, H.J. (eds.), 1984, *Reduction in Science*. Dordrecht.
Bartelborth, T., 1988, *Eine logische Rekonstruktion der klassischen Elektrodynamik*. Frankfurt etc.
Bartelborth, T., 1993, 'Hierarchy versus holism, A structuralist view on general relativity', *Erkenntnis* 39, 383-412.
Bartels, A. & Schmidt, H.J., 1993, 'Two Talks on the Individuation of Events'. Preprint no. 16 of the research group on *Semantical Aspects of Spacetime Theories*, Center for Interdisciplinary Research, University of Bielefeld.
Bourbaki, N., 1968, *Elements of Mathematics, Theory of Sets*, Reading, Mass.
Coleman, R. & Korté, H., 1994, 'A semantic analysis of model and symmetry diffeomorphisms in modern spacetime theories', In, Majer, U. & Schmidt, H.J. (1994), 83-94.
Colodny, R. (ed.), 1986, *From Quarks to Quasars*, Pittsburgh.
Diederich, W., 1989, 'The development of structuralism', *Erkenntnis* 30, 363-386.
Dieudonné, J.A. & Carrell, J.B., 1971, *Invariant Theory, Old and New*, New York etc.
Earman, J., 1989, *World Enough and Space-Time*. Cambridge, Mass.
Earman, J. & Norton, J., 1987, 'What price spacetime substantivalism? The hole story', *British Journal for the Philosophy of Science* 38, 515-525.
Ehlers, J., 1973, 'Survey of general relativity theory', In, Israel, W. (1973), 1-125.
Ehlers, J., 1973, 'The nature and structure of spacetime', In, Mehra, J. (1973), 71-91.
Einstein, A., 1916, 'Die Grundlage der allgemeinen Relativitätstheorie', *Annalen der Physik* 49, 769-822.
Einstein, A., 1949, 'Autobiographical notes', In Schilpp, P.A. (1949), 2-94.
Einstein, A., 1973, *Grundzüge der Relativitätstheorie*, Braunschweig.
Fine, A. & Leplin, J. (eds.), 1989, *PSA 1988, Vol. 2*, East Lansing, Michigan.
Fraassen, B.C.v., 1989, *Laws and Symmetry*, Oxford.
French, P.A. & Uehling, T.E.,Jr. & Wettstein, H.K. (eds.), 1992, *Midwest Studies in Philosophy*, Vol. XVII, The Wittgenstein Legacy, Notre Dame.
Friedman, M., 1983, *Foundations of Space-Time Theories*, Princeton.
Gähde, U., 1983, *T-Theoretizität und Holismus*, Frankfurt, etc.

Gähde, U., 1990, 'On innertheoretical conditions for theoretical terms', *Erkenntnis* 32, 215-233.

Goenner, H., 1984, 'Killing vector fields and the Einstein-Maxwell field equations for null electromagnetic fields', *General Relativity and Gravitation* 16, 795-796.

Goldfarb, W., 1992, 'Wittgenstein on understanding', In, French, P.A. & Uehling, T.E.,Jr. & Wettstein, H.K. (1992), 109-122.

Havas, P., 1964, 'Fourdimensional formulations of Newtonian mechanics and their relations to the special and general theory of relativity', *Reviews of Modern Physics* 35, 938-965.

Hilbert, D., 1964, 'Die Grundlagen der Physik', In, Hilbert, D. (1964), 47-78.

Hilbert, D., 1964, *Hilbertiana*, Darmstadt.

Israel, W. (ed.), 1973, *Astrophysics and Cosmology*, Dordrecht.

Kamlah, A., 1983, 'The significance of physical invariance principles for the measurement of spacetime quantities', In Mayr, D. & Süßmann, G. (1983), 171-194.

Kanitscheider, B., 1987, 'Probleme und Grenzen einer geometrisierten Physik', In, Weingartner, P. & Schurz, G. (1987), 129-144.

Kreisel, G., 1960, 'Wittgenstein's theory and practice of philosophy', *British Journal for the Philosophy of Science* 11, pp. 238-252.

Lukács, B. & Perjés, Z., 1977, 'Time dependent Maxwell fields in stationary geometry', In, Ruffini, B. (1977), 281-288.

Mach, E., 1993, *Die Mechanik, historischkritisch dargestellt*, Leipzig.

Majer, U. & Schmidt, H.J. (eds.), 1994, *Semantical Aspects of Spacetime Theories*, Mannheim.

Manin, Yu.I., 1981, *Mathematics and Physics*, Boston etc.

Mayr, D. & Süssmann, G. (eds.), 1983, *Space, Time, and Mechanics*. Dordrecht.

McKinsey, J.C.C. & Sugar, A.C. & Suppes, P., 1953, 'Axiomatic foundations of classical particle mechanics', *Journal of Rational Mechanics and Analysis* II, 253-272.

McKinsey, J.C.C. & Suppes, P., 1955, 'On the notion of invariance in classical mechanics', *British Journal for the Philosophy of Science* 5, 290-302.

Meetz, K. & Engel, W.L., 1980, *Elektromagnetische Felder*, 1980.

Mehra, J. (ed.), 1973, *The Physicist's Conception of Nature*, Dordrecht.

Mühlhölzer, F., 1988a, 'On objectivity', *Erkenntnis* 28, 185-230.

Mühlhölzer, F., 1988b, 'Ununterscheidbarkeit und intertheoretische Relationen', *Conceptus* XXII, 107-130.

Mühlhölzer, F., 1989, *Objektivität und Erkenntnisfortschritt*, Habilitationsschrift, Universität München.
Mühlhölzer, F., 1994, 'Scientific explanation and equivalent descriptions', In, Salmon, W.C. & Wolters, G. (1994), 119-138.
Niiniluoto, I. & Tuomela, R. (eds.), 1979, *The Logic and Epistemology of Scientific Change*, Acta Philosophica Fennica, 30, Amsterdam.
Norton, J.D., 1989, 'The hole argument', In, Fine, A. & Leplin, J. (1989), 56-64.
Norton, J.D., 1993, 'General covariance and the foundations of general relativity, Eight decades of dispute', *Reports on Progress in Physics* 56, 791-858.
Rosen, G., 1972, 'Galilean invariance and the general covariance of nonrelativistic laws', *American Journal of Physics* 40, 693-687.
Ruffini, B. (ed.), 1977, *Proceedings of the First Marcel Grossmann Meeting on General Relativity*, Amsterdam.
Salmon, W.C. & Wolters, G. (eds.), 1994, *Logic, Language, and the Structure of Scientific Theories*, Pittsburgh and Konstanz.
Scheibe, E., 1981, 'Invariance and covariance', In, Agassi, J. & Cohen, R.S. (1981), 311-331.
Scheibe, E., 1994, 'Canonical invariance in physics', In, Majer, U. & Schmidt, H.J. (1994), 195-201.
Schilpp, P.A. (ed.), 1949, *Albert Einstein, Philosopher-Scientist*, LaSalle.
Schmidt, H.J., 1993, 'The principle of coreference and individuation of spacetime points', In, Bartels, A. & Schmidt, H.J. (1993).
Sneed, J.D., 1971, *The Logical Structure of Mathematical Physics*, Dordrecht.
Sneed, J.D., 1979, 'Theoretization and invariance principles', In, Niiniluoto, I. & Tuomela, R. (1979), 130-178.
Sneed, J.D., 1984, *Reduction, interpretation, and invariance*, In, Balzer, W. & Pearce, D. & Schmidt, H.J. (1984), 95-129.
Stegmüller, W., 1973, *Theorie und Erfahrung, Zweiter Halbband, Theorienstrukturen und Theoriendynamik*, Berlin etc.
Torretti, R., 1983, *Relativity and Geometry*, Oxford etc.
Wald, R.M., 1984, *General Relativity*, Chicago.
Weingartner, P. & Schurz, G. (eds.), 1987, *Logic, Philosophy of Science and Epistemology*, Wien.
Weyl, H., 1946, *The Classical Groups, Their Invariants and Representations*, Princeton.
Weyl, H., 1949, *Philosophy of Mathematics and Natural Science*, Princeton.
Winnie, J.A., 1986, 'Invariants and objectivity', In, Colodny, R. (1986).

Chapter 11: Bridges, Constraints, and Links[1]

C. Ulises Moulines & Marek Polanski

According to the structuralist metatheory, intertheoretical relations are not to be interpreted as relations between (sets of) statements but as relations between (sets of) models. In order to analyze this kind of relation model-theoretically, the concept of a link has been introduced into the structuralist notion of a theory. Typical global relations between theories - like reduction, equivalence, incommensurability, specialization, theoretization, approximation and others with no particular names - are assumed to be constituted by links as the smallest units of intertheoreticity. Links are so to speak the 'cells' of global intertheoretical relations.

Links have to belong to the concept of an empirical theory because it is fundamentally inadequate to view empirical theories as isolated entities. The identity of a given empirical theory may be grasped fully only if one takes its links to other theories into account. (This may be an important difference between pure mathematics and empirical science.)

The idea of introducing links into the general notion of an empirical theory was first put forward in Moulines (1984a). The idea was further developed in Balzer & Moulines & Sneed (1986) and applied to particular examples and epistemological issues in *Architectonic*, especially to deal with T-theoreticity (Ch. II) and to introduce the notion of the most complex unit of science - a theory-holon (Ch. VIII). More recently, Moulines (1992) presents a first attempt at a typology of links. The present essay is essentially a revision and further development of the last article mentioned.

This essay consists of two parts: a) a general discussion of links and their main types; b) an exploration of the formal relationships between the main types of links and also between them and the notions of interpretation and definability by applying metalogical means.

a) *Bridges*

Links are relations between models of different theories. They may be viewed as a particular case of a more general concept - that of a 'bridge' between different models, whether they belong to the same theory or not[2].

[1] We owe Thomas Mormann and José Pedro Ubeda some helpful remarks to a previous version of this essay.

[2] The idea of introducing a general notion of a model-theoretical bridge goes back to U. Gähde and Th. Mormann, independently. Both authors dwell upon these ideas in a more detailed way in their respective contributions to this volume.

Constraints intuitively are the other kind of model-theoretical bridges - those between the models of one and the same theory. Accordingly, we may define the notion of a bridge as follows:
Let $M_p^1, \ldots, M_p^n$ be (different or identical) classes of potential models.
Def.1: β is a bridge over $\langle M_p^1, \ldots, M_p^n \rangle$ iff

(1) $n > 1$

(2) $\emptyset \neq \beta \subseteq M_p^1 \times \ldots \times M_p^n$.

We distinguish between internal and external bridges according to whether they are defined on one and the same class of potential models or not.
Def.2: β is an internal bridge over M_p iff

(1) β is a bridge over $\langle M_p^1, \ldots, M_p^n \rangle$

(2) $\forall i (1 \leq i \leq n \rightarrow M_p^i = M_p)$

Def.3: β is an external bridge over $\langle M_p^1, \ldots, M_p^n \rangle$ iff

(1) β is a bridge over $\langle M_p^1, \ldots, M_p^n \rangle$

(2) $\exists i, j (1 \leq i, j \leq n \wedge M_p^i \neq M_p^j)$.

b) *Constraints*

We have claimed in a) above that, intuitively, constraints should be viewed as internal bridges between the models of one and the same theory. However, from a formal point of view, it is not immediately obvious that the classical notion of a constraint, as usually defined in the structuralistic literature, really corresponds to a sort of bridge. Remind the standard notion of a (transitive) constraint as laid out, for example, in *Architectonic*, p. 47: a transitive constraint C for M_p is a non-vacuous subclass of $Po(M_p)$ which contains all singletons but does not contain the empty set, and which furthermore satisfies the condition:

$$\forall X, Y : X \in C \wedge Y \subseteq X \rightarrow Y \in C.$$

Now, this notion of a constraint has certainly not an apparent relational character, as it should have if it has to come out of a bridge. However, one can define a specific kind of internal bridge (we will call it an n-constraint) which leads in a natural way to a structure (call it a relational constraint) having the properties of the standard notion of a transitive constraint. On the other hand, the new, relational concept of a constraint is somewhat more restrictive than the standard one, in the sense that some 'crazy' classes of sets of potential models, which would formally be constraints in the traditional sense, are ruled out by the new notion. But this is good so because all 'real-life' examples of constraints are of the new kind as well, while those excluded are intuitively unwanted. Let's set up all this formally. First, we still need some technical notation for permutations.

Let μ be a one-one-map of the set $\{1, \ldots, n\}$ onto itself. μ induces a permutation σ on $M_p \times \ldots^{n)} \times M_p$ in the following manner:

$$\forall x_1, \ldots, x_n \in M_p : \sigma(x_1, \ldots, x_n) =: \langle x_{\mu(1)}, \ldots, x_{\mu(n)} \rangle$$

Let's denote the group of all permutations by 'Σ'. We get the new notion of an n-constraint when we add two further conditions to internal bridges as settled in Def.2: general reflexivity and symmetry.

Def.4: γ is an n-constraint on M_p iff
(1) γ is an internal bridge over M_p
(2) $\forall x \in M_p : \langle x, \ldots, x \rangle \in \gamma$ (reflexivity)
(3) $\forall x_1, \ldots, x_n \in M_p \, \forall \sigma \in \Sigma :$
$\langle x_1, \ldots, x_n \rangle \in \gamma \rightarrow \sigma(x_1, \ldots, x_n) \in \gamma$ (symmetry).

On the basis of the present concept of an n-constraint we may introduce a general concept of a relational constraint having the required properties of the traditional notion of a transitive constraint.

Def.5: C is a relational constraint on M_p iff there is an n-constraint γ such that $C := \{X : \emptyset \neq X \subseteq M_p \wedge \forall x_1, \ldots, x_n (x_1 \in X \wedge \ldots \wedge x_n \in X \rightarrow \langle x_1, \ldots, x_n \rangle \in \gamma)\}$

Th.1: If C is a relational constraint (in the sense of Def.5) then C is a transitive constraint, i.e.:
(1) $C \subseteq \mathrm{Po}(M_p)$
(2) $C \neq \emptyset$
(3) $\emptyset \notin C$
(4) $\forall x \in M_p : \{x\} \in C$
(5) $\forall X, Y (X \in C \wedge \emptyset \neq Y \subseteq X \rightarrow Y \in C)$

Proof:

1. is immediate by construction of C.

2. By definition of a bridge,$\gamma \neq \emptyset$. So, take any $\langle x_1, \ldots, x_n \rangle \in \gamma$ and build $X_0 =: \{x_1, \ldots, x_n\} \in C$. So, $C \neq \emptyset$.

3. is immediate by construction of C.

4. Take any $x \in M_p$. Obviously, $\emptyset \neq \{x\}$ and $\{x\} \subseteq M_p$. Moreover, for any $z_1, \ldots, z_n$ with $z_i \in \{x\}$, we have $z_1 = \ldots = z_n = x$ But also, by the reflexivity of γ, we have $\langle z_1, \ldots, z_n \rangle \in \gamma$. Therefore $\{x\} \in C$.

5. Take any X, Y with $\emptyset \neq Y \subseteq X$, and take any $z_1, \ldots, z_n$ with $z_1 \in Y \wedge \ldots \wedge z_n \in Y$. Since $Y \subseteq X$, it follows that $z_1 \in X \wedge \ldots \wedge z_n \in X$; and, by $X \in C, \langle z_1, \ldots, z_n \rangle \in \gamma$.. Therefore, for any $z_1, \ldots, z_n$, if

$z_1 \in Y \wedge \ldots \wedge z_n \in Y$, then $\langle z_1, \ldots, z_n \rangle \in \gamma$. Therefore, $Y \in C$.
Q.E.D.

Thus, it follows from Theorem 1 that the newly defined notion of a constraint has all the essential properties of a constraint in the traditional structuralist sense. A constraint as a bridge always induces a constraint in the original sense. The converse, however, does not hold. We can construe (intuitively awkward) instances of (traditional) transitive constraints which are not relational constraints. Let's produce one such example.[3]
Take any infinite, denumerable subclass of $M_p : X = \{x_1, x_2, x_3, \ldots\}$, and define $C(X) =: \{Y : \emptyset \neq Y \subseteq X \wedge \|Y\| < \aleph_o\}$. It is immediately obvious that $C(X)$ is a transitive constraint in the traditional sense (however awkward it may be). Since $\|X\| = \aleph_o$, note that $X \notin C(X)$. Now, let us see that $C(X)$. is not a relational constraint. Suppose it were. Then, it should come out of an n-constraint γ for a finite, given n; that is,
$\gamma \subseteq X \times \ldots \times X$, n times.
Take any element A of $C(X)$:
$A = \{x_1, \ldots x_n\}$ with $x_i \in X$ for $1 \leq i \leq n$. By hypothesis, $< x_1, \ldots, x_n > \in \gamma$.
Since A has been chosen arbitrarily, it follows that any n-tuple of elements of X is contained in γ, i.e.: $\gamma = X \times \ldots \times X$.
By Def.5, if $C(X)$ actually were the relational constraint constructed out of γ, the last equality would imply that $X \in C(X)$.
But we already had noted that $X \notin C(X)$.
Q.E.D.

c) *Links*
We get links out of bridges whenever all M_p^i are different.

Def.6: λ is a link over$\langle M_p^1, \ldots, M_p^n \rangle$ iff
(1) λ is a bridge over$\langle M_p^1, \ldots, M_p^n \rangle$
(2) $\forall i, j (1 \leq i, j \leq n \wedge i \neq j \rightarrow M_p^i \neq M_p^j)$.

Links are trivial (or superfluous) whenever they don't rule out any actual models, i.e. whenever $M^1 \times \ldots \times M^n \subseteq \lambda$. Such links are uninteresting for a reconstruction of real-life intertheoretical relations. In the following definitions, we assume that all links are non-trivial in this sense.
Further, we'll take only dyadic links into account. Though it may be that some real-life links are n-adic for $n > 2$, the great majority of links for essential intertheoretical relations in empirical science are certainly dyadic.
Let's introduce some simplifying terminology.
Remember that a pair $E^i = \langle M_p^i, M^i \rangle$ is called a 'model-element'.

[3] The idea of this example is due to Th. Mormann.

If we have a link $\lambda \subseteq M_p^i \times M_p^j$, we also write: $E^i \lambda E^j$.
If we have $\langle x^i, x^j \rangle \in \lambda$, we also write: $x^i \lambda x^j$.

d) *Types of links*
In Moulines (1992) the claim was set up that there are two fundamental sorts of links: entailment and determining links. All other links come out of these two types either by putting some further restrictive conditions on them or by combining them. Entailment links are 'global' in the sense that they relate whole classes of structures of different theories; determining links are term-to-term connections. Reduction, equivalence and approximation are clearly made up of special kinds of entailment links; theoretization and other terms-connecting links with no particular label (like the link connecting pressure in hydrodynamics with energy and volume in thermodynamics – see *Architectonic*, p. 135) are determining links. Let's formalize these ideas.

Def.7: λ is an entailment link between E^1 and E^2 iff
$(1) E^1 \lambda E^2 \wedge M^1 \cap D_1(\lambda) \neq \emptyset \wedge M^2 \cap D_2(\lambda) \neq \emptyset$
$(2) \forall x^1 \in M_p^1 \forall x^2 \in M_p^2 (x^1 \lambda x^2 \wedge x^2 \in M^2 \rightarrow x^1 \in M^1)$.

In order to formalize the other fundamental type of links, i.e. determining links, we need some special notation. Suppose any $x^i \in M_p^i$ is an n-tuple.For any j with $1 \leq j \leq n$, we define
$t^{ij} =: \{t : \exists x^i \in M_p^i (t = \prod_j (x^i))\}$,
where $\prod_j$ is the so-called 'j-th projection function' on tuples.
Intuitively, t^{ij} is the so-called 'abstract term' of model-element E^i appearing on the j-th place of its models.
Determining links have to refer to particular abstract terms in a theory.

Def.8: λ is a determining link between E^1 and E^2 for t^{1j} iff:
$(1) E^1 \lambda E^2$
$(2) \forall x_1^1, x_2^1 \in M_p^1 \forall x^2 \in M_p^2 \forall t_1, t_2 \in t^{1j}$
$(x_1^1 \lambda x^2 \wedge x_2^1 \lambda x^2 \wedge t_1 = \prod_j (x_1^1) \wedge t_2 = \prod_j (x_1^2) \rightarrow t_1 = t_2)$.

(It would be more realistic to generalize the consequent of the last conditional into a convenient equivalence relation - e.g. coincidence up to a scale or invariance transformation - but we keep identity to simplify the exposition.)
In principle, the notions of an entailment and a determining link are conceptually independent. Neither one implies the other. However, in Moulines (1992) the question was raised whether some systematic relationship

between both kinds of links may be found in the sense that one might always accompany the other or one can be constructed out of the other in a canonical way. By analyzing the real-life example of the relationship between collision mechanics and Newtonian mechanics, it was there argued that the entailment link between these two theories presupposes (or implies) some specific determining links for velocity and mass. More concretely, we construct the entailment link from collision mechanics to Newtonian mechanics, first, by identifying the corresponding base sets, and secondly, by identifying the values of the mass functions in both theories and by setting the velocity values of collision mechanics equal to the derivative of position with respect to time in Newtonian mechanics. On the other hand, we set up these term-to-term identifications only for those systems which we expect to satisfy Newton's basic laws, from which, in turn, the basic law of collision mechanics (momentum conservation) may be derived. Intuitively speaking, this example suggests a kind of pragmatic 'second-order' equivalence between the global entailment link between the two theories and the corresponding determining links.

To make the generalization from this example more plausible, consider the two most paradigmatic cases of intertheoretical relations: reduction and theoretization. Any reduction is obviously 'made out' of an entailment link. But, to be taken seriously, any reduction must be, first of all, what has been called a case of 'ontological reduction' in the sense of Moulines (1984b); this means that the base sets of the reduced theory must be set-theoretically constructible out of some corresponding base sets of the reducing theory - and such construction procedures are nothing but determining links. Moreover, if the reduced theory has metrical functions among its primitive concepts, their values will be identified with the result of some operations on the values of some of the functions of the reducing theory. Again, these are nothing but determining links.

In some rather complex cases, the claim that an entailment link is always accompanied by determining links should be interpreted in a qualified manner. It doesn't necessarily mean that the accompanying link determines single magnitudes of one of the theories linked by entailment. They may determine some functional composition of several magnitudes. A case in point here may be this. In some traditional formulations of Lagrangian mechanics, several distinct generalized coordinates ω_1 might appear as individual magnitudes. There certainly is an entailment link from Lagrangian to Newtonian mechanics which assures the empirical equivalence of both theories (see *Architectonic*, Ch. VI.5.1). Now, in many models of Lagrangian mechanics, we can't state a determining link for each ω_i. However, we may formally construct a functional composition of the ω_i, call it $\bar{\omega}$, and state a determining link for $\bar{\omega}$ to the Newtonian position function (for the details, see *Architectonic, loc. cit.*). As a matter of fact, in the axiomatization of Lagrangian mechanics provided in *Architectonic*, it is this 'global' $\bar{\omega}$ and not the particular ω_i which appears as a primitive function in the theory's models. Thus, formally, in this more complex case, the close connection between the entailment and the corresponding determining link may be seen in quite the same way as in the simpler cases previously discussed in Moulines (1992). There is some reason to think that the case of Lagrangian mechanics is paradigmatic for other complex examples of a similar sort.[4]

Take now the case of theoretization as a global relationship between two theories. This is clearly a case where we need determining links: the T-non-theoretical terms are to be determined in the underlying theory T' of which T is a theoretization. Now, in principle, we could imagine that we determine the T-non-theoretical terms in an actual model of T by means of the T'-theoretical terms of potential models of T' which are never actual models. But such a determination would not be taken seriously as a case of theoretization. A 'reasonable' theoretization must be such that, whatever we do with the potential models, at any rate the non-theoretical terms in the actual models of T must be determined by terms appearing in the actual models of the underlying theory T'. (Consider the example of the determination of non-theoretical terms of Newtonian mechanics by means of Galilean kinematics.) This means that the determination of T-non-theoretical terms in actual models of T presupposes the appearance of their corresponding terms in the actual models of T', and therefore presupposes the existence of a particular entailment link.

Generally speaking, it is plausible to expect this sort of correspondence between entailment and determining links in any real-life intertheoretical relationship. Even if some counter-examples may be found, they probably would be regarded as rather 'pathological'. At any rate, the assumed generality of this correspondence suggests that it is worthwhile to explore the formal connections between the two kinds of links in a more general, formal setting. More concretely, the question may now be raised whether some general formal conditions may be found such that, when they are given, the correspondence, or even 'equivalence', between both kinds of links, becomes automatically assured, whereas under different general conditions, there would be (at least formally) a substantial difference between

[4] We owe the idea to discuss this case to a hint by Wolfgang Balzer.

an entailment and a determining link. Actually, such formal conditions can be found by employing some general results of model theory. Also, it can be shown that there are some interesting connections between our notions of links and the formal concepts of translatability and definability. This allows for a more complete perspective on intertheoretical relations. This will be shown in the following sections. It is convenient to introduce first some formal terminology.

e) *Formal background*

We will make use of the notion of a type (essentially as introduced in *Architectonic*, Ch.I). Types may be conceived formally as finite sequences of natural numbers, call them σ, σ', τ etc. (of the length s,s',t, etc.). The class symbolized by 'Str(σ)' is the class of all structures of type σ; it therefore contains elements of the sort
$x = \langle A, P_1, \ldots, P_s \rangle$ where $P_i \subseteq A^{\sigma(i)}$, for $i \leq s$.
As in §§c) and d) above, we restrict our attention to the concept of a model-element E as a simplified version of the structuralistic theory-concept. Formally, a model-element may be defined now as a pair $\langle Str(\sigma), M \rangle$, where σ is a type and $M \subseteq Str(\sigma)$.
Furthermore, we assume that M is a Δ -elementary class. This means that M satisfies the following condition. Let Sent(σ), viz. Fm(σ), be the set of all sentences, viz. the set of all formulae, of a first-order language, where the set of extra-logical constants V(σ) only contains s relation symbols $\bar{P}_1, \ldots, \bar{P}_s$ so that each $\bar{P}_i$ (for $i \leq s$) has $\sigma(i)$ argument places. Then, we assume that there is a $\Sigma \subseteq Sent(\sigma)$ such that $M = \mathrm{Mod}(\Sigma)$, where $\mathrm{Mod}(\Sigma) =: \{x \in Str(\sigma) : \forall \alpha \in \Sigma(x \models \sigma)\}$.

If Σ is a singleton, we say that M is just an elementary class and we abbreviate this by: $M \in EC$. If $\sigma' \subseteq Str(\sigma(M))$, then $Red_{\sigma'(M)}$ is the class of σ'-reducts of all elements of M.

With these formal notions in mind let's see now how some general concepts of model theory may be applied to the analysis of links. (We keep essentially the same notions of entailment and determining links as defined in Def.7 and Def.8 with the only inessential difference that we conceive the link λ itself now as a relation $\lambda \subseteq Str(\sigma) \times Str(\tau)$ for two different types σ, τ.

f) *Interpretability and definability*

Two central notions of formal model theory may be taken as the basis to explore the formal nature of determining and entailment links. They are the concept of interpretability of a theory in another theory and the concept of definability of a structure (a model) in another structure (another

model). Both interpretability and definability may be seen in this context as a sort of formal scheme for intertheoretical relations. They are closely related to each other as well.
Two equivalent notions of interpretability may be used in this context. Let's take the traditional concept of a theory as a deductively closed set of sentences or formulae. According to one notion of interpretability, we will say that theory T is interpretable in theory T' iff there is an appropriate common extension $T^\star$ of both T and T'. (The precise explication of this idea will be given below, in the proof of Th. 4.)
There is another, intuitively different, but formally equivalent, notion of interpretability, which is settled in the following definition.

Def.9: T is interpretable in T' iff there are two functions f and g with the following properties:

(1) $D_I(f) = At(L(T))$ and $D_I(g) = Fm(L(T))$, where $At(L(T))$ is the set of all atomic formulae and $Fm(L(T))$ the set of all formulae of the language used to formulate T.

(2) $D_{II}(f) \subseteq D_{II}(g) \subseteq (Fm(L(T')))$

(3) $\forall\delta \in At(L(T)), f(\delta)$ and δ have the same free variables.

(4) (i) $g(\delta) = f(\delta)$ if $\delta \in At(L(T))$
(ii) $g(\neg\delta) = \langle\neg g(\delta)\rangle$
(iii) $g(\langle\delta \wedge \alpha\rangle) = \langle g(\delta) \wedge g(\alpha)\rangle$
(iv) $g(\langle\forall v\delta\rangle) = \langle\forall v g(\delta)\rangle$ for any $\delta, \alpha \in Fm(L(T))$.

(5) $g[T] \subseteq T'$, i.e. $\forall\alpha \in Fm(L(T))(T \vdash \alpha \rightarrow T' \vdash g(\alpha))$

Let's now introduce the notion of definability of a structure in another structure according to the definition provided by Montague (1965).

Def.10: Take any $x \in Str(\sigma)$and $y \in Str(\tau)$ so that $x = \langle A, P_1, \ldots P_s\rangle$ and $y = \langle B, R_1, \ldots, R_t\rangle$.

(1) Let R be any relation on $A, R \subseteq A^n$. We say that R is definable in x iff there is a formula $\varphi(v_1, \ldots, v_n) \in Fm(\sigma)$ such that $v_i \neq v_j$ for any $i, j \leq n$ with $i \neq j$, and $v_1, \ldots v_n$, are all free variables occurring in φ and: for all $\alpha_1, \ldots, \alpha_n \in A$: $\langle\alpha_1, \ldots \alpha_n\rangle \in R$ iff $\models \varphi(v_1, \ldots, v_n)[\alpha_1, \ldots, \alpha_n]$

(2) y is definable in x iff $A = B$ and any R_i with $1 \leq i \leq t$ is definable in x.

In the article above-mentioned, Montague proved the following result:
Th. 2: If T is a finitely axiomatizable theory and T' is a theory, then T is interpretable in T' iff, in any model of T', a model of T is definable.

g) *Formal properties of determining and entailment links*
We now apply the former general results of model theory to our analysis of links. The first thing to notice is that the concept of definability introduced in Def. 10 is closely connected to the notion of a determining link. This is stated in the following theorem.
Th. 3: Let $\lambda \subseteq Str(\sigma) \times Str(\tau), \langle\varphi(v_1, \ldots, v_{\tau(j)})\rangle \in Fm(\sigma)$, $x = \langle A, P_1, \ldots, P_S\rangle \in Str(\sigma)$ and $y = \langle B, R_1, \ldots, R_t\rangle \in Str(\tau)$; further, suppose λ is such that $\langle x, y\rangle \in \lambda$ iff $A = B$ and for any $R_j \subseteq B_{\tau(j)} = A_{\tau(j)}, R_j$ is definable in x through $\varphi(v_1, \ldots, v_{\tau(i)})$.
Then: λ is a determining link for R_j.

Proof: The proof is immediate from Def. 10 and (the reformulated version of) Def. 8.
Next we show that, given the appropriate conditions, it is always possible to construct a relation between two theories which is an entailment link and a determining link at the same time.

Th. 4: Let $E = \langle Str(\sigma), M\rangle$ and $E' = \langle Str(\tau), M'\rangle$ be two model elements such that $V(\sigma) = \{\bar{P}_1, \ldots, \bar{P}_s\}, V(\sigma) = \{\bar{R}_1, \ldots, \bar{R}_t\}$, and $V(\sigma) \cap V(\tau) = \emptyset$. Let $M = Mod(\Sigma)$ and $M' = Mod(\Sigma')$.
If $Cn(\Sigma')$ is interpretable in $Cn(\Sigma)$ then there is a $\lambda \subseteq Str(\sigma) \times Str(\tau)$ so that λ is an entailment link and λ is a determining link.

Proof: Suppose $Cn(\Sigma')$ is interpretable in $Cn(\Sigma)$. Then there is a theory $T^\star$ and a set $\Delta \subseteq Sent(\sigma \cup \tau)$ such that:

(a) $Cn(\Sigma') \subseteq T^\star$ and $Cn(\Sigma) \subseteq T^\star$;

(b) all elements of Δ are theorems of $T^\star$ (i.e.$\Delta \subseteq T^\star$) and they are possible definitions of the extra-logical constants of the set $V(\tau) = \{\bar{R}_1, \ldots, \bar{R}_t\}$, that is, $\Delta = \{\langle\forall v_1, \ldots, v_{\tau(t)}(\bar{R}_1(v_1, \ldots, v_{\tau(t)}) \leftrightarrow \varphi(v_1, \ldots v_{\tau(t)}\rangle, \ldots, \langle\forall v_1, \ldots, v_{\tau(t)}(\bar{R}_t(v_1, \ldots, v_{\tau(t)}) \leftrightarrow \varphi(v_1, \ldots v_{\tau(t)}\rangle\}$ where $\langle\varphi(v_1, \ldots, v_{\tau(i)})\rangle \in Fm(\sigma)$, for $1 \leq i \leq t$;

(c) every extra-logical constant of $V(\tau)$ occurs in one and only one sentence of Δ;

(d) $T^\star = Cn(\Sigma \cup \Delta)$

Now, the models of T^* look as follows: $x^\star = \langle A_1, P_1, \ldots, P_s, R_1, \ldots, R_t \rangle$.

Since $R_1, \ldots, R_t$ are definable in $T^\star$, it follows that the relation Red_σ^{-1} is a function. Let $x^\star$ be the unique extension of x to a model of $T^\star$ (when $x \in M$), that is,

$$\star : M \longmapsto M^\star = Mod(\tau^\star).$$

On the other hand, it also holds that

$$Red_\tau(x^\star) = \langle A, R_1, \ldots, R_t \rangle \in M'.$$

Thus, we have the following diagram:

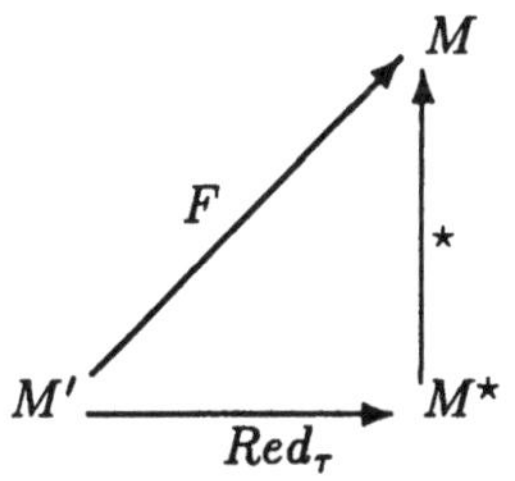

Therefore, we get that, for all $x \in M, F(x) = Red_\tau(x^\star)$.
Clearly, this means that F is an entailment link and a determining link at once.

Q.E.D.

h) *Entailment links and the relation of consequence*
From the structuralist literature on intertheoretical relations we know that the concept of an entailment link was intended as a kind of (model-theoretical) generalization of the concept of consequence. From the very beginning, this was the guiding intuition behind the concept of an entailment link. However, the formal basis for this intuition had not been explored until now. Let us look after the formal conditions under which an entailment link induces the usual consequence relation between sentences of two different theories.

Th. 5: Let $E = \langle Str(\sigma), M \rangle$ and $E' = \langle Str(\tau), M' \rangle$ be two model-elements, and let $\lambda \subseteq Str(\sigma) \times Str(\tau)$ be a link between them. Let's further assume that the following conditions are fulfilled:

(1) λ is an entailment link between E and E';

((2) there is a $\Sigma \subseteq Sent(\sigma)$ such that $D_I(\lambda) = Mod(\Sigma)$,

((3) there is a $\Sigma_1 \subseteq Sent(\sigma)$ such that $M = Mod(\Sigma_1)$;

((4) there is an $\alpha \in Sent(\tau)$ such that $M' = Mod(\alpha)$

((5) $\Sigma \cup \Sigma_1$ is consistent;

((6) for any $x \in Str(\sigma), y \in Str(\tau)$: $\langle x, y\rangle \in \lambda$ iff y is definable in x.

Then: all theorems of T' follow from the theorems of a specialization $T^\star$ of T.

Proof: Construct $T^\star = \langle Str(\sigma), M \cap D_I(\lambda)\rangle = \langle Str(\sigma), M \cap Mod(\Sigma)\rangle$. Clearly, $T^\star$ is a specialization of T. It follows from (1)-(6) that, in any model of $T^\star$, a model of $T' = \langle Str(\tau), M'\rangle$ is definable. It follows from this fact and from Th. 2 above that $Cn(\alpha)$ is interpretable in $Cn(\Sigma \cup \Sigma_1)$. There is also a function g such that $Cn(\Sigma \cup \Sigma) \vdash g(Cn(\alpha))$.

Q.E.D.

A further, related, result is this.

Th.6: Let $E = \langle Str(\sigma), M\rangle, E' = \langle Str(\tau), M'\rangle$ and $\lambda \subseteq Str(\sigma) \times Str(\tau)$ be such that:

(1) there are $\Sigma \subseteq Sent(\sigma))$ and $\Sigma' \subseteq Sent(\tau)$ such that $M = Mod(\Sigma)$; $M' = Mod(\Sigma')$;

((2) λ is an entailment link with the following properties:

((2a) $D_I(\lambda) = M$

((2b) $\forall x, y, y'(\langle x, y\rangle \in \lambda \wedge \langle x, y'\rangle \in \lambda \rightarrow y \equiv y')$[*where* $y \equiv y'$ means that y and y' are elementary equivalent, i.e., for any $\varphi \in Sent(\tau), y \models \varphi$ iff $y' \models \varphi$];

((2c) for any class $Y \subseteq Str(\tau)$ such that $Y \in EC$, there is a class $X \subseteq Str(\sigma)$ such that $X \in EC$ and $\lambda^{-1}[Y] = \{x \in Str(\sigma) : \exists y \in Y(\langle x, y\rangle \in \lambda\} = D_I(\lambda) \cap X$.

Then there is a function $f : Sent(\tau) \longmapsto Sent(\sigma)$ so that

((i) for any $x \in Str(\sigma), y \in Str(\tau)$, and $\varphi \in Sent(\tau)$, if $\langle x, y\rangle \in \lambda$ then $(x \models f(\varphi)$ iff $y \models \varphi))$;

((ii) $\Sigma \models f[\Sigma']$.

Proof: First, we construct f. Take any $\varphi \in Sent(\tau)$. We have $Mod(\varphi) \in EC$. Therefore, according to (2c) above, there is an $X \in EC$ such that $\lambda^{-1}[Mod(\varphi)] = D_I(\lambda) \cap X$.
Consider now the set $trans(\varphi) =: \{\alpha \in Sent(\sigma) : Mod(\alpha) = X\}$.
We pick out an element of $trans(\varphi)$ effectively. Call this element 'ψ'.
We set: $f(\varphi) = \psi$.
We first prove (i).
$\Rightarrow$: Assume $< x, y > \in \lambda$ and $x \models f(\varphi)$;
then, $x \models \psi$ because $f(\varphi) = \psi$, therefore, $x \in Mod(\psi)$;
but, since $< x, y > \in \lambda$, we have $x \in Mod(\psi) \cap D_I(\lambda)$;
since $x \in \lambda^{-1}[Mod(\varphi)]$, there is a $y' \in Str(\tau)$ such that $< x, y' > \in \lambda$ and $y' \models \varphi$;
but, according to (2b), it holds $y \equiv y'$;
therefore, $y \models \varphi$;
thus, we have proven that, if $x \models f(\varphi)$, then $y \models \varphi$, when $< x, y > \in \lambda$.
$\Leftarrow$: This can be proven in a completely analogous way.
Thus, we have proven (i).
Now, we proceed to the proof of (ii).
Take $\alpha' \in \Sigma'$. We have to show that $M \models f(\alpha')$.
It follows from (1) that $M' \subseteq Mod(\alpha')$; since λ is an entailment link, we have $\lambda[M] \subseteq M'$;
therefore, $\lambda[M] \subseteq Mod(\alpha')$,
which implies $M \subseteq \lambda^{-1}[Mod(\alpha')]$;
that is, $\forall x \in M(x \in Mod(f(\alpha'))$;
hence, $M \models f(\alpha')$,
since α' is any element of Σ', we have therefore shown that $M \models f[\Sigma']$;
but this just means that $\Sigma \vdash f[\Sigma']$.

Q.E.D.

Both Theorem 5 and Theorem 6 settle certain conditions under which an entailment link induces the traditional relation of consequence between the corresponding theories (considered as sets of sentences). The important difference between both results, however, consists in the fact that Th. 5 establishes the conditions under which the relation of consequence may ultimately be constructed via a determining link, whereas, in Th. 6, it is not presupposed that λ is a determining link too. Consequently, only in a situation where the conditions of Th. 5 are fulfilled, are we entitled to say that there is a sort of 'equivalence' between the notion of an entailment and of a determining link. Theorem 6 makes clear that, in principle, it is possible to have entailment links without determining links. Whether this is a possibility to be encountered in 'real-life' science, is a question that must be left open here.

References

Balzer, W. & Moulines, C.U. & Sneed, J.D., 1987, *An Architectonic for Science*, Dordrecht.
Balzer, W. & Pearce, D. & Schmidt, H.J. (eds.), 1984, *Reduction in Science*, Dordrecht.
Balzer, W. & Moulines, C.U. & Sneed, J.D., The Structure of Science: Local and Global, In: Barcan Marcus, R. & Dorn, G.J.W. & Weingartner, (1986), 291-306.
Barcan Marcus, R. & Dorn, G.J.W. & Weingartner, P. (eds.), 1986, *Logic, Methodology and Philosophy of Science (Proceedings of the 7th Congress)*, Amsterdam.
Echeverría, J. & Ibarra, A. & Mormann, Th. (eds.), 1992, *The Space of Mathematics*, Berlin/New York.
Montague, R., 1965, Interpretability in terms of models, *Indagationes mathematicae*, 21, 221-261.
Moulines, C.U., 1984, Links, Loops, and the Global Structure of Science, *Philosophia Naturalis*, 21(2-4), (cited as Moulines (1984a)), 254-265.
Moulines, C.U., Ontological Reduction in the Natural Sciences, In: Balzer, W. & Pearce, D. & Schmidt, H.J. (1984), (cited as Moulines (1984b)), 51-70.
Moulines, C.U., Towards a Typology of Intertheoretical Relations, In: Echeverría, J. & Ibarra, A. & Mormann, Th. (eds.) (1992), 403-411.

Chapter 12: A Rigorous Set Theoretical Foundation of the Structuralist Approach

Peter Hinst

1. Introduction

The aim of this essay is to provide a purely set theoretical formulation of the fundamental parts of the structuralist theory of empirical theories, and thus to show that the use of 'informal' set theoretic predicates advocated by Suppes, Stegmüller and others does not create new, set theoretic foundational problems.

The set theoretical language this will be done in, is a version of the von Neumann-Bernays-Gödel-type of language including urelements with modifications mainly due to A. Mostowski, J.L. Kelley and A.P. Morse. It will be called NBGU.

In NBGU quantification over proper classes is possible. But only sets and urelements can be elements of classes. Especially, only sets and urelements can figure as values of functions and therefore of sequences. Due to this fact, the definitions of most fundamental notions deviate more or less from the usual definitions e.g. in Architectonic (1987).

In NBGU quantification over formulas is not possible. As a consequence it was necessary to find substitutes for axioms-as-formulas to be able to define the notion of a structure species.

This essay is based on Architectonic (1987). The notions, their denotations and definitions, contained in this work, are overtaken wherever it was possible. This doesn't mean that the author agrees in all respects to the structuralist theory of empirical theories as laid down in Architectonic (1987) ; on the contrary, he would change a lot if he had to construct such a metatheory on his own responsibility.

The presentation of the material is as short as possible, rather a list of definitions and theorems. Proofs of the theorems are suppressed. For examples and discussions the reader is referred to Architectonic (1987) and the literature offered there.

In Architectonic (1987) definitions and theorems are formulated on different levels of abstraction. The lower level is the structure-species-level, the higher level is the model-level. E.g. definition DI-3(a), p. 10 (structure species) is on the lower level, whereas definition DII-12, p. 79 (theory core) is on the higher level. The definitions and theorems of the lower level only use variables for entities on the level of structure species, but no variables for entities on the level of models, global constraints, global links, i.e. for entities which are defined with the help of notions which belong to the lower level. The definitions and theorems on the higher level use variables for (partial) (potential) models, global constraints and links. In this essay all definitions and theorems are formulated on the lower level. As a consequence, the later a definition or theorem of this essay is located in the development of the theory, the more it deviates from the corresponding definition resp. theorem in Architectonic (1987).

2. The Set-Theoretical Language NBGU

The language we will use to represent the fundamental notions of the structuralist theory of empirical theories will be a version of the set-theoretical language.

We will use italics $a, \ldots, z, A, \ldots, Z, a', a''$, etc. as variables of NBGU. Variables will be used only in connection with variable binding operators as bound variables. As free variables in theorems, definitions and proofs we will use the letters a,...,z, A,...,Z, a', a", etc. called parameters of NBGU. The undefined descriptive constants of NBGU are the two-place predicate constant $\in$ for the relation of membership, the one-place predicate constant Urel for the property to be an urelement, and the variable binding operator $\{|\}$ for class abstraction. Variables, parameters and individual constants are the atomic terms of NBGU. The class of terms of NBGU and the class of formulas of NBGU are simultaneously inductively defined as usual, so that $(\alpha \in \beta)$, Urel(α) are atomic formulas of NBGU, if α, β, are terms of NBGU, and $\{\xi \mid \Gamma\}$ is a term of NBGU, if ξ is a variable of NBGU and Γ is a formula of NBGU. If B, Γ are formulas of NBGU, we use the formulas $\sim B$, (not B) for the negation of $B, (B \& \Gamma), (B$ and $\Gamma)$ for the conjunction of B, Γ, $(B$ or $\Gamma)$ for the adjunction of $B, \Gamma, (B \Longrightarrow \Gamma)$, (if B, then Γ) for the conditional of B, Γ, and $(B \Longleftrightarrow \Gamma), (B$ iff $\Gamma), (B$ if and only if $\Gamma)$ for the biconditional of B, Γ. If ξ is a variable of NBG and B is a formula of NBG, then we use the formulas $\forall \xi B$, for all $\xi : B$ for the universal quantification of B by ξ, and the formulas $\exists \xi B$, for at least one ξ : B, there exist ξ : B for the existential quantification of B by ξ.

The notions of a term/formula being a part of a term/formula, of a variable being free, bound in a term/formula, and the notion of substitution of a term/formula for a term/formula in a term/formula are to be defined as usual. We use Subst(α, β, γ) to denote the result of substituting α for β in γ.

A closed term and a closed formula is a term resp. a formula without a variable being free in it.

We will omit parentheses when they are not necessary to understand terms and formulas uniquely following the usual rules.

If B is a closed formula of NBGU, we will use **Theorem** .-B to state a theorem of NBGU, DEF.-B to express a definition of NBGU.

We presuppose that a notion of proof based on a system of rules of classical logic, e.g. the rules of natural deduction, is established for NBGU, and that the notion of a theorem of NBGU is defined so that a theorem of NBGU is a closed formula of NBGU for which there exists a proof of NBGU. But we will omit all proofs.

In NBGU we can distinguish between classes and urelements. The notion of a class (Cl) is defined by the formula $\forall x\ (\mathrm{Cl}(x) \iff \sim\mathrm{Urel}(x))$.

Classes can be divided in sets (Set) and proper classes (PrCl). These notions are resp. defined by the formulas $\forall x\ (\mathrm{Set}(x) \iff (\mathrm{Cl}(x)\ \&\exists y\ x \in y)$ and $\forall x\ (\mathrm{PrCl}(x) \iff \mathrm{Cl}(x)\ \&\ \sim \exists y x \in y))$.
Clearly, urelements, sets, and proper classes form a trichotomy.
Equality (=) is in NBGU defined by the formula $\forall x\ \forall y(x = y \iff (\forall z(z \in x \iff z \in y)\ \&\ \forall z(x \in z \iff y \in z)))$.

Soon we will list the axioms of NBGU, especially the axiom of extensionality. Then the following formulas of NBGU are theorems of NBGU:

$$\forall x\forall y(\mathrm{Cl}(x)\ \&\ \mathrm{Cl}(y) \Longrightarrow (x = y \iff \forall z(z \in x \iff z \in y)))$$
$$\forall x\forall y(\mathrm{Urel}(x)\ \&\ \mathrm{Urel}(y) \Longrightarrow (x = y \iff \forall z(x \in z \iff y \in z)))$$
$$\forall x\forall y(\mathrm{Urel}(x)\ \&\ \mathrm{Cl}(y) \Longrightarrow x \neq y).$$

We use $(\alpha \neq \beta)$ as an abbreviation of $\sim (\alpha = \beta)$ and $(\alpha \notin \beta)$ as an abbreviation of $\sim (\alpha \in \beta)$.

The class of natural numbers **N** is defined in NBGU by the formula of NBGU $\mathbf{N} = \{x \mid \forall M([\{z \mid z \neq z\} \in M\ \&\ \forall y(y \in M \Longrightarrow \{z \mid z \in y \text{ or } z = y\} \in M)] \Longrightarrow x \in M)\}$.

Now we are able to list the axioms of NBGU:

Axiom of Urelements
$\forall y(\mathrm{Urel}(y) \Longrightarrow \sim \exists x\ x \in y)$.

Axioms of Abstraction
If ξ is a variable of NBGU and Γ is a formula of NBGU and at most ξ is free in Γ, then the following formula is an axiom of NBGU:
$\forall \xi\ (\mathrm{Set}(\xi) \text{ or } \mathrm{Urel}(\xi) \Longrightarrow (\xi \in \{\xi \mid \Gamma\} \iff \Gamma))$.

Axioms of Classes
If ξ is a variable of NBGU and Γ is a formula of NBGU and at most ξ is free in Γ, then the following formula is an axiom of NBGU: $\mathrm{Cl}(\{\xi \mid \Gamma\})$.

Axiom of the Set of Urelements
Set $(\{x \mid \text{Urel}(x)\})$.

Axiom of Subsets
$\forall A \forall B(\text{ Set }(B) \ \& \ \text{Cl}(A) \ \& \ \forall x(x \in A \Longrightarrow x \in B) \Longrightarrow \text{Set}(A))$.

Axiom of Power Sets
$\forall B(\text{Set}(B) \Longrightarrow \text{Set}(\{A \mid \text{Cl}(A) \ \& \ \forall x(x \in A \Longrightarrow x \in B)\}))$.

Axiom of Union Sets
$\forall A(\text{Set}(A) \Longrightarrow \text{Set}(\{x \mid \exists y(y \in A \ \& \ x \in y)\}))$.

Axiom of Infinity
Set$(\mathbf{N})$.

Axioms of Replacement
If ξ, ζ, ζ', η are distinct variables of NBGU and Γ is a formula of NBGU and at most ξ, ζ are free in Γ and ζ' is no part of Γ, then the following formula is an axiom of NBGU: $\forall \eta(\forall \xi \forall \zeta \forall \zeta'(\Gamma \ \& Subst(\zeta', \zeta, \Gamma) \Longrightarrow$
$\Longrightarrow \zeta = \zeta') \ \& \ \text{Set}(\eta) \Longrightarrow \text{Set}(\{\zeta \mid \exists \xi(\xi \in \eta \ \& \ \Gamma)\}))$.

Axiom of Extensionality
$\forall x \forall y(\text{Cl}(x) \ \& \ \forall z(z \in x \Longleftrightarrow z \in y) \Longrightarrow x = y)$.

This class of axioms is not independent. Further axioms could be added, e.g. the Axiom of Choice.

We now give a short outline of the further development of NBGU, which precedes by extensions by definitions.

The subclass-relation is defined by the formula $\forall A \forall B(A \subseteq B \Longleftrightarrow \text{Cl}(A) \ \& \ \text{Cl}(B) \ \& \ \forall x(x \in A \Longrightarrow x \in B))$.
If α is a term of NBGU, Po(α) denotes the power class of α $\{x \mid x \subseteq \alpha\}$. 0 is an individual constant of NBGU and denotes the empty class $\{x \mid x \neq x\}$, which is a set. The individual constant $\mathcal{V}$ denotes the universal class $\{x \mid x = x\}$, which is a proper class. The unordered pair is defined by the formula $\forall x \forall y(\{x, y\} = \{z \mid \text{PrCl}(x) \text{ or } \text{PrCl}(y) \text{ or } z = x \text{ or } z = y\})$.
The singleton of an object is defined by the formula $\forall x(\{x\} = \{x, x\})$.
The union, intersection and difference of two objects is denoted resp. by $(\alpha \cup \beta), (\alpha \cap \beta), (\alpha \backslash \beta)$ and defined as usual. $\bigcap(\alpha)$ denotes the intersection of the class α. The ordered pair is defined by the formula $\forall x \forall y((x, y) = \{\{x\}, \{x, y\}\})$.

The first and the second coordinate of an ordered pair denoted by α are

denoted resp. by pr1(α),pr2(α). The Cartesian product is defined by the formula $\forall A \forall B(A \times B = \{p \mid p$ is an ordered pair & pr1$(p) \in A$ &pr2$(p) \in B\})$.
A relation is a class of ordered pairs. Domain, range and field of a relation are defined by the following formulas:
$\forall R(\text{Dom}(R) = \{x \mid \exists y(y, x) \in R\})$
$\forall R(\text{Ran}(R) = \{y \mid \exists x(y, x) \in R\})$
$\forall R(\text{Fd}(R) = \text{Dom}(R) \cup \text{Ran}(R))$.

The domain of a relation is the class of the second coordinates of the members of that relation, the range the class of the first coordinates of the members of that relation. The image to the left and the image to the right of an object under a relation are defined by the formulas
$\forall R \forall A(R\text{''}A = \{y \mid \exists x(x \in A \;\&\; (y, x) \in R)\})$
$\forall R \forall A(R\text{``}A = \{x \mid \exists y(y \in A \;\&\; (y, x) \in R)\})$.

The domain-restriction, range-restriction and total-restriction are defined as follows:
$\forall R \forall A(R\lceil A = \{p \mid p \in R \;\&\; \text{pr2}(p) \in A\})$
$\forall R \forall A(R\rceil A = \{p \mid p \in R \;\&\; \text{pr1}(p) \in A\})$
$\forall R \forall A(R[A = \{p \mid p \in R \;\&\; \text{pr1}(p) \in A \;\&\; \text{pr2}(p) \in A\})$.
The converse of an object is defined by the formula $\forall R(R^{-1} = \{p \mid (\text{pr2}(p), \text{pr1}(p)) \in R\})$.
The composition of two objects is defined by the formula $\forall R \forall S(R \circ S = \{p \mid \exists z(\text{pr1}(p), z) \in R \;\&\; (z, \text{pr2}(p)) \in S)\})$.
We use the individual constant Id to denote the equality relation: Id $= \{(x, y) \mid x = y\}$.
The property of being single valued to the left resp. single valued to the right resp. one-to-one are defined as follows:
$\forall R(R$ is single-valued to the left $\Longleftrightarrow \forall x \forall y \forall z((x, y) \in R \;\&\; (z, y) \in R \Longrightarrow x = z))$.
$\forall R(R$ is single-valued to the right $\Longleftrightarrow \forall x \forall y \forall z((x, y) \in R \;\&\; (x, z) \in R \Longrightarrow y = z))$.
$\forall R(R$ is one-to-one $\Longleftrightarrow R$ is single valued to left and to the right).

A function is defined as a relation which is single valued to the left. (Therefore the domain of a relation was defined as the class of all second coordinates of the members of that relation.) If α, β are terms of NBGU, the value of α for β is denoted by $\alpha'\beta, \alpha(\beta)$, and α_β. The definition is:

$\forall f \forall x \forall y((f'x) = y \Longleftrightarrow ((f$ is a function & $(y, x) \in f)$ or (f is not a function & $y = \mathcal{V})))$.

The definitions of a mapping, an injection, surjection, bijection from one object to another object are as usual. A function on an object is defined as a function whose domain is that object: $\forall f \forall A(f$ is a function on $A \Longleftrightarrow (f$ is a function & $\mathrm{Dom}(f) = A))$.

The class of all mappings from one object to another object is defined by the formula $\forall A \forall B(A^B = \{f \mid f$ is a function on B & $f'' B \subseteq A\}$.
If ξ, ζ are distinct variables, α a closed term, and β a term of NBGU, and at most ξ is free in β, then the function, whose domain is given by α and whose values are given by β, is defined by the formula $< \beta >_{\xi \in \alpha} = \{(\zeta, \xi) \mid \xi \in \alpha \ \& \ \zeta = \beta\}$. $< \beta >_{\xi \in \alpha}$ denotes the function with domain given by α and values given by β only, if the formula $\forall \xi(\xi \in \alpha \Longrightarrow \mathrm{Set}(\beta)$ or $\mathrm{Urel}(\beta))$ is a theorem of NBGU. Under the same conditions the term $\{\beta\}_{\xi \in \alpha}$ denotes the class of all objects given by β for ξ in α: $\{\beta\}_{\xi \in \alpha} = \{\zeta \mid \exists \xi(\xi \in \alpha \ \& \ \zeta = \beta)\}$. Under the same conditions, the formula $\mathrm{Ran}(< \beta >_{\xi \in \alpha}) = \{\beta\}_{\xi \in \alpha}$ is a theorem of NBGU. Under the same conditions for $\xi, \zeta, \alpha, \beta$ the term $\mathrm{CPr}_{\xi \in \alpha}(\beta)$ denotes the generalized Cartesian product of the objects given by β for ξ in α and is defined by the formula $\mathrm{CPr}_{\xi \in \alpha}(\beta) = \{\zeta \mid \zeta$ is a function on α & $\forall \xi(\xi \in \alpha \Longrightarrow \zeta' \xi \in \beta)\}$.

The above given definition of the set $\mathbf{N}$ of natural numbers can be reformulated: $\mathbf{N} = \{x \mid \forall M([0 \in M \ \& \ \forall y(y \in M \Longrightarrow y \cup \{y\} \in M)] \Longrightarrow x \in M)\}$.

The two-place predicate constant $<$ is used to express the fact that a natural number is smaller than another natural number. The above given definition of the set of natural numbers allows to prove in NBGU that every natural number is just the set of natural numbers smaller than that number, i.e. that the formula $\forall x(x \in \mathbf{N} \Longrightarrow x = \{z \mid z < x\})$ is a theorem of NBGU. The two-place predicate constants $\leq, >, \geq$ are used to express the facts that a natural number is smaller or equal, resp. greater, resp. greater or equal than another natural number.

The notion of a sequence is defined as follows: $\forall f(f$ is a sequence $\Longleftrightarrow$ (f is a function & $\mathrm{Dom}(f) \subseteq \mathbf{N}$ & $\forall x \forall y(x \in \mathrm{Dom}(f)$ & $y < x \Longrightarrow y \in \mathrm{Dom}(f))))$.

A sequence f in A is a sequence with $\mathrm{Ran}(f) \subseteq A$. An infinite sequence f is a sequence with $\mathbf{N} \subseteq \mathrm{Dom}(f)$. A finite sequence f is a sequence with $\sim (\mathbf{N} \subseteq \mathrm{Dom}(f))$. A sequence of length n is a function on n with $n \in \mathbf{N}$. If the finite length of a sequence and the values of that sequence are explicitly known, the following representation of such a sequence is often used:
$\forall x(< x >= \{(x, 0)\}), \forall x \forall y(< x, y >= \{(x, 0), (y, 1)\}), \forall x \forall y \forall z(< x, y, z >= \{(x, 0), (y, 1), (x, 2)\}), etc.$

A denumeration is a sequence which is single valued to the right. The definitions of the notions of a finite resp. an infinite denumeration resp. a denumeration of length n are obvious. A denumeration f of an object A is a denumeration with $\mathrm{Ran}(f) = A$. A strictly monotone sequence of natural numbers is a sequence f with $\mathrm{Ran}(f) \subseteq \mathbf{N}$ & $\forall i \forall j(i, j \in \mathrm{Dom}(f)$ & $i < j \Longrightarrow f_i < f_j)$.

A strictly monotone sequence f in A is a strictly monotone sequence f of natural numbers with $\mathrm{Ran}(f) \subseteq A$. A class A is finite iff there exists a finite denumeration of A. A class is infinite iff it is not finite. A class is denumerably infinite iff there exists an infinite denumeration of A. A n-place relation R is a relation with $n \in \mathbf{N}$ & $R \neq 0$ & $\forall x(x \in \mathrm{Dom}(R) \Longrightarrow$ x is a sequence of length $n)$. A n-place function$f$ is a n-place relation which is single-valued to the left. A n-place predicate P is a class with $n \in \mathbf{N}$ & $P \neq 0$ & $\forall x(x \in P \Longrightarrow x$ is a sequence of length $n)$.

If α, β denote in NBGU natural numbers, than we use $(\alpha + \beta)$, $(\alpha \cdot \beta)$, $(\alpha \div \beta)$ to denote the sum, the product, and the difference of the numbers denoted by α, β. The difference is defined for all pairs of natural numbers and the following formula is a theorem of NBGU: $\forall x \forall y(x, y \in \mathbf{N}$ & $x < y \Longrightarrow (x \div y) = 0)$. If α, β are terms of NBGU, then the concatenation $\alpha^\wedge\beta$ is defined by the following formula: $\forall u \forall v(u, v$ are finite sequences $\Longrightarrow (u^\wedge v)$ is a sequence of length $\mathrm{Dom}(u) + \mathrm{Dom}(v)$ & $\forall i(i \in \mathrm{Dom}(u) \Longrightarrow (u^\wedge v)_i = u_i)$ & $\forall i(i \in \mathrm{Dom}(v) \Longrightarrow (u^\wedge v)_{\mathrm{Dom}(u)+i} = v_i))$ & $\forall u \forall v(u$ or v is not a finite sequence $\Longrightarrow (u^\wedge v) = \mathcal{V})$.

The definitions of the notions of a reflexive, irreflexive, symmetric, asymmetric, antisymmetric, transitive, connected relation are as usual. The notions of a partial resp. total ordering are defined by the following formulas:

$\forall R(R$ is a partial ordering $\Longleftrightarrow$ $(R$ is a relation & R is antisymmetric & R is transitive$))$

$\forall R(R$ is a total ordering $\Longleftrightarrow$ $(R$ is a partial ordering & R is connected$))$.

The notions of an upper bound and a supremum are defined as follows:

$\forall u \forall R \forall A(u$ is an upper bound for A wrt.$R \Longleftrightarrow \forall v(v \in A$ & $v \neq u \Longrightarrow (v, u) \in R))$,

$\forall u \forall R \forall A$ (u is a supremum for A wrt. $R \Longleftrightarrow$ (u is an upper bound for A wrt. R & $\forall u'(u'$ is an upper bound for A wrt. R & $u' \neq u \Longrightarrow (u, u') \in R)))$, $\forall u \forall R \forall A(\mathrm{Sup}_R(A) = u \Longleftrightarrow$ ([R is antisymmetric & u is a supremum for A wrt. R] or [$\sim$ (R is antisymmetric & u is a supremum for A wrt. R) & $u = \mathcal{V}$])).

The notions of a lower bound and an infimum are defined correspondingly. An equivalence relation is a relation which is symmetric and transitive. The notion of an equivalence class is defined as follows:

$\forall u \forall R(A$ is an equivalence class wrt. $R \Longleftrightarrow (R$ is an equivalence relation & $\exists u\, u \in A$ & $\forall u(u \in A \Longrightarrow \forall v(v \in A \Longleftrightarrow (v, u) \in R)))$.

The following formula is a theorem of NBGU: $\forall A \forall R \forall x(A$ is an equivalence

class wrt. R & $x \in A \Longrightarrow A = R\text{"}\{x\}$).
Values of sequences have to be sets or urelements. Sometimes one would like to construct sequences the values of which are proper classes, what is impossible. In these cases one can use quasi sequences. This notion is defined by the following formula:
$\forall f$(f is a quasi sequence iff f is a relation & Dom(f) $\subseteq \mathbf{N}$).

An infinite quasi sequence f is a quasi sequence such that $\sim \exists u$ u is an upper bound for Dom(f) wrt. $<$. A finite quasi sequence is a quasi sequence such that $\exists u$ u is an upper bound for Dom(f)wrt. $<$. If f is a quasi sequence & $n \in$ Dom(f), then $f\text{"}\{n\}$ is the quasi term of f with index n. We denote it by $f\text{'}n$, i.e. we define $\forall f \forall n(f\text{'}n = f\text{"}\{n\})$. If A is a proper class and $n \in \mathbf{N}$, then $A = (A \times \{n\})\text{"}\{n\}$; therefore proper classes can be construed as quasi terms of quasi sequences. If f is a quasi sequence & (Set(n) or Urel(n)) & $n \notin$ Dom(f), then $f\text{'}n = 0$. If f is a quasi sequence & $n \in$ Dom(f), $then f\text{'}n \neq 0$.

3. Structure types and structures of a type

We now define the first foundational notions of the structuralist theory of empirical theories: the notions of a structure type and a structure of a type. To that end we need some auxiliary notions, namely the notions of an echelon scheme, an echelon and an echelon class.

We start with the notion of an echelon scheme. The following definition differs a little from the original definition of Bourbaki. We first give the definition and comment then on the deviation.

2.1 Definition . - For all L: L is an echelon scheme iff the following conditions are satisfied:

(0) L is a finite sequence & $0 \in$ Dom(L) &

(1) for all n: if $n \in$ Dom(L), then

- **a)** $L_n \in \mathbf{N}$, or
- **b)** $\exists j(j < n$ & $L_n = <0, j>$, or
- **c)** $\exists j \exists k(j, k < n$ & $L_n = <1, j, k>$), or
- **d)** $\exists j$(j is a finite sequence in n & $0 \in$ Dom(j) & $L_n = <2>^\wedge j$).

An echelon scheme is used to construct echelon classes. Condition (1) a) serves to indicate the basic classes of the construction; condition (1) b) is used to construct the power class of a class, condition (1) c) is used to construct the Cartesian product of two classes. These conditions are well known from Bourbaki's definition of the notion of an echelon scheme. The above mentioned deviation concerns condition (1) d), which serves to construct the generalized Cartesian product. Why this condition is established will be explained after the definition of the notion of an echelon class.

If L is an echelon scheme and $n \in \text{Dom}(L)$, then L_n is a natural number or a finite sequence of natural numbers. Both kinds of entities are sets, therefore echelon schemes can be defined by listing the values without running a risk of producing contradictions. Echelon schemes themselves are finite sequences and therefore sets.

If L is an echelon scheme, then L is a finite sequence. Then $\mathbf{N} \cap \text{Ran}(L)$ is a finite subclass of $\mathbf{N}$; then there exists a supremum in $\mathbf{N} \cap \text{Ran}(L)$ wrt. to the relation smaller then between natural numbers. This fact motivates the following definition:

2.2 Definition . - For all L, k : L is an echelon scheme for at most k classes iff L is an echelon scheme & $\text{Sup}_<(\mathbf{N} \cap \text{Ran}(L)) < k$.

Next we define the notion of the echelon wrt. an echelon scheme over a sequence of classes:

2.3 Definition . - For all L, D: (1) If L is an echelon scheme & D is a sequence & $\text{Sup}_<((\mathbf{N} \cap \text{Ran}(L)) \in \text{Dom}(D)$, then:

(i) $\text{Ech}(L, D)$ is a function on $\text{Dom}(L)$ &

(ii) for all n: if $n \in \text{Dom}(L)$, then

a) if $L_n \in \mathbf{N}$, then $\text{Ech}(L, D)_n = D_{L_n}$ &

b) $\forall j(j \leq n \& L_n =< 0, j > \Longrightarrow \text{Ech}(L, D)_n = \text{Po}(\text{Ech}(L, D)_j))\&$

c) $\forall j \forall k(j, k < n \ \& \ L_n =< 1, j, k > \Longrightarrow \text{Ech}(L, D)_n = (\text{Ech}(L, D)_j \times \text{Ech}(L, D)_k))$ &

d) $\forall j(j$ is a finite sequence in n & $0 \in \text{Dom}(j)$ & $L_n =< 2 >^\wedge j \Longrightarrow \text{Ech}(L, D)_n = \text{CPr}_{i \in \text{Dom}(j)}(\text{Ech}(L, D)_{j_i}))$.

(2) If L is not an echelon scheme or D is not a sequence or $\text{Sup}_<(\mathbf{N} \cap \text{Ran}(L)) \notin \text{Dom}(D)$, then $\text{Ech}(L, D) = \mathcal{V}$.

2.4 Definition . - For all L, D : the echelon wrt. L over $D = \text{Ech}(L, D)$.

If the presuppositions of 2.3 (1) are fulfilled for L, D, then each value of Ech (L, D) is a value of the sequence D, hence a set or an urelement, or is a power class of a set or an urelement, hence a set, or is the Cartesian product of two sets or urelements, hence a set, or is a generalized Cartesian product $\mathrm{CPr}_{i\in\alpha}(\beta)$, where α denotes a finite class of natural numbers and therefore a set, and β denotes for each $i \in \alpha$ a set or urelement, and therefore is a set. Consequently, Ech (L, D) is a well defined function and gives no rise to a contradiction.

If L is an echelon scheme, D is a sequence and $\mathrm{Sup}_{<}(\mathbf{N} \cap \mathrm{Ran}(L)) \in \mathrm{Dom}(L)$, then $\mathrm{Dom}(\mathrm{Ech}(L, D)) = \mathrm{Dom}(L)$ and $\mathrm{Dom}(L) \dot{-} 1$ is the supremum of $\mathrm{Dom}(L) wrt. <$. Then $\mathrm{Ech}(L, D)_{\mathrm{Dom}(L)\dot{-}1}$ is the last member of the sequence $\mathrm{Ech}(L, D)$. This class is called the echelon class wrt. L over D:

2.5 Definition . - For all L, D : $\mathrm{EchCl}(L, D) = \mathrm{Ech}(L, D)_{\mathrm{Dom}(L)\dot{-}1}$.

2.6 Definition . - For all L, D: the echelon class wrt. L over D = $\mathrm{EchCl}(L, D)$.

Echelon classes are used to typify the basic relations of (potential) models of a theory. If R is such a basic relation, the fact that R is a relation between elements of a class A and a class B could be expressed by $R \in \mathrm{Po}(A \times B)$ with $\mathrm{Po}(A \times B) = \mathrm{EchCl}(< 0, 1, < 1, 0, 1 >, < 0, 2 >>, < A, B >)$. That R is a n-place predicate on classes $A_0, \ldots, A_{n\dot{-}1}$ could be expressed by $R \in \mathrm{Po}(\mathrm{CPr}_{i\in n}(A_i))$ with $\mathrm{Po}(\mathrm{CPr}_{i\in n}(A_i)) =$
$\mathrm{EchCl}(< 0, 1, \ldots, n \dot{-} 1, < 2 >^\wedge < 0, \ldots, n \dot{-} 1 >, < 0, n >>, < A_i >_{i\in n})$. That R is a n-place relation (function) between elements of a class B and sequences of elements of classes $A_0, \ldots, A_{n\dot{-}1}$ could be expressed by $R \in \mathrm{Po}(B \times \mathrm{CPr}_{i\in n}(A_i))$ with $\mathrm{Po}(B \times \mathrm{CPr}_{i\in n}(A_i)) = \mathrm{EchCl}(0, 1, \ldots, n, < 2 >^\wedge < 1, \ldots, n >, < 1, 0, n + 1 >, < 0, n + 2 >>, < B, A_0, \ldots, A_{n\dot{-}1} >)$.

The motivation for handling n-place predicates and relations by means of condition (1) d) in the definition 2.1 of the notion of an echelon scheme is as follows: in most text books on set theory the ordered pair of a, b is written $< a, b >$ and defined by the formula

$$< a, b >= \{\{a\}, \{a, b\}\}.$$

The notion of an ordered n-tupel ($n \in \mathbf{N}$ is defined recursively by the pair of formulas
$< x >= x$
$< x_0, \ldots, x_n, x_{n+1} >=<< x_0, \ldots, x_n >, x_{n+1} >$ $(n \in \mathbf{N})$.

Here x has to be a sequence as defined above in Sec. 1. That means, the notion of a sequence (and the set of natural numbers) has already to be at hand, when the above definition of an n-tupel is made. But then the whole definition is superfluous, because the notion of a sequence of length n can do the job. Consistently, the notions of a n-place predicate and a n-place relation are to be defined as above in Sec. 1, and not as sets of ordered n-tupels of the form $< \ldots << x_0, x_1 >, x_2 >, \ldots, x_{n \div 1} >$. But then condition (1)d) in definition 2.1 is necessary to be able to construct echelon classes which could be used to typify n-place predicates and relations. We continue with the definition of the notion of a structure type:

2.7 Definition . - For all t : t is a structure type iff the following conditions are satisfied:

(0) t is a finite sequence & $2 \in \mathrm{Dom}(t)$ &

(1) $t_0 \in \mathbf{N} \,\&\, t_1 \in \mathbf{N} \,\&\, \sim (t_0 = 0 \,\&\, t_1 = 0)$ &

(2) $\forall i (i \in \mathrm{Dom}(t) \,\&\, i > 1 \Longrightarrow t_i$ is an echelon scheme for at most $t_0 + t_1$ classes).

If t is a structure type, then t_0 is called the number of principal base sets, t_1 is called the number of auxiliary base sets, the number $\mathrm{Dom}(t) \div 2$ is called the number of structure terms of t.

Condition (1) of 2.7 deviates a little from the orthodox condition concerning the number of base sets: now it is possible that the number of principal base sets is 0, which makes it possible to subsume theories with no principal base sets. This little liberalization goes back to a proposal by W. Balzer. The condition that $2 \in \mathrm{Dom}(t)$ will guarantee that a theory contains at least one structure term.

A structure type comprises complete information about the form of all set theoretical structures of that type: how many principal base sets, how many auxiliary base sets, how many structure terms such a structure contains and of what type each structure term is. We now define the notion of a structure of a type:

2.8 Definition . - For all x, t : x is a structure of type t iff the following conditions are satisfied:

(0) t is a structure type &

(1) x is a sequence of length $t_0 + t_1 + \mathrm{Dom}(t) \div 2$ &

(2) $\forall i (i < t_0 + t_1 \Longrightarrow \mathrm{Cl}(x_i))$ &

(3) $\forall i (i < \mathrm{Dom}(t) \div 2 \Longrightarrow x_{i+t_0+t_1} \in \mathrm{EchCl}(t_{i+2}, x))$.

If x is a structure of type t, then t is called the type of $x, x_0, \ldots, x_{t_{0+1}}$ are called the principal base sets of x, $x_{t_0}, \ldots, x_{t_0+t_1 \div 1}$ are called the auxiliary base sets of x, and $x_{t_0}, x_{t_1}, \ldots, x_{t_0+t_1+\mathrm{Dom}(t) \div 2}$ are called the structure terms of x.

Condition (2) of definition 2.8 excludes urelements from the base sets. Condition (3) stipulates the types of the structure terms; mostly structure terms are relations or predicates, i.e. the typifying echelon set is a power class of a Cartesian product or a power class of a generalized Cartesian product.

Since structures of a type t are defined as sequences, only sets can function as principal and auxiliary base sets. In all known examples of empirical theories this condition is satisfied. If the situation should change one day, structures could be construed as quasi sequences.
We introduce a term for the class of all structures of a given type:

2.9 Definition . - For all t : $\mathbf{Str}(t) = \{x \mid x$ is a structure of type $t\}$.

The notion of a set theoretical predicate can also be defined now:

2.10 Definition . - For all P, t : P is a set theoretical predicate of type t iff $P \subseteq \mathbf{Str}(t)$.

4. Structure species and structures of a species

In this section we will reconstruct in a rigorously set theoretical manner the notions of a structure species and a structure of a species. For that we need the notion of what Bourbaki calls a transportable predicate. To define that notion we need some auxiliary notions which are to be defined first.

3.1 Definition . - For all f : $\mathrm{pot}(f) = \{(x, y) \mid x \in \mathrm{Po}(\mathrm{Dom}(f))$ & $y = f"x\}$.

3.2 Definition . - For all f, g : $(f \otimes g) = \{(y, x) \mid x \in (\mathrm{Dom}(f) \times \mathrm{Dom}(g))$ & $y = (f'\mathrm{pr1}(x), g'\mathrm{pr2}(x))\}$.

3.3 Definition scheme for NBGU. For all $\xi, \zeta, \eta, \alpha, \beta$ if ξ, ζ, η are distinct variables of NBGU and α is a closed term of NBGU and β is a term of NBGU, which contains at most the variable ξ free, then the following formula of NBGU shall be an axiom of NBGU: $\mathrm{cpr}_{\xi \in \alpha} = \{(\eta, \zeta) \mid \zeta \in \mathrm{CPr}_{\xi \in \alpha}(\mathrm{Dom}(\beta)$ & $\eta = < \beta(\zeta_\xi) >_{\xi \in \alpha}\}$.

If the term β denotes for each $\xi \in \alpha$ a function, then $\text{cpr}_{\xi\in\alpha}(\beta)$ denotes a function with domain $\text{CPr}_{\xi\in\alpha}(\text{Dom}(\beta))$. If $\zeta \in \text{CPr}_{\xi\in\alpha}(\text{Dom}(\beta))$, then for $\xi \in \alpha, \zeta_\xi \in \text{Dom}(\beta)$ and $\beta(\zeta_\xi)$ is the value of β for ζ_ξ, and $< \beta(\zeta_\xi) >_{\xi\in\alpha}$ is the 'sequence' of these values, i.e. $< \beta(\zeta_\xi) >_{\xi\in\alpha} \in \text{CPr}_{\xi\in\alpha}(\text{Ran}(\beta))$ So, if for each $\xi \in \alpha, \beta$ is a mapping from $\text{Dom}(\beta)$ to $\text{Ran}(\beta)$, then $\text{CPr}_{\xi\in\alpha}(\beta)$ is a mapping from $\text{CPr}_{\xi\in\alpha}(\text{Dom}(\beta))$ to $\text{CPr}_{\xi\in\alpha}(\text{Ran}(\beta))$.
Obviously, the class of all the axioms given in 3.3 is a definition of cpr.

The constants pot, $\otimes$, cpr will now be used to define the notion of *the construction of the canonical extension of a sequence of functions wrt. an echelon scheme*:

3.4 Definition . - For all L, f :

(1) If L is an echelon scheme & f is a sequence of length $1 + \text{Sup}_<(\mathbf{N} \cap \text{Ran}(L))$ & $\forall i(i \in \text{Dom}(f) \Rightarrow f_i$ is a function), then:

- **(i)** $\{f/L\}$ is a function on $\text{Dom}(L)$ &
- **(ii)** for all n: if $n \in \text{Dom}(L)$, then
  - **a)** if $L_n \in \mathbf{N}$, then $\{f/L\}_n = f_{L_i}$ &
  - **b)** $\forall j(j < n \;\&\; L_n =< 0, j > \Rightarrow \{f/L\}_n = \text{Po}(\{f/L\}_j)$ &
  - **c)** $\forall j \forall k(j, k < n \;\&\; L_n =< 1, j, k > \Rightarrow \{f/L\}_n = (\{f/L\}_j \otimes \{f/L\}_k))$ &
  - **d)** $\forall j(j$ is a finite sequence in n & $0 \in \text{Dom}(j)$ & $L_n =< 2 >^\wedge j \Rightarrow \{f/L\}_n = \text{cpr}_{i\in\text{Dom}(j)}(\{f/L\}_{j_i}))$.

(2) If L is not an echelon scheme or f is not a sequence of length $1 + \text{Sup}_<(\mathbf{N} \cap \text{Ran}(L))$ or $\sim \forall i(i \in \text{Dom}(f) \Rightarrow f_i$ is a function), then $\{f/L\} = \mathcal{V}$.

3.5 Definition . - For all L, f: *the construction of the canonical extension of f wrt. L* $= \{f/L\}$.

If the presuppositions of 3.4 are fulfilled for L, f, then we have one of the following four cases:
Case a): The value of $\{f/L\}$ is a value of the sequence f of functions and therefore a set.
Case b): The value of $\{f/L\}$ is pot(h) for some function h being a set. But $\text{pot}(h) \subseteq \text{Po}(\text{Ran}(h)) \times \text{Po}(\text{Dom}(h))$ and the latter is a set, since h is a set. Hence pot(h) is a set.
Case c): The value of $\{f/L\}$ is $g \otimes h$ for some functions g, h being sets. But $g \otimes h \subseteq (\text{Ran}(g) \times \text{Ran}(h)) \times (\text{Dom}(g) \times \text{Dom}(h))$ and the latter is a set, since g, h are sets. Hence $g \otimes h$ is a set.

Case d): The value of $\{f/L\}$ is $\mathrm{cpr}_{i\in \mathrm{Dom}(j)}(h_{j_i})$ with Dom(j) being a finite class of natural numbers and therefore a set and h_{j_i} being a function which is a set for each $i \in$ Dom(j).
But $\mathrm{cpr}_{i\in \mathrm{Dom}(j)}(h_{j_i}) \subseteq \mathrm{CPr}_{i\in \mathrm{Dom}(j)}(\mathrm{Ran}(h_{j_i}) \times \mathrm{CPr}_{i\in \mathrm{Dom}(j)}(\mathrm{Ran}(h_{j_i}))$ and the latter is a set, since Dom(j) is a set and for each $i \in$ Dom(j), h_{j_i} is a set. Hence $\mathrm{CPr}_{i\in \mathrm{Dom}(j)}(h_{j_i})$ is a set.

Since the values of a construction of the canonical extension of a sequence of functions wrt. an echelon scheme are sets, such a construction can be defined by listing their values without running a risk of producing contradictions. Furthermore, such constructions are finite sequences and therefore sets.

If L is an echelon scheme and f is a sequence of length $1 + \mathrm{Sup}_<(\mathbf{N} \cap \mathrm{Ran}(L))$, then $\mathrm{Dom}(\{f/L\}) = \mathrm{Dom}(L)$ and $\mathrm{Dom}(L) \div 1$ is the supremum of Dom(L) wrt. $<$. Then $\{f/L\}_{\mathrm{Dom}(L)\div 1}$ is the last member of the sequence $\{f/L\}$. This class is called *the canonical extension of f wrt. L*.

3.6 Definition . - For all $L, f : [f/L] = \{f/L\}_{\mathrm{Dom}(L)\div 1}$

3.7 Definition . - For all L, f: the canonical extension of f wrt. $L = [f/L]$.

We note a theorem about canonical extensions:

3.8 Theorem .- For all L, f, A, B : if L is an echelon scheme and f, A, B are sequences of length $1 + \mathrm{Sup}_<(\mathbf{N} \cap \mathrm{Ran}(L))$ and $\forall i(i \in \mathrm{Dom}(f) \Rightarrow$ f_i is a mapping / injection / surjection / bijection from A_i to B_i, then $[f/L]$ is a mapping / injection / surjection / bijection from EchCl(L, A) to EchCl(L, B).
Transportable predicates or *structure predicates* are set theoretical predicates which are closed under isomorphism. We define an appropriate notion of isomorphism:

3.9 Definition . - For all f, x, y, t : f is an isomorphism from x to y of type t iff the following conditions are satisfied:

(0) x, y are structures of type t & $\forall i(t_0 \leq i < t_0 + t_1 \Rightarrow x_i = y_i)$ &

(1) f is a sequence of length t_0 & $\forall i(i < t_0 \Rightarrow f_i$ is a bijection from x_i to $y_i)$ &

(2) $\forall j(j \in \mathrm{Dom}(t) \div 2 \Rightarrow [(f^\wedge < \mathrm{Id}\lceil x_{t_0+i} >_{i\in \mathrm{Dom}(t_1)})/t_{j+2}](x_{t_0+t_1+j})$.

Condition (0) claims that the auxiliary base sets of isomorphic structures are identical. Condition (1) claims that the isomorphism is a sequence of bijections from the principal base sets of the first structure to the principal base sets of the second structure. And condition (3) claims that the structure terms of the second structure are pictures of the corresponding structure terms of the first structure under the canonical extension of the sequence of the bijections between the principal base sets and the identity functions between the auxiliary base sets wrt. the appropriate echelon schemes.
Structures are isomorphic if there exists an isomorphism from the one to the other:

3.10 Definition . - For all x, y, t: structures x, y of type t are isomorphic iff $\exists f$ f is an isomorphism from x to y of type t.

Now we are able to define the notion of a transportable predicate:

3.11 Definition . - For all P, t : P is a transportable predicate of type t iff P is a set theoretical predicate of type t & $\forall x \forall y(x \in P$ & the structures x, y of type t are isomorphic $\Rightarrow y \in P)$.

Our next aim is to define the notion of a *structure species*. This definition requires some preliminary remarks, because a rigorous set theoretical definition of that notion has to deviate from the usual definitions, which refer to formulas.
Many different theories (in the structuralist view) have the same structure type. E.g. all theories dealing with a relation on a class without using an auxiliary base set have the structure type $< 1, 0, < 0, < 1, 0, 0 >, < 0, 1 >>>$. A structure of that type is a sequence $< A, R >$which answers the condition $R \in \text{Po}(A \times A)$. Examples of theories of this type are the theory of equivalence relations on a set, the theory of partial orderings on a set and the theory of progressions. Different theories of the same structure type differ from each other by the axioms. The notion of a structure species shall express these differences made by the axioms of the theories. Therefore a structure species should be a structure type with additional information about what axioms the theory contains. But axioms in the usual sense are formulas and in NBGU there is no possibility to speak generally about formulas, since NBGU has no formula-variables and therefore no formula quantifiers. Hence formulas have to be substituted by set theoretical entities which could take the place of axioms of a theory.

As substitutes for formulas we choose predicates. If Γ is a formula of

NBGU with the free variables $\xi_0, \ldots, \xi_{n \div 1}$, we take the predicate $\{< \xi_0, \ldots, \xi_{n \div 1} > | \Gamma\}$ as a substitute for Γ.

Such axioms-as-predicates in general cannot be proved to be sets. As a consequence axioms-as-predicates cannot be values of a sequence. We therefore define structure species as quasi sequences.

We deviate from the standard definition of the notion of a structure species still in another respect: We incorporate into the notion of a structure species equivalence relations so that a structure species gives for each index of a structure term of the structures of the type of the structure species an equivalence relation the field of which contains all structure terms with that index. These equivalence relations will be used to determine which of the structure terms are theoretical and which are non- theoretical (see Sec. 4 definitions 4.3, 4.4, and 4.5).

3.12 Definition . - For all s : s is a *structure species* iff the following conditions are met:

(0) s is a quasi sequence & $3 \subseteq \mathrm{Dom}(s)$ &

(1) s'0 is a structure type &

(2) s'1 is a finite quasi sequence & $\mathrm{Dom}(s\text{'}1) \neq 0$ & $\forall i(i \in \mathrm{Dom}(s\text{'}1) \Rightarrow$ $(s\text{'}1)\text{'}i$ is a transportable predicate of type s'0) &

(3) s'2 is a quasi sequence & $\mathrm{Dom}(s\text{'}2) = \{(s\text{'}0)_0 + (s\text{'}0)_1 + i\}$ where $i \in \mathrm{Dom}(s\text{'}0) \div 2$
& $\forall j(j \in \mathrm{Dom}(s\text{'}2) \Rightarrow (s\text{'}2)\text{'}j$ is an equivalence relation &
$\{x_j \mid x \in \mathrm{Str}(s\text{'}0)\} \subseteq \mathrm{Fd}((s\text{'}2)\text{'}j))$.

The term with index 0 of a structure species is a structure type and the term with index 1 is a quasi sequence of transportable predicates of the type given by the term with index 0. These predicates are the axioms-as-predicates which serve as substitutes for the axioms-as-formulas. The term with index 2 is a quasi sequence of the above mentioned equivalence relations.

3.13 Definition . - For all s : **type**$(s) = s$'0

3.14 Definition . - For all s : the structure type of s = **type**(s).

3.15 Definition . - For all s, P : P is an axiom of s iff $\exists i(i \in \mathrm{Dom}(s\text{'}1)$ & $P = (s\text{'}1)\text{'}i)$.

3.16 Theorem . - For all s, P: if s is a structure species and P is an axiom of s, then P is a transportable predicate of type $\mathbf{type}(s)$.

3.17 Definition . - For all j, s : $\mathbf{E}(j,s) = (s\text{'}2)\text{'}j$.

3.18 Definition . - For all j, s : the equivalence relation with index j of $s = \mathbf{E}(j,s)$.

3.19 Theorem . - For all s, j :if s is a structure species & $j \in \{(s\text{'}0)_0 + (s\text{'}0)_1 + i\}$ where $i \in \mathrm{Dom}(s\text{'}0) \div 2$, then $\mathbf{E}(j,s)$ is an equivalence relation & $\{x_j \mid x \in \mathbf{Str}(s\text{'}0)\} \subseteq \mathrm{Fd}(\mathbf{E}(j,s))$.

Next we define the notion of a structure of a structure species:

3.20 Definition . - For all s, x : x is a structure of species s iff s is a structure species & $\forall P(P$ is an axiom of $s \Rightarrow x \in P)$.

A structure of a species is a structure which satisfies all axioms of that structure species. The class of all structures of a species is called the class of models of that structure species:

3.21 Definition . - For all s: $\mathbf{M}(s) = \{x \mid x$ is a structure of species s $\}$.

3.22 Definition . - For all s, x : x is a model of s iff $x \in \mathbf{M}(s)$.

The potential models of a structure species are the structures of the type of that structure species:

3.23 Definition . - For all s : $\mathrm{M_p}(s) = \mathbf{Str}(\ \mathbf{type}(s))$.

3.24 Definition . - For all s, x : x is a potential model of s iff $x \in \mathrm{M_p}(s)$.

3.25 Theorem . - For all s: if s is a structure species, then $\mathbf{M}(s)$ and $\mathrm{M_p}(s)$ are set theoretic predicates of type $\mathbf{type}(s)$ and $\mathbf{M}(s) \subseteq \mathrm{M_p}(s)$.

As the foregoing definitions show, we have not defined the notion of a typification. The reason is that this notion is not necessary for the definition of the notion of a structure species. Especially, it is not necessary to distinguish in the class of axioms-as-predicates between typifications and other kinds of axioms to build up the structuralist theory of empirical theories. In definition 3.12 nothing is said about the content of the axioms-as-predicates. The axioms somehow have to guarantee that the structures which satisfy the axioms are structures of the structure type of the structure species. They do not have to be divided in two classes, the typifications (or characterizations) and the other axioms ('laws').

On the other hand, one might try to characterize kinds of axioms like typifications, characterizations and laws. But no effort will be made in this essay in that direction.

5. Theoretical terms and partial potential models

The aim of this section is to explicate the notions of a theoretical resp. a non-theoretical term in a purely set theoretic way. The consequence of this is that no reference to formulas is allowed. Obviously only what is called in Architectonic (1987) a formal criterion of theoreticity can be formulated in purely set theoretic terms.
We begin with an auxiliary notion:

4.1 Definition . - For all $x, x', j : x \simeq_j x'$ iff $\forall i(i \in \mathrm{Dom}(x)$ & $i \neq j \Rightarrow x_i = x'_i)$.

4.2 Definition . - For all $x, x', j : x : x, x'$ differ at most wrt. j iff $x \simeq_j x'$.

The idea underlying the formal criterion of theoreticity is that theoretical terms are structure terms which are determined by the models of a structure species up to an equivalence relation. This idea is made precise with the help of the notion of an admissible method of determination for a structure term index up to an equivalence relation:

4.3 Definition . - For all M, j, s : M is an admissible method of determination for j in s iff the following conditions are met:

(0) s is a structure species & $j \in \{(s\text{'}0)_0 + (s\text{'}0)_1 + i\}$ where $i \in \mathrm{Dom}(s\text{'}0) \div 2$ &

(1) $M \subseteq \mathbf{M}(s)$ &

(2) $\forall x \forall x'(x, x' \in M$ & $x \simeq_j x' \Rightarrow (x_j, x'_j) \in \mathbf{E}(j, s))$ &

(3) $\forall x \forall x'(x \in M$ & $x \simeq_j x'$&$x' \in \mathbf{M}(s) \Rightarrow x' \in M)$.

In condition (0) $\{(s\text{'}0)_0 + (s\text{'}0)_1 + i\}$ where $i \in \mathrm{Dom}(s\text{'}0) \div 2$ is the set of indices of the structure terms of the structures of species s.

The formal criterion of theoreticity can now be defined as follows:

4.4 Definition . - For all R, x, s : R is a theoretical term in x of s iff $\exists M \exists j(M$ is an admissible method of determination for j in s & $x \in M$ & $R = x_j)$.

4.5 Definition . - For all R, x, s : R is a non-theoretical term in x of s iff $x \in \mathrm{M_p}(s)$ & $R \in \mathrm{Ran}(x\lceil\{(s\text{'}0)_0 + (s\text{'}0)_1 + i\}$ where $i \in \mathrm{Dom}(s\text{'}0) \div 2)$ & R is not a theoretical term in x of s.

The partial potential models of a structure species are obtained from the potential models by omitting the theoretical terms. To define the notion of a partial potential model we first define the notion of an indices function for the non-theoretical terms:

4.6 Definition . - For all f, s : f is the indices function for the non-theoretical terms of s iff the following conditions are satisfied:

(0) s is a structure species &

(1) f is a strictly monotone sequence in $\{(s\text{'}0)_0 + (s\text{'}0)_1 + i\}$ where $i \in \mathrm{Dom}(s\text{'}0) \div 2$ &

(2) $\mathrm{Ran}(f) = \{j \mid\sim \exists M M$ is an admissible method of determination for j in $s\}$.

4.7 Theorem .- For all s : if s is a structure species, then there exists exactly one indices function for the non-theoretical terms of s.

Theorem 4.7 justifies the following definition:

4.8 Definition . - For all s:

(1) If s is a structure species, then **if**(s) is the indices function for the non-theoretical terms of s.

(2) If s is not a structure species, then **if**(s) $= \mathcal{V}$.

The notion of a partial potential model now can be defined easily:

4.9 Definition . - For all z, s : z is a partial potential model of s iff s is a structure species & $\exists x(x \in \mathrm{M_p}(s)$ &
$z = (x\lceil((s\text{'}0)_0 + (s\text{'}0)_1))^\wedge(x\lceil\mathrm{Ran}(\mathbf{if}(s))))$.

4.10 Definition . - For all s : $\mathrm{M_{pp}}(s) = \{z \mid z$ is a partial potential model of $s\}$.

In definition 4.9 $(x\lceil((s'0)_0+(s'0)_1))$ is the sequence of the base sets occurring in x, and $x\lceil\mathrm{Ran}(\mathbf{if}(s))$ is the sequence of the non-theoretical terms occurring in x. A partial potential model is therefore obtained from a potential model by omitting the theoretical terms.

Given a structure species s, the class of potential models of s is uniquely determined and therefore the class of partial potential models is uniquely determined. Therefore we can define an operation on the class of all structure species which assigns to each structure species a mapping from the class of potential models to the class of partial potential models as follows:

4.11 Definition . - For all s :

(1) If s is a structure species, then $\mathbf{r}(s)$ is a function on $\mathrm{M_p}(s)$ & $\forall x(x \in \mathrm{Dom}(\mathbf{r}(s)) \Rightarrow \mathbf{r}(s)(x) = (x\lceil((s'0)_0 + s'0)_1))^\wedge(x\lceil\mathrm{Ran}(\mathbf{if}(s))))$.

(2) If s is not a structure species, then $\mathbf{r}(s) = \mathcal{V}$.

4.12 Theorem .- For all s : if s is a structure species, then $\mathbf{r}(s)\text{"}\mathrm{M_p}(s) = \mathrm{M_{pp}}(s)$.

6. Theory cores and application operators

Structure species are only one component of the theoretical part of a theory. Two other components are necessary: constraints and links. The complete theoretical part of an empirical theory is called a theory core and comprises a structure species, constraints and links.
The definition of a constraint will be taken over nearly unchanged from Architectonic, p. 47:

5.1 Definition . - For all $C, s : C$ is a constraint for s iff s is a structure species & $C \subseteq \mathrm{Po}(\mathrm{M_p}(s))$ & $C \neq 0$ & $0 \notin C$ & $\forall x(x \in \mathrm{M_p}(s) \Rightarrow \{x\} \in C)$.

5.2 Definition . - For all $C, s : C$ is a transitive constraint for s iff C is a constraint for s & $\forall x \forall y(y \neq 0$ & $y \subseteq x$ & $x \in C \Rightarrow y \in C)$.

If s is a structure species and $\mathrm{M_p}(s)$ is a proper class, then $\mathrm{Po}(\mathrm{M_p}(s))$ does not contain every subclass of $\mathrm{M_p}(s)$ as an element, because $\mathrm{Po}(\mathrm{M_p}(s))$ contains only the subclasses of $\mathrm{M_p}(s)$ which are sets. If this will turn out to

be an unacceptable limitation, the purely set theoretical formulation of the structuralist theory of empirical theories would become rather complicated. We leave this question for further investigation.
In contrast to constraints, which single out classes of potential models of one and the same structure species, links are relations between potential models of different structure species:

5.3 Definition . - For all s, s', L : L is an abstract link for s, s' iff s, s' are structure species & $L \subseteq \mathrm{M_p}(s) \times \mathrm{M_p}(s')$.
In general a link relates not potential models as a whole, but some of the structure terms of the potential models. To make this explicit, we define the notion of a concrete link:

5.4 Definition . - For all L, i, i', s, s' : L is a concrete link for i, i' from s, s' iff the following conditions are met:

(0) s, s' are structure species & i is a strictly monotone sequence in $\{(s'0)_0+(s'0)_1+j\}$ where $j \in \mathrm{Dom}(s'0) \div 2$ & i' is a strictly monotone sequence in $\{(s'0)_0 + (s'0)_1 + j\}$ where $j \in \mathrm{Dom}(s'0) \div 2$ &

(1) $L \subseteq \{t \mid \exists x \exists x'(x \in \mathrm{M_p}(s) \;\&\; x' \in \mathrm{M_p}(s') \;\&\; t =< x, x\lceil \mathrm{Ran}(i), x', x'\lceil \mathrm{Ran}(i') >)\}$.

Let L be a concrete link for i, i' from s, s'. Then $\{(s'0)_0+(s'0)_1+j\}$ where $j \in \mathrm{Dom}(s'0) \div 2$ is the set of indices of the structure terms of the structures of species s and i is a (strictly monotone) sequence of such indices. If x is a structure of species s, then $x\lceil \mathrm{Ran}(i)$ is the sequence (not in the technical sense) of the structure terms of x indicated by i. A concrete link therefore is a relation between structures of species s and their structure terms indicated by i on the one hand and structures of species s' and their structure terms indicated by i' on the other hand.

5.5 Definition . - For all L, s, s' : L is a concrete link for s, s' iff $\exists i \exists i' L$ is a concrete link for i, i' from s, s'.

The following theorem shows how a concrete link can be used to single out a subset of the potential models of a structure species:

5.6 Theorem .- For all L, s, s' : if L is a concrete link for s, s', then $\{x \mid \exists x' \exists u \exists u' < x, u, x', u' >\in L\} \subseteq \mathrm{M_p}(s)$.

We now have the ingredients for the definition of the notion of a theory core. This definition widely deviates from the definition given in Architectonic, p. 79 to meet the requirements of the theory of definitions and of the decision made in Sec. 0 concerning the level of abstraction underlying the definitions.

5.7 Definition . - For all K : K is a theory core iff the following conditions are met:

(0) K is a structure species &

(1) K'3 is a finite quasi sequence & $\forall i(i \in \text{Dom}(K\text{'}3) \Rightarrow (K\text{'}3)\text{'}i$ is a constraint for K) &

(2) K'4 is a finite quasi sequence & $\forall i(i \in \text{Dom}(K\text{'}4) \Rightarrow (K\text{'}4)\text{'}i$ is a structure species) &

(3) K'5 is a quasi sequence &
$\text{Dom}(K\text{'}5) = \text{Dom}(K\text{'}4)$ & $\forall i(i \in \text{Dom}(K\text{'}5) \Rightarrow (K\text{'}5)\text{'}i$ is a concrete link for $K\lceil 3, (K\text{'}4)\text{'}i)$.

Let K be a theory core. Then condition (0) says that K'0 is a structure type and K'1 is a finite quasi sequence of transportable predicates of type K'0. Condition (1) gives the finite quasi sequence of constraints regarded as necessary for K. Condition (2) gives the finite quasi sequence of structure species K is considered to be linked with, and condition (3) gives a link for the structure species K'3 and each structure species given by K'4. Since the same structure species can appear several times in $K'4$, the possibility is left open that K is linked by several links with the same other structure species.

The foregoing introduced terms **type**, $\mathbf{M}$, $\mathbf{M_p}$, **if**, $\mathbf{M_{pp}}$ are likewise applicable to structure species and theory cores. Furthermore, the notions of a global constraint and a global link can be defined as follows:

5.8 Definition . - For all K : $\mathbf{GC}(K) = \{A \mid \forall i(i \in \text{Dom}(K\text{'}3) \Rightarrow A \in \text{Dom}(K\text{'}3)\text{'}i)\}$.

5.9 Definition . - For all K : the global constraint of $K = \mathbf{GC}(K)$.

5.10 Definition . - For all K : $\mathbf{GL}(K) = \{x \mid \forall i(i \in \text{Dom}(K\text{'}5) \Rightarrow \exists u \exists x' \exists u' < x, u, x', u' > \in (K\text{'}5)\text{'}i)\}$.

5.11 Definition . - For all K : the global link of $K = \mathbf{GL}(K)$.

5.12 Theorem .- For all K : if K is a theory core, then if $3 \in \mathrm{Dom}(K)$, then $\mathbf{GC}(K) \in \mathrm{Po}(\mathrm{M_p}(K))$, and if $3 \notin \mathrm{Dom}(K)$, then $\mathbf{GC}(K) = \mathcal{V}$.

5.13 Theorem .- For all K : if K is a theory core, then if $5 \in \mathrm{Dom}(K)$, then $\mathbf{GL}(K) \in \mathrm{M_p}(K)$, and if $5 \notin \mathrm{Dom}(K)$, then $\mathbf{GL}(K) = \mathcal{V}$.

5.14 Theorem .- For all K : if K is a theory core & $\forall i(i \in \mathrm{Dom}(K\text{'}3) \Rightarrow (K\text{'}3)\text{'}i$ is a transitive constraint for K), then $\forall A \forall B(A \neq 0 \;\&\; A \subseteq B \;\&\; B \in \mathbf{GC}(K) \Rightarrow A \in \mathbf{GC}(K))$.

In connection with theory cores three notions of content can be defined. The definitions of the notion of isolated content and theoretical content can be overtaken from Architectonic (1987), p. 82:

5.15 Definition . - For all K : $\mathbf{cn}(K) = \mathbf{M}(K) \cap \mathbf{GL}(K)$.

5.16 Definition . - For all K : the isolated content of $K = \mathbf{cn}(K)$.

5.17 Definition . - For all K : $\mathbf{Cn_{th}}(K) = \mathrm{Po}(\mathbf{M}(K)) \cap \mathbf{GC}(K) \cap \mathrm{Po}(\mathbf{GL}(K))$.

5.18 Definition . - For all K : the theoretical content of $K = \mathbf{Cn_{th}}(K)$.

With regard to the theoretical content of K a remark corresponding to the remark made after 5.2 can be made: the theoretical content of a theory core contains only sets of models of K, and the question arises, whether this is an unacceptable limitation.

If K is a theory core, the elements of the isolated content of K are structures of type $\mathbf{type}(K)$, and the elements of the theoretical content of K are sets of structures of type $\mathbf{type}(K)$. To get the empirical content of K we have to omit the theoretical terms from the theoretical content of K. For this purpose we have introduced in Sec. 4 the operation $\mathbf{r}$. Since every theory core is a structure species, $\mathbf{r}$ is defined for theory cores, and if K is a theory core, $\mathbf{r}(K)$ is a function on $\mathrm{M_p}(K)$, i.e. $\mathbf{Str}(\mathbf{type}(K))$, and if $A \subseteq \mathrm{M_p}(K)$, then $\mathbf{r}(K)\text{"}A$ is the class of partial potential models which can be obtained from the elements of A by omitting the theoretical terms. Since $\mathbf{cn}(K) \subseteq \mathrm{M_p}(K), \mathbf{r}(K)\text{"}\mathbf{cn}(K)$ is the isolated empirical content of K. No separate term is introduced in Architectonic for this sort of content.

To get the empirical content contained in $\mathbf{Cn_{th}}(K)$ by the same method we have got the empirical content contained in $\mathbf{cn}(K)$, we need a 'strip-off-the-theoretical-terms' function, which could be applied to a class of classes of potential models of K in the same way $\mathbf{r}(K)$ was applied to $\mathbf{cn}(K)$. This function can be defined in a canonical way as follows:

5.19 Definition . - For all $s : \bar{\mathbf{r}}(s) = \mathrm{pot}(\mathbf{r}(s))$.

5.20 Theorem .- For all s : if s is a structure species, then $\bar{\mathbf{r}}(s)$ is a function on $\mathrm{Po}(\mathrm{M_p}(s))$ & $\forall A(A \in \mathrm{Dom}(\bar{\mathbf{r}}(s)) \Rightarrow \bar{\mathbf{r}}(s)(A) = \mathbf{r}(s)\text{"}A)$ & $\mathrm{Ran}(\bar{\mathbf{r}}(s)) \subseteq \mathrm{Po}(\mathrm{M_{pp}}(s))$.

We now can proceed as follows: since a theory core is a structure species, $\bar{\mathbf{r}}$ is defined for theory cores, and if K is a theory core, $\bar{\mathbf{r}}(K)$ is a function on $\mathrm{Po}(\mathrm{M_p}(K))$, and if $X \subseteq \mathrm{Po}(\mathrm{M_p}(K)), \bar{\mathbf{r}}(K)\text{"}X$ is the class of all classes of partial potential models which can be obtained from the elements of the elements of X by omitting the theoretical terms. Since $\mathbf{Cn_{th}}(K) \subseteq \mathrm{Po}(\mathrm{M_p}(K)), \bar{\mathbf{r}}(K)\text{"}\mathbf{Cn_{th}}(K)$ is the empirical content of K, which is given a separate denotation:

5.21 Definition . - For all K : $\mathbf{Cn}(\mathrm{K}) = \bar{\mathbf{r}}(K)\text{"}\mathbf{Cn_{th}}(K)$.

5.22 Theorem .- For all K : if K is a theory core, then $\mathbf{Cn}(K) \subseteq \mathrm{Po}(\mathrm{M_{pp}}(K))$.

6. Theory elements

If we add to a theory core a set of intended applications we get a theory element:

6.1 Definition . - For all T : T is an idealized theory element iff the following conditions are satisfied: (0) T is a theory core & (1) $T\text{'}6 \subseteq \mathrm{M_{pp}}(T)$ & $\mathrm{Set}(T\text{'}6)$.

The above defined terms **type**, **M**, $\mathbf{M_p}$, **if**, $\mathbf{M_{pp}}$, **GL**, **GC**, **cn**, $\mathbf{Cn_{th}}$, **r**, $\bar{\mathbf{r}}$ are likewise applicable to theory cores and theory elements. We further introduce the notion of the intended applications of a theory element:

6.2 Definition . - For all $T : \mathbf{I}(T) = T\text{'}6$.

6.3 Definition . - For all T, x : x is an intended application of T iff $x \in \mathbf{I}(T)$.

In Architectonic, p. 91 the notion of an idealized empirical claim is defined. But this notion cannot be defined by purely set theoretical means, because the idealized empirical claim is a formula and in a set theoretical language like NBGU there exist no formula variables. But we can define a notion which reflects the fact that the empirical claim of a theory element is true :

6.4 Definition . - For all T : T is a successful idealized theory element iff T is an idealized theory element & $\mathbf{I}(T) \in \mathbf{Cn}(T)$.

7. Theory nets

The frame for the structuralist reconstruction of an advanced empirical theory is a theory net. A theory net is a class of theory elements ordered by a relation of specialization. In order to meet the conditions concerning the level of abstraction laid down in sec. 0 and the general conditions for definitions we have to depart much from the definition of the notion of a specialization relation given in Architectonic, p. 170.

7.1 Definition . - For all T', T : T' is an idealized specialization of T iff the following conditions are met:

(0) T', T are idealized theory elements & $\mathrm{Dom}(T') = \mathrm{Dom}(T)$ &

(1) $\mathbf{type}(T') = \mathbf{type}\,(T)$ &

(2) $T\text{'}2 = (T')\text{'}2$ &

(3) $\forall i(i \in \{1,3,4,5\} \Rightarrow \mathrm{Dom}(T\text{'}i) \subseteq \mathrm{Dom}((T')\text{'}i)$ & $\forall j(j \in \mathrm{Dom}(T\text{'}i) \Rightarrow (T\text{'}i)\text{'}j = ((T')\text{'}i)\text{'}j))$ &

(4) $\mathbf{I}(T') \subseteq \mathbf{I}(T)$ &

(5) $\forall M \forall j(M$ is an admissible method of determination for j in $T \Rightarrow M \subseteq \mathbf{M}(T'))$ &

(6) $\forall M \forall j(M$ is an admissible method of determination for j in $T' \Rightarrow \forall x \forall x'(x \in M \;\&\; x \simeq_j x' \;\&\; x' \in \mathbf{M}(T) \Longrightarrow x' \in \mathbf{M}(T')))$ &

(7) $\forall i(i \in \mathrm{Dom}(T) \;\&\; i > 6 \Rightarrow (T')\text{'}i = T\text{'}i)$.

7.2 Definition . - For all T', T : $(T'\mathbf{Spec}T)$ iff T' is an idealized specialization of T.

Spec is a predicate constant of NBGU, not an individual constant denoting a relation (class of ordered pairs). Since theory elements are specializations of each other, and theory elements in general are not sets, it would be inadequate to introduce such a relation (class of ordered pairs).

We now can state (and prove) some theorems which partly are constituents of the definition of the specialization relation given in Architectonic, p. 170.

7.3 Theorem .- For all T', T: if $(T'\mathbf{Spec}T)$, then $\mathrm{M_p}(T') = \mathrm{M_p}(T)$.

7.4 Theorem .- For all T', T : if $(T'\mathbf{Spec}T)$, then $\forall M \forall j(M$ is an admissible method of determination for j in T' iff M is an admissible method of determination for j in T).

7.5 Theorem .- For all T', T: **if** $(T'\mathbf{Spec}T)$, then **if** $(T') =$ **if**(T).

7.6 Theorem .- For all T', T : if $(T'\mathbf{Spec}T)$, then $\mathrm{M_{pp}}(T') = \mathrm{M_{pp}}(T)$.

7.7 Theorem .- For all T', T: if $(T'\mathbf{Spec}T)$, then $\mathbf{GC}(T') \subseteq \mathbf{GC}(T)$ & $\mathbf{GL}(T') \subseteq \mathbf{GL}(T)$.

The next two theorems are concerned with the content of idealized theory elements related by the relation of specialization.

7.8 Theorem .- For all T', T: if $(T'\mathbf{Spec}T)$, then $\mathbf{Cn}(T') \subseteq \mathbf{Cn}(T)$.

7.9 Theorem .- For all T', T: if $(T'\mathbf{Spec}T)$ and $\forall i(i \in \mathrm{Dom}(T'3) \Rightarrow$ $(T'3)'i$ is a transitive constraint for T) & $\mathbf{I}(T) \in \mathbf{Cn}(T)$, then $\mathbf{I}(T') \in \mathbf{Cn}(T)$.

The relation of specialization is reflexive, antisymmetric and transitive:

7.10 Theorem .-

(1) For all T: if T is an idealized theory element, then $(T\mathbf{Spec}T)$.

(2) For all T', T : if $(T'\mathbf{Spec}T)$ & $T' \neq T$, then $\sim (T\mathbf{Spec}T')$.

(3) For all T'', T', T: if $(T''\mathbf{Spec}T')$ & $(T'\mathbf{Spec}T)$, then $(T'\mathbf{Spec}T)$.

Next we will define the notion of a theory net and in doing that, we will depart from the definition given in Architectonic, p. 172 in several respects. In Architectonic, the notion of a theory net is defined as a sequence of length 2 of which the first member is a non-empty and finite class of idealized theory elements, and the second member is the specialization relation (class of ordered pairs) restricted to that class. But then the explicit mentioning of the relation is redundant and the definition can be simplified to saying that a theory net is a non- empty finite class of theory elements. Because theory elements in general are not sets they cannot be elements of a class. Therefore we have to construct theory nets as quasi sequences of theory elements. Furthermore we will define theory nets as what is called in Architectonic a connected theory net, since till now there are no realistic examples of theory nets which are not connected ones.

7.11 Definition . - For all N : N is an idealized theory net iff the following conditions are met:

(0) N is a finite quasi sequence & $\mathrm{Dom}(N) \neq 0$ &

(1) $\forall i(i \in \mathrm{Dom}(N) \Rightarrow N'i$ is an idealized theory element) &

(2) $\forall m \forall n(m, n \in \mathrm{Dom}(N) \Rightarrow \exists j(j$ is a finite sequence in $\mathrm{Dom}(N)$ & $j_0 = m$ & $j_{\mathrm{Dom}(j) \div 1} = n$ & $\forall i(i \in \mathrm{Dom}(j) \div 1$ $\Rightarrow ((N'(j_i)\mathbf{Spec}(N'(j_{i+1})$ or $(N'(j_{i+1})\mathbf{Spec}N'(j_i))))))$.

7.12 Theorem .- For all N : if N is an idealized theory net, then $\forall m \forall n(m, n \in \mathrm{Dom}(N) \Rightarrow \mathrm{M_p}(N'm) = \mathrm{M_p}(N'n))\&\mathrm{M_{pp}}(N'm) =$ $= \mathrm{M_{pp}}(N'n))$.

Theory nets contain maximal theory elements. We first define this notion and then formulate a corresponding theorem.

7.13 Definition . - For all T, N : T is a maximal theory element in N iff N is an idealized theory net & $\exists i(i \in \mathrm{Dom}(N)$ & $T = N'i$ & $\forall j(j \in \mathrm{Dom}(N)$ & $N'j \neq N'i \Rightarrow\sim (N'i\mathbf{Spec}N'j)))$.

7.14 Theorem .- For all N: if N is an idealized theory net, then $\exists T$ T is a maximal theory element in N.

For the class of maximal elements of an idealized theory net which is - following 7.14 - not empty, we will introduce a separate denomination.

7.15 Definition . - For all N : $\mathbf{B}(N) = \{(x, i) \mid (x, i) \in N$ & $N'i$ is a maximal theory element in $N\}$.

7.16 Definition . - For all N : the basis of $N = \mathbf{B}(N)$.

7.17 Definition . - For all N, T : T is a basic theory element of N iff $\exists i(i \in \mathrm{Dom}(\mathbf{B}(N))$ & $T = \mathbf{B}(N)\text{'}i)$.

7.18 Theorem .- For all N : if N is an idealized theory net, then $\exists T$ T is a basic theory element of N.

In general, the basis of a realistic example of a theory net contains exactly one basic theory element of N. Theory nets with this characteristic are called theory trees in Architectonic. But this denomination seems to be inadequate, because it is not in accordance with the usual use of the term 'tree': in trees the class of predecessors of an element forms a chain, but this is not so in theory nets with a single maximal element. (If T'**Spec**T), then T is thought of to be the predecessor of T'.) Therefore we use the term 'single based'.

7.19 Definition . - For all N : N is a single based idealized theory net iff N is an idealized theory net & $\exists T \forall T'(T'$ is a basic theory element of $N \Longleftrightarrow T' = T)$.

A single based idealized theory net N has exactly one maximal theory element. Then $\mathrm{Ran}(\mathbf{B}(N))$ is that element. We introduce a term for it:

7.20 Definition . - For all N : $\mathbf{b}(N) = \mathrm{Ran}(\mathbf{B}(N))$.

7.21 Definition . - For all N : the basic theory element of $N = \mathbf{b}(N)$.
For the basic theory element we note three theorems:

7.22 Theorem .- For all N: if N is a single based idealized theory net, then $\exists i(i \in \mathrm{Dom}(\mathbf{B}(N))$ & $(\mathbf{B}(N) = (\mathbf{b}(N) \times \{i\}))$ and $\mathbf{b}(N)$ is a maximal theory element in N and $\forall T(T$ is a maximal theory element in $N \Rightarrow T = \mathbf{B}(N))$.

7.23 Theorem .- For all N: if N is a single based idealized theory net, then $\forall i(i \in \mathrm{Dom}(N) \Rightarrow (N\text{'}i\ \mathbf{Spec}\ \mathbf{b}(N)))$.

7.24 Theorem . For all N : if N is a single based idealized theory net, then $\forall i(i \in \mathrm{Dom}(N) \Rightarrow \mathbf{I}(N\text{'}i) \subseteq \mathbf{I}(\mathbf{b}(N)))$.

The ordering relation **Spec** in the field of idealized theory elements gives rise to corresponding ordering relations in the field of idealized theory cores and the field of intended applications of the theory elements.

7.25 Definition . - For all K', K : K' is a core specialization of K iff the following conditions are met:

(0) K', K are theory cores & $\mathrm{Dom}(K') = \mathrm{Dom}(K)$ &

(1) $\mathbf{type}\ (K') = \mathbf{type}\ (K)$ &

(2) $K\text{'}2 = (K')\text{'}2$ &

(3) $\forall i(i \in \{1,3,4,5\} \rightarrow \mathrm{Dom}(K\text{'}i) \subseteq \mathrm{Dom}((K')\text{'}i)$ & $\forall j(j \in \mathrm{Dom}(K\text{'}i) \Rightarrow (K\text{'}i)\text{'}j = (K')\text{'}i)\text{'}j)$ &

(4) $\forall M \forall j(M$ is an admissible method of determination for j in $K \Rightarrow M \subset \mathbf{M}(K'))$ &

(5) $\forall M \forall j(M$ is an admissible method of determination for j in $K' \Rightarrow \forall x \forall x'(x \in M$ & $x \simeq_j x'$ & $x' \in \mathbf{M}(K) \Rightarrow x' \in \mathbf{M}(K')))$ &

(6) $\forall i(i \in \mathrm{Dom}(K)$ & $i > 5 \Rightarrow (K')\text{'}i = K\text{'}i)$.

7.26 Definition . - For all K', K : $(K'\mathbf{CSpec}K)$ iff K' is a core specialization of K.

The relation of core specialization is reflexive, antisymmetric and transitive:

7.27 Theorem .

(1) For all K : if K is a theory core, then $(K\mathbf{CSpec}K)$.

(2) For all K', K : if $((K'\mathbf{CSpec}K)$ & $K' \neq K$, then $\sim (K\mathbf{CSpec}K')$.

(3) For all K'', K', K : if $(K''\mathbf{CSpec}K')$ & $(K'\mathbf{CSpec}K)$, then $(K''\mathbf{CSpec}K)$.

An idealized theory net induces a core net in the following way:

7.28 Definition . - For all N : $\mathbf{CNet}(N) = \{(p,j) \mid j \in \mathrm{Dom}(N)$ & $p \in (N\text{'}j)\lceil 6\}$.

7.29 Definition . - For all N :the core net associated with $N = \mathbf{CNet}(N)$.

The content of the next theorem are some fundamental properties of core nets :

7.30 Theorem .- For all N: if N is an idealized theory net, then:

(0) $\mathbf{CNet}(N)$ is a finite quasi sequence & $\mathrm{Dom}(\mathbf{CNet}(N)) = \mathrm{Dom}(N)$ &

(1) $\forall j(j \in \mathrm{Dom}(\mathbf{CNet}(N)) \Rightarrow \mathbf{CNet}(N\text{'}j) = (N'j)\lceil 6)$ &

(2) $\forall j(j \in \mathrm{Dom}(\mathbf{CNet}(N)) \Rightarrow \mathbf{CNet}(N)\text{'}j$ is a theory core) &

(3) $\forall m \forall n(m, n \in \mathrm{Dom}(\mathbf{CNet}(N))$ & $(N\text{'}m\mathbf{Spec}N\text{'}n) \Rightarrow$
$\mathbf{CNet}(N)\text{'}m\mathbf{CspecCnet}(N)\text{'}n))$.

In the same way an idealized theory net induces a core net it induces an intended application net:

7.31 Definition . - For all N : $\mathbf{INet}(N) = \{(z, j) \mid j \in \mathrm{Dom}(N)$ & $z \in (N\text{'}j)\text{'}6\}$.

7.32 Definition . - For all N : the intended application net associated with N = **INet**(N).

The following theorem lists some fundamental properties of intended application nets:

7.33 Theorem .- For all N : if N is an idealized theory net, then:

(0) $\mathbf{INet}(N)$ is a finite quasi sequence & $\mathrm{Dom}(\mathbf{INet}(N)) \subseteq \mathrm{Dom}(N)$
& if $\forall i(i \in \mathrm{Dom}(N) \Rightarrow 6 \in \mathrm{Dom}(N\text{'}i)$, then
$\mathrm{Dom}(N) \subseteq \mathrm{Dom}(\mathbf{INet}(N))$ &

(1) $\forall j(j \in \mathrm{Dom}(N) \Rightarrow \mathbf{INet}(N)\text{'}j))$ &

(2) $\forall m \forall n(m, n \in \mathrm{Dom}(\mathbf{INet}(N))$ & $(N\text{'}m\mathbf{Spec}N\text{'}n) \Rightarrow$
$\mathbf{INet}(N)\text{'}m \subseteq \mathbf{INet}(N)\text{'}n))$.

As 7.33(2) shows, the order-relation in an intended application net is simply the subclass relation.

In Architectonic, p. 177 ff. three notions of the empirical claim of a theory net are discussed. For the same reason as for theory elements, these notions cannot be defined in purely set theoretical terms. But as for theory elements according notions of successful theory nets can be defined. We give only the definition of the most simple case:

7.34 Definition . - For all N : N is a successful idealized theory net iff N is an idealized theory net & $\forall i(i \in \mathrm{Dom}(N) \Rightarrow \mathbf{I}(N\text{'}i) \in \mathbf{Cn}(N\text{'}i))$.

8. References

Balzer, W. & Moulines, C.U. & Sneed, J.D., 1987, *An Architectonic for Science*, Dordrecht.
Bourbaki, N., 1970, *Elèments de mathèmatiques, Thèorie des ensembles*, Paris.
van Dalen, D. & Doets, H.C. & de Swart, H., 1978, *Sets: Naive, Axiomatic and Applied*, Oxford.
Gödel, K., 1940, *The consistency of the axiom of choice and of the Generalized Continuum Hypothesis with the Axioms of Set Theory*, Princeton.
Hatcher, W.S., 1968, *Foundations of Mathematics*, Philadelphia.
Kelley, J.L., 1955, *General Topology*, Princeton.
Lemmon, E.J., 1968, *Introduction to Axiomatic Set Theory*, London.
Levy, A., 1979, *Basic Set Theory*, Berlin.
Rubin, J.E., 1967, *Set Theory for the Mathematician*, San Francisco.

Chapter 13: Categorial Structuralism

Thomas Mormann

1. Introduction

The idea of using concepts of category theory for the structuralist approach has been in the air for some time, but it cannot be said to have gained too great a popularity in the structuralist community. This is somewhat surprising: in recent years structuralism has concentrated, besides the more traditional reconstruction of theories, on the elucidation of the global structure of empirical science: on the other hand, category theory has turned out to be a very useful conceptual framework for elucidating the global structure of mathematics. Thus, remembering Stegmüller's characterisation of 'structuralism as a possible analogue of the Bourbaki programme' (cf. Stegmüller 1973) one might conjecture that the structuralist reconstruction of empirical science could follow an analogous track to the category theoretical reconstruction of mathematics. To take some steps towards the realisation of this conjecture is the aim of this paper. Let us start with the following quite intuitive characterisation of a category:

'A category may be thought of in the first instance as a universe for a particular kind of mathematical discourse. Such a universe is determined by specifying a certain kind of 'object', and a certain kind of 'arrow' that links different objects.' Goldblatt (1977, p.1)

Thus, if we want to cast the structuralist approach in the framework of category theory, we have to outline a category or, more realistically, several categories - where the structuralist discourse about empirical science takes place. However, in order to bring structuralism into a fruitful contact with category theory it is not sufficient just to show that the structuralist concepts can be reformulated in the language of category theory: even an ordinary set might be considered as a (rather trivial) category. What we have to show is that the structuralist approach gives rise to interesting categories. Of course, there is no general agreement about what is an interesting category. However, the following general criterion for interesting categories probably can be unanimously accepted: an interesting category is a category that allows for certain universal constructions relevant and genuine for the discourse which is to be described in the framework of category theory.

The outline of this paper is as follows: in section 2 we recall the concepts of constraints and links crucial for the structuralist approach. We provide a new characterisation of constraints in terms of reflexive and symmetric relations. Section 3 provides a compilation of the category theoretical concepts necessary for the following. In section 4 category theory is applied to a reformulation of the structuralist concepts of constraints and links. In section 5 we discuss several categories that may be used as 'universes of discourse' for structuralism. In section 6 we introduce the category LTC of linked theory cores wherein the global structure of empirical theories can be represesented as a net of linked theory elements.

2. Constraints and Links.

The traditional structuralist concept of constraint (cf. Sneed 1971, Balzer & Moulines & Sneed 1987) is quite abstract. Its fate in the development of the structuralist approach has been ambivalent: on the one hand it has been hailed as one of the characteristic features of structuralism providing it with its distinctive flavour, on the other hand, many authors consider constraints as a not-so-welcomed complication of the formal apparatus. Hence they often try to get rid of it as soon as possible. In this paper I'd like to join the former group, i.e., I'll consider constraints as an essential ingredient of the structuralist approach.

(2.1) Definition. Let X be a set and PX its power set. A (transitive) constraint C for X is non-empty subset of PX satisfying the following requirements:

(1) $(x)(x \in X \Rightarrow \{x\} \in C)$

(2) $\emptyset \neq Y \subseteq Y' \in C \Rightarrow Y \in C.$

Intuitively, constraints express relations of compatibility between different models of a theory element. That is, if two (or more) models belong to an element of a constraint they may be used for a consistent description of some part of reality. This interpretation renders plausible the two requirements for constraints defined above: (1) each model is consistent with itself, and (2) if some elements of a consistent set of models are removed the resulting set of models is still consistent. The concept of constraint given in (2.1) is extremely general, it comprises a lot of constraints nobody ever would seriously regard as 'real' constraints for structuralist theory-elements. Hence it is a rather pleasing fact that in reality all structuralist constraints are of a much more special kind. As I will explain in a moment, all constraints important for structuralist considerations are reflexive and symmetric relations in the following sense:

(2.2) Definition. Let X be a set. Denote its n-th Cartesian product by $X \times \ldots \times X$ (n-times). Let s be an n-permutation, i.e. s is a bijective map of $\{1, 2, \ldots, n\}$ onto itself. A subset $R \subseteq X \times \ldots \times X$ is a symmetric and reflexive relation iff the following two requirements are satisfied:

(i) $(x)(x \in X \Rightarrow (x, \ldots, x) \in R)$

(ii) for all n-permutations s: $(x_1, \ldots, x_n) \in R \Leftrightarrow (x_{s(1)}, \ldots, x_{s(n)}) \in R$.

(2.3) Lemma. Let $R \subseteq X \times \ldots \times X$ be a reflexive and n-symmetric relation of X, R gives rise to a constraint $C_R(X)$ on X in the sense of (2.1) in the following way:
$C_R(X) := \{Y \mid \emptyset \neq Y \subseteq X \wedge \forall x_1, \ldots, x_n(\{x_1, \ldots, x_n\} \subseteq Y \Rightarrow (x_1, \ldots, x_n) \in R)\}$.

(2.4) Definition. A constraint on X that is generated by an n-symmetric relation on X is called an n-constraint.

The interesting fact about n-constraints is not that they can be generated by symmetric and reflexive n-relations, but rather that all 'real-life'-constraints are indeed n-constraints. This assertion cannot be rigorously proved, of course, rather it is a more or less well-confirmed empirical fact. As evidence we now show that some of the most common types of constraints are indeed 2- and 3-constraints.

(2.5) Example. The identity constraint $C^{=,=}(m)$ of mass in CPM is defined as follows: $X \in C^{=,=}(m)$ iff $(x)(y)(\{x, y\} \subseteq X \Rightarrow (p)(p \in P_x \cap P_y \Rightarrow m_x(p) = m_y(p)))$.

Now define a reflexive and symmetric relation R_{id} on M_p (CPM) as follows: $R_{id}xy$ iff $(p)(p \in P_x \cap P_y \Rightarrow m_x(p) = m_y(p)))$ Then, obviously, the identity constraint $C^{=,=}(m)$ is just C_{Rid}, i.e. the identity constraint of the mass function in CPM is a 2-constraint.

(2.6) Example. The extensivity constraint $C^{<0.+>}(m)$ of mass in CPM is defined as follows: $X \in C^{(0.+)}(m)$ iff $(x)(y)(z)(x, y, z \in X \Rightarrow$
$\Rightarrow (p_i)(p_j)(p_k)(p_i \in P_x \wedge p_j \in P_y \wedge p_k \in P_z \wedge p_k = p_i o p_j \Rightarrow m_z(p_k) = m_x(p) + m_y(p)))$

Now define a ternary reflexive and symmetric relation R_x on $M_p(CPM)$ as follows:

$R_x xyz$ iff $(p_i)(p_j)(p_k)(p_i \in P_x \wedge p_j \in P_y \wedge$
$p_k \in P_z \wedge p_k = p_i o p_j \Rightarrow m_z(p_k) = m_x(p_i) + m_y(p_j)))$ or
$(p_i)(p_j)(p_k)(p_i \in P_z \wedge p_j \in P_y \wedge$

$p_k \in P_x \wedge p_k = p_i o p_j \Rightarrow m_x(p_k) = m_z(p_i) + m_y(p_j)))$ or
$(p_i)(p_j)(p_k)(p_i \in P_x \wedge p_j \in P_z \wedge$
$p_k \in P_y \wedge p_k = p_i o p_j \Rightarrow m_y(p_k) = m_z(p_i) + m_y(pj))).$

The relation R_{ex} is a ternary reflexive and symmetric relation on $M_p(CPM)$ with $C_{Rex} = C^{(0.+)}(m)$. Hence, the extensionality constraint of CPM is a 3-constraint.

Stegmüller (1986) claimed that any structuralist constraint is a 2-constraint. This is wrong: As is easily seen, extensionality constraints are genuine 3-constraints which cannot be realised as 2-constraints. However, Stegmüller got it almost right: scrutinising the existing structuralist reconstructions I haven't found one constraint of higher order than 2 or 3. Hence, it seems to be a reasonably confirmed conjecture that all 'real' constraints are 2- or 3-constraints. Of course, for any n one can easily construct n-constraints that are not (n-1)-constraints, and moreover there are constraints that are not n-constraints for any n whatsoever.

A formal advantage of the traditional definition of a constraint is that it allows the combination of constraints by intersection: if C and C' are constraints the set theoretical intersection $C \cap C'$ is a constraint as well. This fact enables us to consider a single global constraint for a theory element. This procedure cannot be applied directly to n-constraints. Of course, if we have two reflexive and symmetric n-relations R and R', its intersection $R \cap R'$ again is a reflexive and symmetric n-relation. However, how we can combine an n-relation and an $(n+1)$ relation? This is not directly possible, but can be achieved by the trick of 'blowing up' the n-relation. We define this process of blowing up only for 2-relations but it should be obvious how to do it for any $n \geq 2$:

(2.7) Definition and Lemma. Let $R^2 \subseteq X \times X$ be a reflexive and symmetric relation on X. Then R^2 can be blown up to a reflexive and symmetric relation $R^3 \subseteq X \times X \times X$ such that the constraints C_{R^2} and C_{R^3} induced by R^2 and R^3, coincide. R^3 is defined as follows: R^3xyz iff R^2xy and R^2yz and R^2zx.

Lemma 2.7) enables us to avoid some awkward technical complications in the following way: assume we have a set X with some constraints $C_1, \ldots, C_m$. Let us further assume that all these constraints C_i are n-constraints for some n_i. Then (2.7) entitles us to the simplifying assumption that n_i are equal to each other. Thus we can assume that X is constrained by a single n-relation. Actually, the situation for the structuralist is even simpler yet: as I said above, all (transitive) constraints are actually 2- or 3-constraints. Hence, by applying lemma (2.7) we are entitled to assume that all structuralist theory elements are endowed with

3-constraints arising from ternary reflexive and symmetric relations.

Now let us come to the concept of a structuralist link which may be seen as an external generalisation of the concept of constraint. A link is to be seen as the most basic model-theoretic unit to deal with intertheoretical relations (cf. Moulines 1992, p.18):

(2.8) Definition. A link L between sets X and Y is a relation $L \subseteq X \times Y$.

Seen from this perspective, a constraint, or more precisely a 2-constraint is nothing but a special internal link. Needless to say, this definition of a link is too broad to be really useful, further restrictions have to be imposed on 'real-life' links.

3. Basics of Category Theory.

In this section we recall the fundamental notions of category theory to be used in the following. In no way can this section be considered as an independent introduction to category theory, it is simply a compilation of the relevant material mainly taken from Goldblatt (1978), McLarty (1992) and others. The reader not acquainted with category theory should consult these sources for further information and motivation.

(3.1) Definition. A category C is given by the following data:

(i) A collection ob(C) of C-objects denoted by X, Y, Z and so on.

(ii) For each ordered pair of objects (X, Y) a set $C(X, Y)$ of C-morphisms from X to Y. Morphisms may also be called maps or arrows.

(iii) Operations assigning to each C-morphism f a C-object *dom* f (the domain of f) and a C-object *cod* f (the codomain or target of f). If $X = dom\ f$ and $Y = cod\ f$ we display this as $f : X \rightarrow Y$ or $X \xrightarrow{f} Y$. The collection of C-morphisms whose domain is f and whose domain is Y is denoted by $C(X, Y)$.

(iv) An operation assigning to each pair (g, f) of C-morphisms with *dom* $g = cod\ g$, a C-morphism $g \circ f$, the composite of f and g having *dom* $(g, f) = dom\ f$ and $cod(g, f) = cod\ g$, such that the following condition obtains: Given $f : X \rightarrow Y, g : Y \rightarrow Z, h : Z \rightarrow U$ of C-objects and C-morphisms the following law of associativity holds: $(h \circ (g \circ f)) = ((h \circ g) \circ f)$.

(v) An assignment to each C-object X of a C-morphism $id_x : X \rightarrow X$ called the identity morphism of X such that the following holds: for any C-morphisms $f : Y \to X$ and $g : X \to Z$ $id_x \circ f = f$ and $g \circ id_x = g$.

If there is no danger of confusion we speak of objects and morphisms (arrows) instead of C-objects and C-morphisms. Examples of categories abound in mathematics and elsewhere (cf. Goldblatt (1978) Mac Lane (1971), Adamek et a. (1990), Pitt et al. (1986)).

(3.2) Examples.

(1) The category SET of sets:

- **(i)** SET-objects: sets.
- **(ii)** SET-morphisms: set theoretical maps $f : X \to Y$.

(2) The category TOP of topological spaces:

- **(i)** TOP-objects: topological spaces
- **(ii)** TOP-morphisms: continuous maps $f : X \to Y$.

(3) The GRP of groups:

- **(i)** GRP-objects: groups.
- **(ii)** GRP-morphism: group-homomorphism f: $G \to H$.

(4) The category POS of partially ordered sets:

- **(i)** POS-objects: partially ordered sets.
- **(ii)** POS-morphisms: order-preserving maps f: $P \to Q$.

One should note that a category is not determined by its objects, there may be categories with exactly the same objects but different morphisms. In the following we'll explicitly deal with such a case: we start with a category that has the 'right' objects but whose morphisms are less than optimal for our purposes. Hence, later we'll replace them by other morphisms thereby moving to another category with the same objects.

In a sense which I do not want to make precise here, all these categories mentioned above are rather similar to each other. The concept of a category, however, comprises a lot of other creatures that may not come immediately to mind if one concentrates on the standard examples above:

(3.3) Examples.

(1) Let G be a group. G may be conceived of as a category in the following way:

 (i) G-objects: the unit element e of G.

 (ii) G-morphisms: the elements of G.

(2) Let W be a topological space. W may be conceived of as a category in the following way:

 (i) W-objects: open subsets of W.

 (ii) W-morphisms: inclusion maps between open sets.

(3) Let X be a set. X may be conceived of as a category in the following way:

 (i) X-objects: elements of X.

 (ii) X-morphisms: id_x for $x \varepsilon X$.

The category X has no morphisms except identity morphisms. It is called a discrete category.

The external orientation of category theory aiming at relations rather than 'things in themselves' renders it natural not to consider categories in isolation but to take into account relations between several categories as well. These relations are called functors:

(3.4) Definition. Let C and D be categories. A functor F from C to D is a map that assigns

(1) to each C-object X a D-object $F(X)$.

(2) to each C-morphism f: $X \rightarrow Y$ a D-morphism $F(f) : F(X) \rightarrow F(Y)$ such that $F(id_x) = id_{F(x)}$ and $F(g \circ f) = F(g) \circ F(f)$, whenever $g \circ f$ is defined.

A functor F from C to D is denoted by $F : C \Rightarrow D$.

(3.5) Definition. (i) A functor $F : C \Rightarrow D$ is faithful iff the following holds: for any pair of C-objects, for any C-morphisms $f, g : X \rightarrow Y$ if $F(f) = F(g)$ then $f = g$. (ii) A functor $F : C \rightarrow D$ is full iff the following holds: for any C-objects X and Y and D-morphism $g : F(X) \rightarrow F(Y)$ there is some $f : X \rightarrow Y$ with $F(f) = g$.

Examples of functors abound, see the above mentioned references. In particular, the following examples are important:

3.6. Examples. Let C be one of the categories TOP, GRP or POS. There is a canonical 'forgetful' functor $F_c : C \Longrightarrow SET$ which assigns to each C-object its underlying set and to each C-map $f : X \Longrightarrow Y$ its underlying set theoretical map. Forgetful functors are faithful but not full.

In no way are all functors forgetful functors. Some go into quite a different direction as the following example shows:

(3.7) Example. The functor $U :\Longrightarrow TOP$ assigns to each set X the topological space on X with the discrete topology. Each set theoretical map $f : X \longrightarrow Y$ is continuous with respect to the discrete topologies on X and Y. Hence U is indeed a functor that is (in some sense to be specified) an inverse of the forgetful functor $V : TOP \Longrightarrow SET$. U is full and faithful. Actually, much more can be said about the pair of functors U and V. As is well known they are adjoint to each other (cf. Goldblatt 1978 or Mc Larthy 1992).

The examples show that the concept of a category is an extremely general one, almost everything can be made a category in a more or less artificial way. Hence, the point one has to drive home in a category theoretical reformulating of some realm is not that something can be rendered a category, rather, it should be an interesting category. An interesting category should not simply give a static description of some realm, rather it should set the stage for doing something interesting with the stuff available, i.e. construing new objects and out of old ones. An interesting category is a kind of arsenal where one finds material and tools to build up new things. Expressed in less picturesque language this means a good category theoretical framework should enable us to carry through interesting universal constructions. In the following I'll recall some of the most basic universal constructions that are useful for the structuralist approach.

(3.8) Definition. Let C be a category. An object 0 is initial in C if for every C-object X there is one and only one arrow $0 \longrightarrow X$. Dually, an object 1 is a terminal object of C if for every C-object Y there is one and only one arrow $Y \longrightarrow 1$. An element of a C-object X is a C-morphism $1 \rightarrow X$.

(3.9) Lemma. In the category SET the empty set is an initial object, and any singleton is a terminal object. Similarly in TOP. In GRP the trivial group with one element serves as an initial and a terminal object.

It is easily seen that SET-elements X can be identified with elements in the ordinary set theoretical sense. In general, however, C-objects need not have elements at all.

(3.10) Definition. Let C be a category. A product of the C-objects X and Y is a C-object $X \times Y$ with a pair $(p_X : X \times Y \longrightarrow X, p_Y : X \times Y \longrightarrow Y)$ of C-morphisms such that for object $X \times Y$ together any pair of C-morphisms of the form $f : Z \longrightarrow X, g : Z \longrightarrow Y$ there is exactly one arrow $(f, g) : Z \longrightarrow X \times Y$ making the following diagram commute:

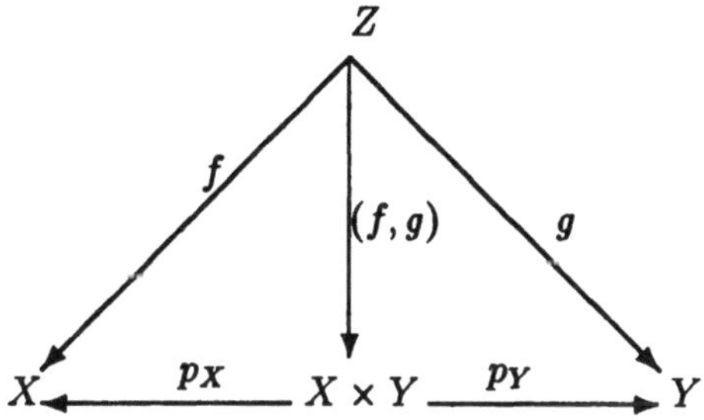

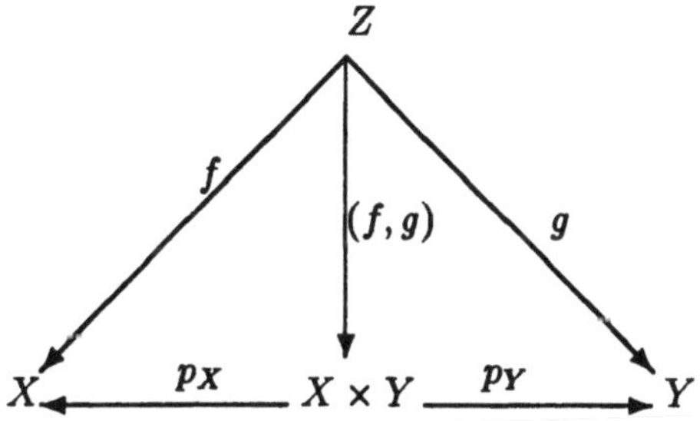

As is easily proved for the categories SET, TOP, GRP and many others this definition just amounts to the well-known set theoretical definition of the Cartesian product of (structured) sets.

(3.11) Definition. Let C be a category. A pullback of a pair $(f : X \longrightarrow Z \longleftarrow Y : g)$ consists of three morphisms $f' : P \rightarrow X, g' : P \rightarrow Y$ such that the following conditions are satisfied:

(i) The following diagram commutes

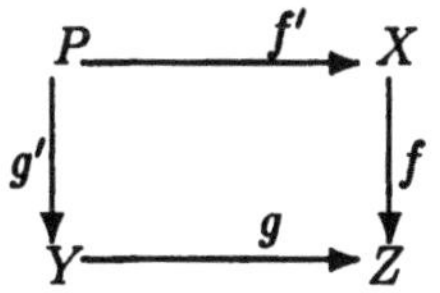

(ii) Whenever $Y \xleftarrow{h} Q \xrightarrow{j} X$ such that $g \circ h = h \circ j$ then

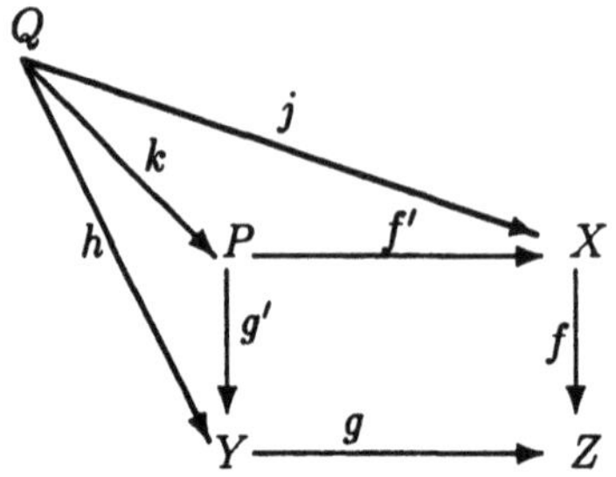

there is exactly one C-arrow k: $Q \longrightarrow P$ such that $h = g' \circ k$ and $j = f' \circ k$.

(3.12) Definition. Let C be a category.

(i) A C-morphism $f : X \longrightarrow Y$ is a monomorphism iff for any pair g: $U \longrightarrow X$ and h: $U \longrightarrow X$ the equality $f \circ g = f \circ h$ implies $g = h$.

(ii) *A* C-morphism $f : X \longrightarrow Y$ is an epimorphism iff for any pair $g : Y \longrightarrow V$ and $h : Y \longrightarrow V$ the equality $g \circ f = h \circ f$ implies $g = h$.

(iii) *A* C-morphism $f : X \longrightarrow Y$ is an isomorphism if there is a C-morphism $g : Y \longrightarrow X$ such that $g \circ f = id_X$ and $f \circ g = id_Y$.

Every isomorphism is a monomorphism and an epimorphism. In general, however, the converse is not true. There are categories with morphisms that are mono and epi without being iso. Categories where mono and epi implies iso are called balanced. SET is balanced. Some of the categories to be considered in the following, to wit, constrained categories, are definitively not balanced. For instance, the category of similarity structures SET(2) is not balanced.

(3.13) Definition. Let C be a category. A subobject of a C-object X is a monic C-morphism $f : Y \longrightarrow X$.

The class $Sub(X)$ of subobjects of X has a natural preorder: if $f : Y \longrightarrow X$ and $f' : Y' \longrightarrow X$ be subobjects of X define $Y \leq Y'$ iff f factorises in the following way:

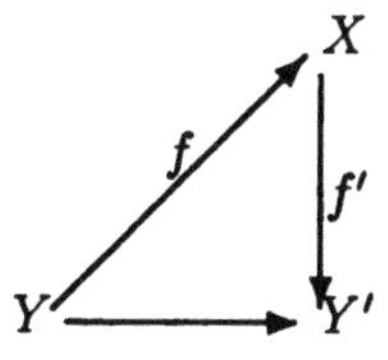

Obviously, id: $X \to X$ is a maximal subobject of X.

(3.14) Definition. A C-morphism is $f : X \longrightarrow Y$ surjective if the smallest subobject of Y it factorises through is $id_y : Y \longrightarrow Y$. This means if we have a commutative diagram

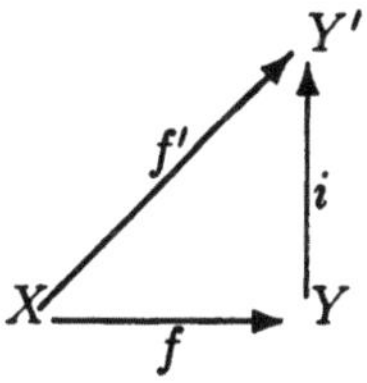

$i : Y' \longrightarrow Y$ is isomorphic to Y.

(3.15) Lemma. A surjective morphism $f : X \longrightarrow Y$ is an epimorphism. In SET, the concepts of epimorphism and surjective map coincide. In general, however, this is not the case.

(3.16) Definition. Let C be a category with products. A relation of C-objects X and Y is a subobject $R \longrightarrow X \times Y$.

It may be useful to compare this definition of relation with the set theoretical one: while in set theory a relation between sets refers to the internal structure of the sets involved, (3.16) provides a completely external characterisation.

4. Constraints and Links in the Framework of Category Theory

After theses general remarks on category theory let us now apply the framework of category theory to a new generalisation of the concept of constraints. Let C be a category with (finite) products. We now embark on the task to define for C constraints that are to be considered as category theoretical generalisations of the set theoretical constraints we dealt with in section 2. For ease of presentation we only deal with 2-constraints. It will be obvious, however, how to treat the general case of n-constraints, $n \geq 2$. According to (2.3) a 2-constraint on a set X can be characterised as a reflexive and symmetric subset $C \subseteq X \times X$. In category theory, the concepts 'subset', 'reflexive' and 'symmetric' are not readily available, rather we have to reconstruct them in terms of morphisms. The counterpart of 'subset' in a category C is of course 'subobject'. Hence, a constraint is to be regarded as a monomorphism i: $C \longrightarrow X \times X$. The only concepts left to explain are 'reflexive' and 'symmetric'. For this purpose we need the following two lemmas:

(4.1) Definition. Let C be a category with products. Let X be a C-object. Then there is a unique C-morphism (the 'diagonal') $D : X \longrightarrow X \times X$ defined by the following diagram:

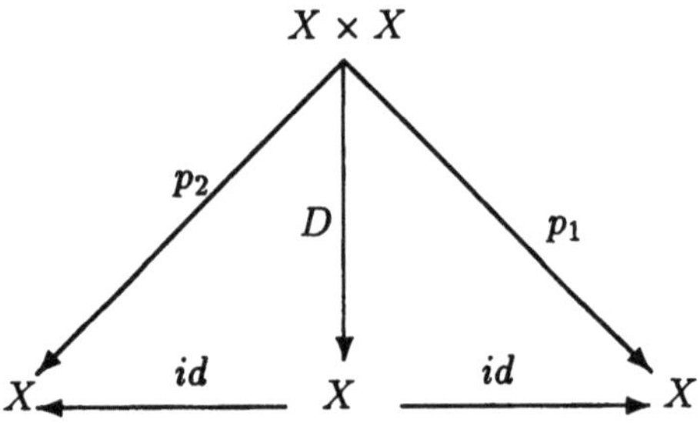

In SET, D is the map $D : X \longrightarrow X \times X$ defined by $D(x) := (x, x)$.

(4.2) Definition. Let C be a category with products. Let X be a C-object. There is a unique C-map (the 'twist') $T : X \times X \longrightarrow X \times X$ defined by the following diagram:

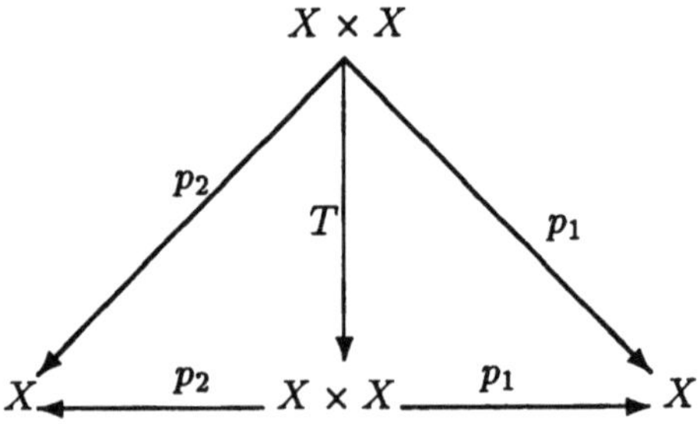

In $SET, T : X \times X \longrightarrow X \times X$ is just the familiar map $T(x, y) = T(y, x)$.

After these preparatory remarks we are ready to define constraints for categories C with products:

(4.3) Definition. Let C be a category with products. Let X be a C-

object. A C-subobject $i : K \longrightarrow X \times X$ is a 2-constraint on X iff the following holds: (i) The diagonal $D : X \longrightarrow X \times X$ factorizes through K, that is, there is a C-map $D_K : K \longrightarrow K$ such that the following diagram commutes:

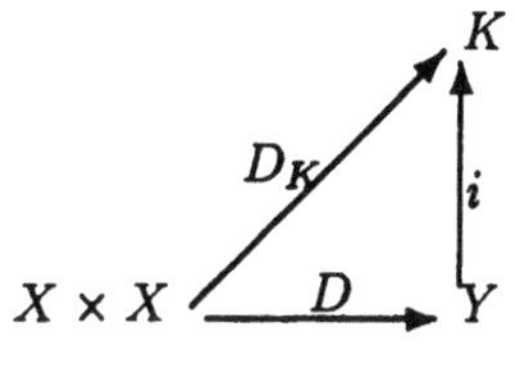

(ii) The twist $T : X \times X \longrightarrow X \times X$ factorizes through a morphism $T_K : K \longrightarrow X \times X$ such that the following diagram commutes:

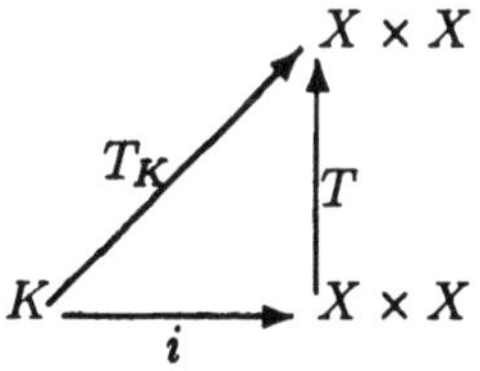

As is easily seen, in SET these conditions boil down to the familiar requirements of reflexivity and symmetry. Moreover, it is obvious, how to generalise these concepts to n-ary reflexivity and symmetry.

Taking stock we may say that we have been able to defined the structuralist notion of constraint independently from any set theoretical considerations for a quite general class of categories, to wit, categories with products. This definition is, as it should be from the vantage point of category theory, a completely external definition, i.e. it is independent from any consideration of the internal structure of the objects involved. The category theoretic concept of constraint can be used to define the concept of a constraint category:

(4.4) Definition and Lemma. Let C be category with products. The 2-constraint-category $C(2)$ of C is defined as follows:

(i) $C(2)$-objects: $(X, K, i), X$ and K are C-objects, and $i : K \longrightarrow X \times X$ is a reflexive and symmetric monomorphism.

(ii) $C(2)$-morphisms $(f, f_K) : (X, K, i) \longrightarrow (X', K', i'), f : X \longrightarrow X'$ and $f_K : K \longrightarrow K'$ being C-morphisms such that the following diagram commutes:

$$\begin{array}{ccc} K & \xrightarrow{f_k} & K' \\ \downarrow i & & \downarrow i' \\ X \times X & \xrightarrow{f \times f} & X' \times X' \end{array}$$

$C(2)$ is a usually non-full subcategory of C. More precisely, there is a forgetful functor $V : C(2) \Longrightarrow C$ and an adjoint functor $U : C \Longrightarrow C(2)$ defined by mapping a C-object X to the $C(2)$-object $(X, X \times X, i)$ where i is the trivial constraint defined by id: $X \times X \longrightarrow X \times X$. Moreover, we get the following pleasing result:

(4.5) Proposition. If C is a finitely complete category, i.e. has a terminal object and pullbacks, $C(2)$ (and more generally $C(n)$) is a finitely complete category as well.

Proof: Some diagram chasing of pullback diagrams.
An interesting example of a constraint category $C(2)$ that deserves special attention is the category $SET(2)$:

(4.6) Example. The category $SET(2)$ is the category of similarity structures. The objects of $SET(2)$ are sets $(S, \sim)$ endowed with a reflexive and symmetric relation $\sim \subseteq S \times S$. The morphisms of $SET(2)$ are set theoretical maps f: $S \longrightarrow S'$ preserving the similarity structures, i.e. maps satisfying for all $x, y \epsilon S$ the requirement $x \sim y \Longrightarrow f(x) \sim f(y)$.

The example (4.6) may be used to point out that the parallelity between C and $C(2)$ is limited: although the finite completeness of C implies the finite completeness of $C(2)$ not all structural features of C may be inherited by $C(2)$. For instance, SET is a topos, but $SET(2)$ is not (cf. Adamek et al. 1990). However, as long as we are content with finite completeness there are no problems concerning constraints provided our base categories C are finitely complete. If this is the case general category theoretical reasoning provides us with ready-made finitely complete constraint categories $C(2)$, (or, more generally, with finitely complete categories $C(n)$). In this way we save a lot of repetitive work. Without the category theoretical apparatus we would have to repeat the proof that constrained entities are as well-behaved as non-constrained ones for each case separately. Now let us come to what may be called an external generalisation of constraints, to wit, links (cf. Balzer & Moulines & Sneed, 1987). Set theoretically links are relations between models of not necessarily the same theory elements, i.e. a link is a relation $L \subseteq M_p \times M_p$.

Definition (4.7). Let C be a finitely complete category.

(i) A link in C is a subobject $f : L \longrightarrow X \times Y$ of the C-objects X and Y.

(ii) A constrained link in $C(n)$ is a subobject $f : L \longrightarrow X \times Y$ of the $C(n)$-objects X and Y.

Quite similarly as in the case of constraints, the notion of a link is much too general to be really useful. One has to single out specific examples of interesting links. First steps in this direction has been undertaken in Moulines (1992) where two essentially different classes of links are distinguished, the entailment links and the determining links. In this paper we'll only deal with the latter. They may be thought of as kinds of 'macroanalytic' relations between theory-elements. The interpreting links, on the other hand, are of a more 'microanalytic' kind, hence they do not fall into the realm proper of a category theoretical reconstruction.

(4.8) Definition. Let C be a finitely complete category. Let $i_{X'} : X' \longrightarrow X$ and $i_{Y'} : Y' \longrightarrow Y$ be subobjects of X and Y, respectively. An entailment link L from X to Y $rel(X', Y')$ is a link between X and Y such that there is a C-morphism e: $P_{X'} \Longrightarrow P_{Y'}$ making the following pullback diagram commutative:

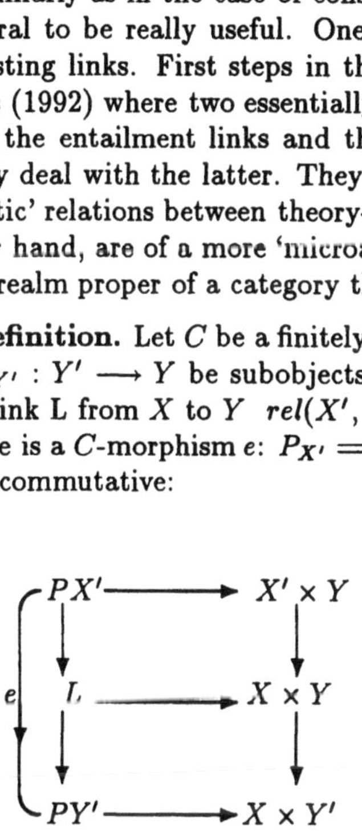

The concept of an entailment link makes sense for any finitely complete category. In particular, if C is finitely complete we are entitled to speak of entailment links in the constraint category $C(n)$ without further ado. We haven't made any reference to sets, elements, etc. That is to say, definition (4.8) is a really external genuinely category theoretical characterisation of this concept. In the next section we will show that (4.8) is indeed the correct category theoretical generalisation of the classical concept of structuralism link. This is done by showing that in the special case of the category TC of formal theory cores it leads to the traditional set theoretical concept of an entailment link.

5. Convenient Categories for Structuralism

After these preparations we are ready to embark on the essential task of this paper, to wit, defining convenient categories that can be used as the stage of the structuralist discourse: Or, to be somewhat more modest, that can be used for at least some of its important aspects, in particular for its 'macro-structures' such as theory-nets and holons. As is well known the structuralist reconstruction of the global structure of science is based on theory-elements and intertheoretical relations between them. Thus, as a convenient set theoretical starting point for a category-theoretical reformulating of the structuralist approach I propose what is known as 'formal theory cores' (cf. Moulines 1992). They are defined as follows:

(5.1) Definition and Lemma. The category TC of formal theory cores is given as follows:

(i) TC-objects: formal theory cores (M_p, M): M_p is a class of potential models and M a class of actual models, $M \subseteq M_p$.

(ii) TC-morphism: $f : (M_p, M) \longrightarrow (M'_p, M)$; f a set theoretical map with $f(M) \subseteq M'$.

Obviously, TC is a category. If we consider the pair (1,1) as a terminal object it is to prove the following:

(5.2) Lemma. The category TC of formal theory cores is a finitely complete category.

The proof is obvious. It depends on the fact that TC-objects actually are sets. For example, the TC-products of TC-objects (M_p, M) and $(M'p, M')$ is just $(M_p \times M'_p, M \times M')$.

(5.3) Corollary. The constraint categories $TC(n)$ are finitely complete categories.

Taking TC (or $TC(n)$) as our base now we are in a position to define further structuralist categories. As is well known, for many purposes the adequate reconstruction of a theory requires two levels: the theoretical level, and the non-theoretical level. This distinction can be rendered precise in the category theoretical framework as follows:

(5.4) Definition and Lemma. The category BTC of core bundles is defined as follows:

(i) BTC-objects: TC-morphisms $f : (M_p, M) \longrightarrow (N_p, N)$, f epimorphism.

(ii) BTC-morphisms: if $f : (M_p, M) \longrightarrow (N_p, N)$ and g: $(M'p, M') \longrightarrow (N'_p, N')$ are BTC-objects a BTC-morphism with domain f and codomain g is a pair (k, h) of TC-morphisms k: $(M_p, M) \longrightarrow (M'_p, M')$ and $h : (N_p, N) \longrightarrow (N'_p, N')$ such that the following diagram commutes:

$$\begin{array}{ccc} (M_p, M) & \xrightarrow{\ k\ } & M'_p, M \\ f\big\downarrow & & \big\downarrow g \\ (N_p, N) & \xrightarrow{\ h\ } & (N'_p, N') \end{array}$$

Obviously, BTC is a category. The terminal object of BTC is the trivial bundle $(1,1) \Longrightarrow (1,1)$. The categories $BTC(n)$ of constraint core bundles are defined in the obvious way. The objects of BTC (or $BTC(n)$) are also structuralist bundles.

(5.5) Lemma. The category BTC is a finitely complete category.

(5.6) Corollary. The categories $BTC(n)$ are finitely complete.

The proofs of these assertions do not depend on the specific structure of TC or $TC(n)$ rather they are carried through by category theoretical 'general abstract nonsense'.

The categories BTC and $BTC(n)$ of structuralist bundles provide a convenient frame for discussing the structuralistically crucial concept of 'empirical claim' in quite general bundle theoretical terms (cf. Mormann 1985).

(5.7) Definition. Let $r : E \longrightarrow B$ be a structuralist bundle, i.e. an BTC- or $BTC(n)$-object. A section of r is a TC or $TC(n)$-morphism $s : B \longrightarrow E$ such that $s \circ r : B \longrightarrow B$ is id_B.

Although bundles $E \longrightarrow B$ in SET always possess sections for trivial reasons, this is not the case for other categories. For example, usually it is quite difficult to find out whether a TOP-bundle $E \longrightarrow B$ has a section or not (cf. Mormann 1985). Similarly, there are constrained bundles $E \longrightarrow B$ that do not possess a section. Thus, for most categories, the 'section problem' is a difficult problem. In structuralist terms, the section problem is just the problem whether the empirical claim for a theory element holds or not:

(5.8) Lemma. Let $E = (M_p, M, C_p)$ a formal theory core, $B = (M_{pp}, I,$

C_{pp}) and $r : M_{pp} \longrightarrow M_p$ the reduction map Then the empirical claim of the structuralist theory element $T := (M_p, M, M_{pp}, I, C)$ holds iff the bundle $E \longrightarrow B$ has a section.

Now let us show how the concept of links is reformulated in the framework of structuralist categories such as $TC, TC(n)$ etc. Let us start with the category TC of formal theory cores. Let (M_p, M) and (M'_p, M_p) be two formal theory cores. According to the definition (4.7) a link is a subobject of $(M_p, M'_p, M \times M')$, i.e. $L = (L_p, L)$ with $L \subseteq L_p$ and $L_p \subseteq M_p \times M'_p$ and $L \subseteq M \times M'$.

Consider the following set theoretical pullback diagram of TC-objects

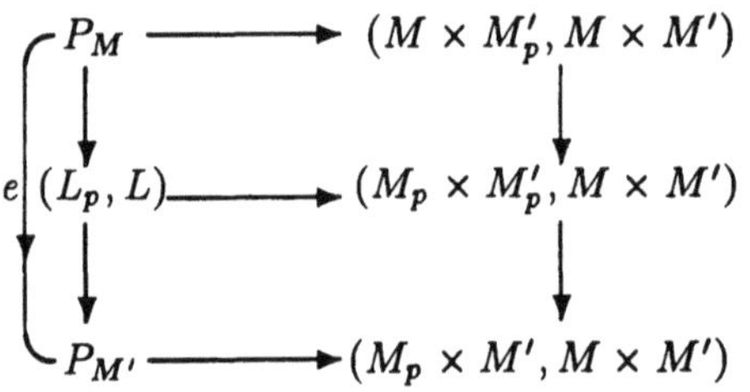

Then it is a simple set theoretical exercise to show that a map e: $P_M \longrightarrow P_{M'}$ iff is an entailment link in the traditional sense, i.e. iff it satisfies the following condition:$(x)(x')((x, x')\varepsilon L_p$ and $x \varepsilon M \longrightarrow x'\varepsilon M')$. Since we remain in the realm of finitely complete categories, for $TC(n)$ and $BTC(n)$ we can talk about entailment links without any further ado. The categories TC and $TC(n)$ are the stage where we can rehearse most of the standard structuralist concepts such as specialisation, theoretization and so on:

(5.9) Definition. Let X, Y and Z be TC or $TC(n)$ objects.

(i) A specialisation of X is a monomorphism s: $Y \longrightarrow X$.

(ii) A theoretisation of X is a surjection t: $Z \longrightarrow X$.

The problem of how to conceptualise structuralist reductions in terms of structure preserving maps, i.e. in terms of morphisms, has been discussed in some detail in Mormann (1988). One of the results of this discussion has been that there are quite a lot of different, not necessarily compatible structuralist reduction concepts. Without going into the rather messy details I mention just one result:

(5.10) Definition. An anomaly-explaining reduction of the BTC bundle $r : E \longrightarrow B$ by the BTC-bundle $r' : E' \longrightarrow B'$ is an epimorphism $(f_E, f_B) : (E', r', B') \longrightarrow (E, r, B)$ as displayed in the following commutative diagram:

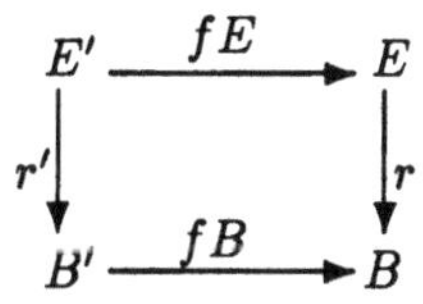

6. Theory-Holons as Functors

According to the structuralist approach, theory-holons are the most complex units which constitute the global structure of empirical science (cf. Balzer & Moulines & Sneed 1987, Moulines 1992). Intuitively, a theory-holon H is a (finite) set of theory elements (theory cores) that are connected by links. Thus, if links could be considered as morphisms of a category a theory-holon H would be nothing but a diagram of that category. Now, a (finite) diagram of a category C is just a functor $F : I \Longrightarrow C$, I some finite category, for instance a directed graph (cf. (3.3)(3)). Happily, there is indeed a category whose objects are theory cores, and whose morphisms are links. And, moreover, this type of category can be nicely described in genuinely category theoretical terms. Let us take SET as the point of departure. We construct a category $LSET$ that has the same objects as SET, and whose morphisms are links, i.e. relations between SET-objects. Intuitively, the construction of $LSET$ is quite obvious. Consider the following diagram in SET:

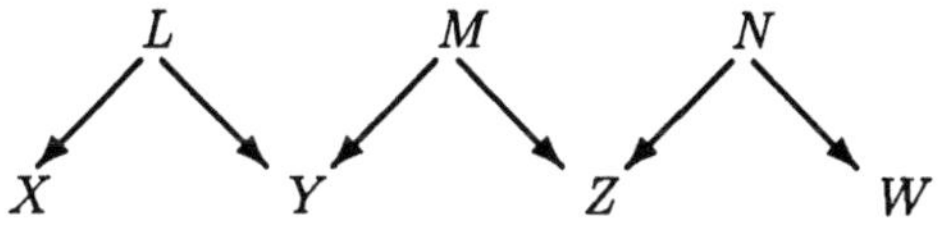

Define the concatenation of L and M as the set theoretical product of the relations L and M, i.e. $(x, z)\varepsilon LM$ iff $\exists y((x, y)\varepsilon L$ and $(y, z)\varepsilon M)$. In category theoretical terms this means the product LM of the relation L

and M is the pullback of the diagram

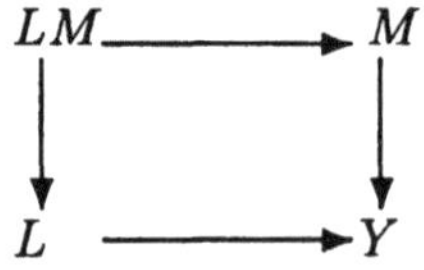

Ordinary morphisms can be rendered 'link-morphism', i.e. morphisms in the 'link category' simply by replacing a morphism $f : X \longrightarrow Y$ by its graph:

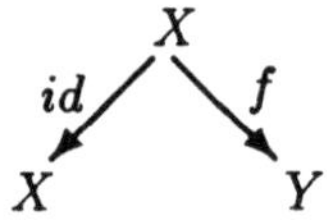

Hence, if there is a 'link-category' LC of C replacing C by LC could be conceived of as a functor. However, there is a technical difficulty, not for all categories C a link category LC can be constructed. The reason is as follows: If links L, M and N are to be considered as morphisms $L : X \longrightarrow Y, M : Y \longrightarrow Z$ and $N : Z \longrightarrow W$ the concatenation has to be associative, i.e. $(LM)N = L(MN)$ This is, in general, not the case (cf. Freyd Scedrov 1992, 1.569, p.83). We have to assume that C is a regular finitely complete category (cf. McLarty 1990, p. 242):

(6.2) Definition. A finitely complete category C is regular iff it satisfies the following requirements:

(i) All surjections are stable, i.e. for every surjection $f : X \rightarrow Y$ and for every s: $Y' \rightarrow Y$ the map $f' : P \rightarrow Y'$ in the following pullback diagram is surjective:

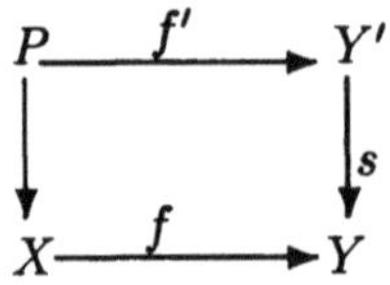

(ii) Every C-morphism $f : X \to Y$ has a surjective image, i.e. f factorises through a surjective f' in the following way

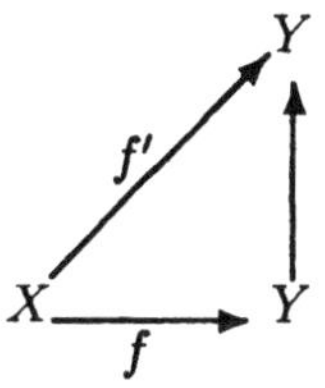

(6.3) Lemma. The categories SET, $SET(n)$ TC, and $TC(n)$ are regular.

Proof: Obvious.

Hence, we finally are able to define a category taking into account the structuralistically crucial concepts of constraints and links as follows:

(6.4) Definition and Lemma. The category $LTC(n)$ is defined as follows and has the following properties:

(1) $LTC(n)$ objects: n-constrained theory cores.

(2) $LTC(n)$-morphisms: n-constrained links between n-constrained theory cores.

(3) $LTC(n)$ is a finitely complete category.

If we like we could even step one level higher defining the category $LBTC(n)$ of linked constrained bundles in the obvious way.

(6.5) Theorem. A theory-holon H in the sense of structuralism is a functor $H : J \Longrightarrow LTC(n)$ (or $H : J \Longrightarrow LBTC(n)$), J being a finite category.

Let us take stock of what we have achieved so far: according to structuralist theory of scièence, the reconstruction of the global features of empirical science is best carried through in terms of theory-holons. As has been shown a theory-holon is a (connected) diagram in the category $LTC(n)$ (or $LBTC(n)$). This may be taken as an evidence that category theory indeed provides a congenial conceptual framework for the elucidation of the global features of empirical science as they are dealt with in the structuralist approach.

7. *References*

Adamek, J. & Herrlich, H. & Strecker, G., 1990, *Abstract and Concrete Categories*, New York.
Arbib, M.A. & Manes, E.G., 1975, *Arrows, Structures and Functors*, The Categorial Imperative, New York.
Balzer, W, & Moulines, C.U. & Sneed, J.D., 1987, *An Architectonic for Science*, Dordrecht.
Freyd, P.J. & Scedrov, A., 1990, *Categories, Allegories*, Amsterdam.
Goldblatt, R.: Topoi, 1978, *The Categorial Analysis of Logic, Amsterdam.*
McLarty, C., 1992, *Elementary Categories, Elementary Toposes*, Oxford.
Mormann, T., 1985, 'Topologische Aspekte strukturalistischer Rekonstruktionen', *Erkenntnis* 21, 319-359.
Mormann, T., 1988, 'Structuralist Reduction Concepts as Structure-Preserving Maps', *Synthese* 77, 215-250.
Moulines, C.U., 1992, 'Structures, Links, and Holons', in H. Westmeyer (ed.), *The Structuralist Program in Psychology*, 13-26.
Moulines, C.U., 'Towards a Typology of Intertheoretical Relations', in J. Echeverria, A. Ibarra, T. Mormann (eds) *The Space of Mathematics*, Berlin, New York, 403-411.
Pitt, D. & Abramsky, S. & Poigné A. & Rydehard, D., 1986, *Category Theory and Computer Programming*, Lecture Notes in Computer Science vol. 240, Berlin, New York.
Sneed, J.D., 1971, *The Logical Structure of Mathematical Physics*, Dordrecht.
Stegmüller, W., 1973, *The Structuralist View of Theories: A Possible Analogue of the Bourbaki Programme in Physical Science*, Berlin, Heidelberg.
Stegmüller, W., 1986, *Theorie und Erfahrung, Die Entwicklung des neuen Strukturalismus seit 1973*, Band II(3), Berlin, Heidelberg.
Westmeyer, H., 1992, *The Structuralist Program in Psychology*: Foundations and Applications (ed.), Bern.

Name Index

Subject Index[1]

[1]Key terms of structuralism as explained in the introductory chapter are not contained in this subject index

Logic, Action, and Information

Essays on Logic in Philosophy and Artificial Intelligence

Edited by André Fuhrmann and Hans Rott

1996. Large-octavo. X, 476 pages. Cloth. ISBN 3-11-013994-4

Formal Theories of reasoning about knowledge and action.

Contents: *Action:* J. Czelakowski, Elements of Formal Action Theory – K. Segerberg, The Delta Operator at Three Levels of Analysis · *Belief Change:* S. O. Hansson, Hidden Structures of Belief – W. Rabinowicz, Stable Revision, or is Preservation Worth Preserving? – M. D. Ryan, Belief Revision and Ordered Theory Presentations – E. Weydert, Doxastic Normality Logic: A Qualitative Probabilistic Modal Framework for Defaults and Belief · *Conditionals and Non-Monotonicity:* G. Crocco/L. Fariñas del Cerro, Counterfactuals: Foundations for Nonmonotonic Inferences. Sequent systems – L. Fariñas del Cerro/A. Herzig, Revisions, Updates, and Interference – St. Lindström, The Ramsey Test and the Indexicality of Conditionals: A Proposed Resolution of Gärdenfors' Paradox – K. Schlechta, Some Completeness Results for Classical Preferential Models · *Non-Classical Logic and Logic Programming:* J. Dix, Semantics of Logic Programs: Their Intuitions and Formal Properties. An Overview – D. M. Gabbay/H. Wansing, What is Negation in a System? Negation in Structured Consequence Relations – E. Orlowska, Relational Environment for Semigroup Logics – D. Pearce, Answer Sets and Constructive Logic. Part I: Monotonic Databases – G. Wagner, Logics Based on Knowledge Representation Systems – C. Witteveen, Belief Revision in Truth Maintenance.

Walter de Gruyter · Berlin · New York

Bei Fragen zur Produktsicherheit wenden Sie sich bitte an:
If you have any questions regarding product safety,
please contact:

Walter de Gruyter GmbH
Genthiner Straße 13
10785 Berlin
productsafety@degruyterbrill.com